Peasant farming in
Muscovy

Peasant farming in Muscovy

R. E. F. SMITH

PROFESSOR OF RUSSIAN, UNIVERSITY OF
BIRMINGHAM

CAMBRIDGE UNIVERSITY PRESS

CAMBRIDGE

LONDON · NEW YORK · MELBOURNE

Published by the Syndics of the Cambridge University Press
The Pitt Building, Trumpington Street, Cambridge CB2 1RP
Bentley House, 200 Euston Road, London NW1 2DB
32 East 57th Street, New York, NY 10022, USA
296 Beaconsfield Parade, Middle Park, Melbourne 3206, Australia

First published 1977

Printed in Great Britain
by W & J Mackay Limited, Chatham

Library of Congress Cataloguing in Publication Data
Smith, Robert E. F.
Peasant farming in Muscovy.

Bibliography: p.
Includes index.
1. Peasantry – Russia – History.
2. Agriculture – Russia – History. 3. Russia – Rural conditions.
I. Title.
HD1511.R9S6   338.1'0947   75–23843
ISBN 0 521 20912 9

Contents

Illustrations

PLATES (between pages 196 and 197)

FIGURES

MAPS

Tables

Acknowledgements

I am most grateful to colleagues and friends who have helped me by reading and commenting on parts of this work. They made numerous criticisms and suggestions, all of which I have carefully considered and most of which have resulted in changes and, I hope, improvements in the text. In particular I should like to thank Gus Alef, L. V. Danilova, John Fennell, Carsten Goehrke, Sergei Kashtanov, A. L. Khoroshkevich, V. T. Pashuto, A. L. Shapiro and his colleagues, and Aleksandr Zimin for their generous help. Others who contributed much include G. R. Barker, J. Goody, R. H. Hilton, Joan Thirsk, the late Daniel Thorner, Mrs M. Perrie, G. S. Smith and C. Thomas. A special word of thanks is due to Michael Pursglove who persistently tracked down missing peasants with an assiduity worthy of a Trinity monastery steward; he has made many of the calculations as accurate as such things are ever likely to be. Finally, Elizabeth Bowden of the Cambridge University Press has worked scrupulously to improve my text and to reduce my ignorance; no one could have tackled these enormous tasks more skilfully.

NOTE ON UNITS

The publisher suggested that, throughout the text, modern equivalents be restricted to metric units. For those who are not familiar with this system, the following table of equivalents may be helpful.

WEIGHT

1 pud = 16.38 kg = 36.11 lb
1 pound (Russian) = 410 g = 14.44 oz

LENGTH

1 versta = 1.067 km = 0.663 mile
1 sazhen = 2.134 m = 7 feet

AREA

1 desyatina = 1.0925 ha = 2.7 acre
1 chet = $\frac{1}{2}$ desyatina

The chetvert or chet was also used as a grain measure. Until the early seventeenth century the chet contained 4 puds of rye or $2\frac{1}{2}$ puds of oats.

Beloozero
Novgorod
Volkhov
Lovat'
Kholm
Toropets
Map 3
Western Dvina
Smolensk
Vyaz'ma
Volokolamsk
Moscow
Map 2
Volga
Klyaz'ma
Oka
Ryazan'
Bryansk
Nizhnii
Novgorod
Vyatka
Kazan'
Map 4
Kama
URAL MOUNTAINS
Kiev
Dnepr
Don
Astrakhan'
0
300
km

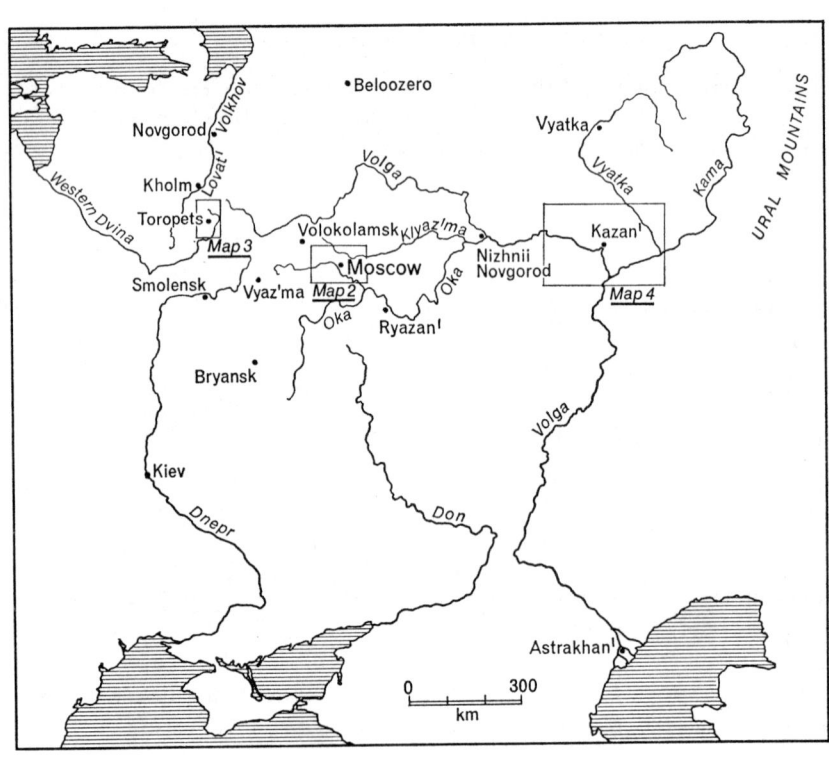

Map 1. Muscovy.

Introduction

Materials for the history of farming in European Russia before the mid seventeenth century originate almost exclusively from princely or monastic sources; this gives rise to considerable difficulties. It might be thought that Soviet historians operating within a Marxist conceptual framework, and therefore paying considerable attention to the economic basis of society and to the primary producers, would have made substantial correctives to offset the limitations and distortions due to the nature of the materials. Yet, for example, Soviet medieval archaeology has so far paid little attention to the countryside; nor have Soviet historians devoted much attention to technology. There have been some Soviet articles on tillage implements, but the latest serious Russian monograph on the main implement was published in Vyatka as long ago as 1908. In fact, neither Soviet literature nor western works on the history of farming before Peter the Great appear to make adequate allowance for the reality of the Russia which lies concealed behind the documents.

It may seem impertinent for an Englishman to attempt to introduce correctives to the generally accepted picture of farming in Russia in the period between the Mongol invasions and the mid seventeenth century (lack of earlier data compels us to concentrate on the period from the fifteenth century). Perhaps, however, the very remoteness of Russia from England and the different experience of our own agricultural history may contribute something new. The main elements with which I am concerned are two.

First, the overwhelming predominance of documentary sources, and their nature, means that attention has been largely focussed on monastic estates, on heritable estates or on estates held by service. In fact, however, much of Russian farming as a system of production was peasant farming organised in small household, mostly family, units; these farms were not in all cases organised into estates of any type. Indeed, for much of the period, even on the estates themselves, similar family farms may have been the basic production units, the manors and estates being largely management units. The micro-economies deserve greater attention than they seem to have received.

Second, many works, particularly those of west European and American historians, generalise somewhat too readily and fail to pay adequate attention to the regional differences which exist within European Russia.

I

In this work, therefore, I hope to be able to shift the focus of attention from the estate organisation to the peasant production unit. This means paying more attention than has hitherto been done to the physical environment and the relatively simple technology available to peasants, that is to factors which existed despite the changes taking place in the macroeconomy and in social structure in this period. By looking at the peasant unit from the consumer as well as from the producer angle, it becomes possible to relate the small production unit of the peasant family farm to the specific physical environment around it. The importance of the forest is then likely to be seen more clearly, not as some sort of minor appendix to farming activity, but rather as the essence which sustained the whole amalgam of farming and gathering in which the peasants were engaged. The specific forms, however, change over time, especially as forest is cleared, and also according to location, both physical and social.

The focus of this study is peasant farming. The peasants who provide the chief interest in our efforts to offset the bias implicit in our sources are those who lived independently of immediate lords. The 'black' peasants were those who were directly subject to the prince's administration; they long continued to be the category which lived, to a great extent, virtually independently; they thus represent an extreme type, but gradually they diminished in numbers as their lands became subject to lords. Peasants on estates of lords, both lay and ecclesiastical, had a different history; they were the immediate objects of a long process of enserfment. In fact, in the sixteenth century the documents often use the term 'peasant' (*krest'yanin*) in the sense of a serf. Our main interest lies with the independent peasant, not with those who were enserfed; but the independent peasant was not found everywhere in the period under review, so we are forced to consider, at least in part, evidence derived from the estates of various types with their dependent peasants. Moreover, the extreme type of the black peasant was not always and everywhere converted immediately into the fully enserfed dependent; there were many intermediate social categories of peasant. For us, then, the term 'peasant' is not used in the Russian sense of a dependent social category, but rather in the broad economic sense of someone making a living directly from the land.

Finally, apart from an attempt to offset the bias of our sources, I hope to be able to convey to English historians a limited impression of some types of the material available in Russian. For this reason I have tried to include rather more primary material in translation (especially in appendices) than might normally be expected.

The first part of the work considers the various branches of activity of a generalised, hypothetical, isolated farm unit leading up to estimates of the technical indices relating to the main elements of production and consumption. There then follows an examination of the evidence dealing with peasant farming in three areas: the Moscow uezd in the centre; Toropets on the western borders of Muscovy; and Kazan' on the eastern frontier. These all roughly lie on the fifty-sixth parallel but in differing climatic, vegetational and cultural environments. A concluding chapter attempts to generalise on the basis of the evidence adduced from these three areas.

The examination of these areas should allow something to be said of how the mass of the population made their living from the land in pre-Petrine Russia. Consideration of the physical environment, the technology and the labour of the small-scale peasant production units applied to this environment involves such matters as forest clearance, the size and nature of settlements, the gradual spread of the three-field system and the possible existence and nature of communes. In this way, and with the help of the input–output calculations for an average peasant farm, we can tentatively propose estimates of surpluses, thus contributing to the discussion about the level of incidence of obligations. Perhaps, too, we may speculate about how far technical limits acted as a brake on economic accumulation and development. The general concern, therefore, is to explain something of how the Russian peasant made his living and, by implication, how far the peasant nature of Russia in this period was itself a major factor determining the slow rate of economic growth.

PART I

THE ELEMENTS OF THE PEASANT HOUSEHOLD

Many Russian charters of the fourteenth and fifteenth centuries contain a three-fold formula 'Wherever the sokha, scythe and axe have gone'; about a hundred such phrases occur in documents of the Trinity monastery of St Sergius alone and the total number runs into several hundreds.[1] Some versions are modified, sometimes substituting 'plough' for 'sokha', sometimes abbreviating the formula by omitting the first element. What did the formula mean? I take it to indicate the traditionally established bounds of the rural settlement, whether waste, hamlet, village or manorial settlement. Given the relatively low density of population and the continuing existence of much forest land, it was likely that, for most parts of the area, such bounds would not be precisely defined. Cultivation was generally extensive; intensive cultivation and precise bounds would only come about with increasing pressure of population on land.[2] In a court case about 1492, Ivan Onisimov, a peasant, claimed he had 'cut the wild forest and tilled that clearing for twelve years as far as the water from Korchmitovo hamlet; and beyond the water, sirs, you can see for yourselves now, the forest is wild to Korchmitovo and axe has not met with axe'.[3] It seems probable that for several centuries after the thirteenth-century Mongol invasions axe had not met with axe in much of European Russia. Incidentally, the various ways in which the land–labour ratio was adjusted were, as will be seen, central to the operation of the peasant economy. Many documents make clear the meaning of the three-fold formula by adding a phrase to the effect that the bounds were those that had been long established.

One Beloozero charter opens with the following words: 'Now I, prince Mikhail Andreevich, have exchanged with Oleshka Afans'ev his Ukhtoma lot and additionally granted Terent'ev land, with everything, with woods, wastes, hay fields, fisheries and all appurtenances and whatever has been subject to that Ukhtoma lot of Oleshka from of old, as it was under Oleshka, wherever the scythe and axe and sokha have gone...'[4] A purchase deed of the early fifteenth century shows the formula applied to a variety of settlements and types of property: 'Now I, prince Dmitrei Yurevich, have bought from Fetin'ya, wife of Ivan Yurevich, her villages in Bezhets Verkh: Priseki village with all its hamlets, with the hamlets beyond the Mologa and the wastes, and the manorial village of Vorob'evo with the hamlets and the wastes, the meadows, rivers, weirs, traps, forest, banks and other appurtenances, and with everything that has been subject to Priseki village and to Vorob'evo manorial settlement, as it was under Fetin'ya and under her son Dmitri, wherever the sokha went, wherever the axe went,

---

[1] Gorskii, *Ocherki*, 39. The sokha was a forked tillage implement, usually drawn by a horse. Further information is given below.

[2] Cp. Mavrodin, *Obrazovanie*, 17.   [3] *ASEI*, II, no. 287.

[4] *ASEI*, II, no. 233, 5 September 1476.

wherever the scythe went.'[1] The villages of Priseki and Vorob'evo were very large.[2] The sequence of villages and their dependent hamlets with their wastes in the surrounding forest is typical, but other documents show the formula applied to the smallest units, for example, when forest land was being opened up for cultivation. At the same period, a cellarer of the Trinity monastery of St Sergius 'bought from Dubrova Ramen'ev the Molitvinski waste and the clearances: Ratmertsevo and Dubrovka, Krugly, Sedelnichi, Pankovy and the two Yanov sites and the Podvezny site and the other Dubrovka, wherever his plough and sokha went and the scythe and the axe, and with everything that was subject to that, with the forest and everything.'[3]

By the fifteenth century then, this formula had evidently become conventional and could, therefore, be modified without substantially affecting its meaning. It applied to estates and to villages with the hamlets, wastes and clearances in the forest which were administratively subordinate to them; it indicated in functional terms their effective limits.

The formula, however, tells us something more. First, although the elements are three implements, it is clear that they stand for the areas over which the characteristic activities performed with such implements were carried out. Gorskii is certainly right to point out that the first element had no real content as regards the particular implement being used in any particular area.[4] It does not tell us whether the plough or the sokha was being used. This element indicates the limits of the arable or cornfield. The implements and techniques used for that cultivation are separate questions which will be dealt with below. Even though the formula seems mainly to occur in areas with small settlements and much forest it seems mistaken to limit its application to slash and burn techniques, as has been suggested by Zimin.[5] The scythe represents the hayfields, a particularly important aspect of farming in Russia because of the crucial part played by feed for livestock which had to be stalled for half the year.[6] The axe represents the forest, but probably not any unclaimed forest, only that which was traditionally exploited by a tenement, an estate or some other unit.

Second, although examples of the formula may be found relating to anything from a clearance or waste to a manorial settlement, this should not be taken to indicate that the activities represented by the implements necessarily were carried out by the members of a settlement in common. In fact, the application of the formula to wastes, which might not always have a house on them, shows that the subject is not properly the settlement, but rather the production and consumption unit.

Third, the frequent occurrence of the formula, together with the fact that it came to be a convention, indicates its importance; this impression is reinforced by considering the activities which it represented. Arable and livestock were the twin bases of farming in Europe and the forest provided many supplements to meet the needs of man and beast. Together these three elements concisely and vividly outline the activities of the holding as a production and consumption unit; the arable and livestock sectors, as well as gathering and extractive industry. In the centuries dealt with here, the impression-

---

[1] *ASEI*, I, no. 163, before 1440.   [2] *Ibid.* p. 603.

[3] *ASEI*, I, no. 18, 1410–27. Could the name Ramen'ev itself be related to *ramenie*, which meant forest at the edge of the tilled land? The two Dubrovkas may have been established by Dubrova himself, judging by the similarity of the names.

[4] *Ocherki*, 39–40 and in *Ocherki russkoi kul'tury*, ch. I, 58.

[5] *Ist. SSSR* (1962), 4.172. Cp. *PRP*, III.64.   [6] See p. 43 below.

istic description given by the formula was valid both for the isolated peasant tenement or hamlet of perhaps two or three households and for the manorial estate based on a number of such units. After the fifteenth century the use of the formula declined; it was sometimes replaced by indications of bounds in sixteenth-century documents; 'when these bounds became stable, they were no longer specially mentioned in the charters' save when bounds were newly established.[1] This change may be associated with the more widespread emergence of three-course rotations on regularly worked fields; the reality, however, remained.

The farm unit was a cell; like biological cells, these units varied in function, but in general it seems reasonable to assume, at least as a tentative hypothesis, that specialised production units were rare. Variations arose from a range of causes, for example from the physical and the social environment; but such variations were only modified versions of what was essentially a basic unit. It was these units that formed the body of most communes, manors, estates and, ultimately, of the state itself.

To start with, then, a static analysis of this unit may be made by considering the three elements mentioned in the formula. It will be necessary to add a fourth element, not mentioned in this connection in the sources. The family or household, like the land off which it lived, was assumed by the sources and needed no mention; but we have to be more explicit, not least because historical family patterns differ in western and eastern Europe. The triple formula reflects the male-dominated world from which it comes; the work of women within the household, including crafts and trades, risks being neglected if we work only within the framework of the formula. When all four elements of the farm unit have been considered, it should then be possible to construct an approximate model to indicate the limits of its production and consumption.

---

[1] *PRP*, IV.75–6. The formula was not found in areas, such as Opol'e, which had few woods and had long been relatively densely settled.

I

Tillage implements: the arable land

A formal classification of implements may help us discern any regularities of distribution in time and space.[1] Moreover attention to terms used for parts of implements seems likely to be more rewarding than the often sterile arguments about terms used for whole implements. Terms of the latter type are, it is true, more frequent in our sources, but we may not be able to rely on *literati* comprehending technical niceties, even if they recorded them correctly. Moreover the indeterminate nature of terms such as *sokha* and *plug* (plough) is increased by the fact that they were also used as measures of land area and as units of taxation. On the other hand, the oral tradition, if it could only be recovered, seems more likely to reflect the terminology of the users. In another context it has been remarked that 'it is by no means uncommon for very ordinary words to remain latent for long periods' and it may not be too much to hope that we may sometimes be able to bring about that 'reemergence out of the obscurity of talk into the light of literature'.[2] such recovery, if achieved for the Russian material, would no doubt provide its own problems, but at the same time, it might help us towards a deeper understanding of the material.

Once a classification of implements and their parts has been established, it becomes necessary to account for any regularities observed. It is here, unfortunately, that what appear to be sectional interests sometimes hinder the development of further understanding. Ethnographers, who at least have taken the trouble to collect material in the field, material which, owing to land reform, mechanisation and a high degree of social mobility, is becoming rare and may be disappearing, have sometimes been accused of ignoring environmental factors.[3] Soil, climate, social and economic conditions, as well as cultural traditions must, of course, be considered in attempting to account for the distributions that are observed. But, at the moment, we lack adequate details for the history of the sokha in Russia. As Novikov rightly observes, 'Unfortunately, the history of Russian tillage implements has not been studied in the past, and nowadays research on this subject (which might give

---

[1] A classification scheme for sokhas will, it is hoped, be published elsewhere.
[2] Sir Ernest Gowers, *Modern English usage*, 524. Ya. A. Sprinchak has cited a few examples of dialect survival of words not in literary Russian; *Voprosy leksikologii*, 56–7.
[3] E.g. Debets, Olderogge & Potekhin, *SE* (1957), 1.163; Novikov, *SE* (1963), 2.99.

most useful results) is almost impossible; the sokha, *saban* and *kosulya,* widely used in Russia even at the turn of the century, are now museum rarities...'[1] Thus, interpretation is hindered. There remains a large element of speculation about the history of Russian implement types.

In a strict sense in modern usage the term *sokha* is restricted to implements which have a split share-beam, or two share-beams or tines and so reflects the idea of forking which the term itself implies.[2] However, this is a recent convention, and is not uniformly observed. In the Russian literature, moreover, *sokha* may be used to indicate any of the three main types of cultivating implements; those which merely scratched the soil, working like harrows (*sokhi cherkayushchie*); those which scratched the soil, but also swept the loosened soil to the side (*sokhi pashushie*); and those which cut and turned a slice.[3] Thus, the term might cover implements from the simplest drags, used as cultivators, to ploughs. It is this lack of precision which, in part, underlies the use of *sokha* and *plug* (plough) without any differentiation in the examples of the formula cited by Gorskii.[4] At the same time, Gorskii is right to stress that 'the question of what the sokha and the plough were in the period under consideration (fourteenth–fifteenth centuries) is extremely complex; this is primarily to be explained by the want of sufficient material, documentary and archaeological, with which to judge'.[5]

An interesting attempt to supplement the evidence hitherto available has been made by Gorskii, using the miniatures illustrating Russian manuscripts of the period.[6] Gorskii goes to some trouble to establish that these illustrations reflect Russian reality, even though, in the case of the eight illustrations from the Illuminated Collation (*Litsevoi svod*), the themes are biblical or mythological and, in the case of the others, only one has a Russian subject. In this he is certainly convincing; there seems no reason to doubt that these miniatures reflect Russian reality.[7] This does not mean, however, that we can agree with Gorskii that 'the sokha irons on the miniatures in the Chronograph correspond to the real "bar-point" sokha irons found in archaeological excavations

---

[1] *VIMK* (1961), 1.47. The *saban* was a light plough, the *kosulya* a soleless plough. Evidence on nineteenth- and twentieth-century materials is to be found in Bezhkovich *i dr., Khozyaistvo i byt* and in Kushner *i dr., Russkie,* 1.42–4, 53–4.

[2] For accounts in English of early Russian scratch ploughs, see Smith, *Origins,* pp. 47f; Chernetsov, *Tools and tillage,* II.1.34–50.

[3] Zelenin, *Russkaya sokha,* 10–12.

[4] *Ocherki,* 40 referring to *ASEI,* II, no. 307.

[5] *Ocherki.* See also his survey in *Ocherki russkoi kul'tury,* ch. 1, 58f.

[6] 'Drevnerusskaya sokha po miniatyuram Litsevogo letopisnogo svoda XVI v.', *Istoriko-arkheologicheskii sbornik* (1962), 339–51; 'Pochvoobrabatyvayushchie orudiya po dannym drevnerusskikh miniatyur XVI–XVII vv.', *MISKh,* VI.15–35; 'Eshche odno izobrazhenie sokhi XVI veka', *VMU, seriya IX* (1963), 3.17–22. See also Gorskii's survey of evidence about thirteenth- and fifteenth-century sokhas in *Ocherki russkoi kul'tury,* ch. 1, 58–75.

[7] Gorskii, in *Istoriko-arkheologicheskii sbornik* (1962), 340–6; *MISKh,* VI.23.

and referred to the fifteenth–sixteenth centuries'.[1] In fact, if we make an estimate of the length of the irons shown in the illustrations, we can see that most do not seem to approximate to the commonest length of those found by excavation.

The illustrations show implement irons roughly estimated to range in length from 15 to 50 cm (the wide range is largely due to the difficulties of making such estimates, and no doubt partly because of artistic licence). Excavated ones are within the range 18–32 cm; most, apparently, being 28 cm in length. Gorskii has himself pointed to a tendency for size to increase over time, giving the following ranges of length:[2]

> seventh–tenth centuries, 15–20 cm
> tenth–thirteenth centuries, 20–25/30 cm
> thirteenth–sixteenth centuries, 28–40 cm

However, quite apart from the question of size, the excavated irons are of a different type. While the miniature illustrations show what appear to be elongated iron cones, bar-point shares, varying somewhat in the extent to which their tips are curved forward and in the way in which the socket terminates, the archaeological finds are mostly pen-shaped broad wedges, the socket not fully closing round the tine of the share-beam. The excavated implement parts which most closely resemble those illustrated are the smaller ones with sockets half their total length or more, and almost totally embracing the tine. None of these appears to date from the fifteenth or sixteenth century; most are much earlier.

It seems difficult to understand why there should be this discrepancy between finds and illustrations. It might be due to the existence of a convention among the illustrators. Certainly, there are conventional usages in these miniatures. Many of these have been dealt with by Professor Artsikhovskii, who has listed, for example, the colours used to indicate objects of specific materials.[3] There also appear to have been some conventional designs, such, for example, as the two miniatures from a seventeenth century manuscript, one showing Adam, the other Cain. Both represent a tillage scene, the elements of which are the same, even though they differ in detail.[4] But the variations noted in the curve of the irons and especially in the way in which the upper end of the socket terminates militate against such an explanation. These seem more likely to be due to observation; had the irons been indicated merely conventionally, one might have expected this to be done by a straight line marking off the tip of the tine. The consistency of usage noted by Artsikhovskii also seems to argue against a lack of skill on the part of the illus-

[1] *MISKh*, VI.23; Kochin, *Sel'skoe khozyaistvo*, 71, similarly relates a pair of sokha irons found in Moscow with those in the miniatures.
[2] In *Ocherki russkoi kul'tury*, ch. 1, 66–7. Cp. Chernetsov, *Tools and tillage* II.1.37, 43.
[3] *Drevnerusskie miniyatyury*.   [4] *MISKh*, VI, ris. 5, 6 (s. 32/3).

trators. The estimates of size of any part shown may well be inaccurate, but, even ignoring size, the other variations remain to be explained.

There is also the possibility that the monks who made these illustrations were familiar with a series of implements which differed somewhat from those to be found on non-monastic lands. There seems, at the moment, to be no evidence to support or refute such a supposition. Perhaps the most probable explanation is that the early type of small sokha iron, represented by the finds from Polotsk and from other small Belorussian fortified settlements of the period before and immediately after the thirteenth century Mongol invasions, long continued in use in such locations, while the other excavated parts for which we have details are from larger settlements such as Novgorod and Moscow. It should be stressed above all that many, probably the vast majority, of the implements in use in the period immediately after the Mongol invasions may not have been shod with iron which was presumably mainly reserved for winged shares which had to be of metal. Moreover, such asymmetric shares are likely to have been used on implements entering the soil less steeply than the nearly vertical tines of some implements with sock-like irons shown in the miniatures; the former would be used to make a furrow, the latter would merely scratch the soil.[1]

The miniatures show 29 implements: 19 sokhas, 5 sokha harrows (see p. 25), 3 harrows, 2 ristle ploughs, but no plough. One representation of a wheeled plough (with neither beam nor sole and drawn by three horses abreast) is known, probably dating from the 13th century.[2] Thus, to judge from this particular form of evidence, the sokha was the commonest implement. In these illustrations, it had no board to break up clods or turn soil to the side; the tines were set at a steep angle to the ground and its irons were conical socks, sometimes curved forwards. The coulter, if used, was mounted as an independent implement. More important were a group of harrows and harrow-like implements.

This illustrative material should be supplemented by what little archae-logical evidence of implements there is for the period after the Mongol invasions. Unfortunately, such material continues to be poorly published in many Soviet works; insufficient details are given in the text, illustrations appear without any indication of scale and sometimes it is not possible to be sure which particular find is being referred to.[3] The number of finds of implement parts from the Russian middle ages in general remains small; I have been able to trace a minimum of sixty-seven thirteenth-century or

---

[1] See Levasheva, *TrGIM*, 32.36.
[2] *Radzivilovskaya ili Kenigsbergskaya letopis'*, 1.9. See also Chernetsov, *Tools and tillage*, II.1.36–9; he also illustrates a seventeenth-century tillage scene showing a wheeled implement 'even less realistic and understandable than the Radzivill chronicle drawing' (pp. 39, 42).
[3] See Gorskii's complaint in *Ocherki russkoi kul'tury*, ch. 1, 62, n. 191.

later implement irons; only a minority of these can be classified on the basis of the published reports. This is partly due to the lack of attention to mediaeval archaeology (with the notable exception of Novgorod); it is, therefore, especially important that finds should be published with the fullest possible details. The same holds true, with greater force, for finds from the countryside. In part, but only in part, there are few finds because implements were not usually shod with iron and have therefore perished without trace.

TABLE I. *Implement types found in miniature illustrations* (numbers)

	Date	Ristle Ploughs	Sokhas	Sokha-harrows	Harrows	Source*
1	First half of or mid 16th century		I			GIM, Muz.sobr., no. 358, l.51v
2	First half of or mid 16th century		I			l.68
3	First half of or mid 16th century		2		I	l.99v
4	First half of or mid 16th century		I			l.123
5	First half of or mid 16th century		I			l.141
6	First half of or mid 16th century		I			l.590
7	First half of or mid 16th century		I			l.599
8	First half of or mid 16th century		I			l.1011
9	c. 1550s		2			BAN (Leningrad) 17.17.9, l.1452v
10	Late 16th century			I		GBL, f.98, no.202, l.57
11	Early 17th century			I		l.58v
12	Unpublished			I		l.58
13	Unpublished			I		l.59
14	Early 17th century		I		I	GBL, f.173, no. 103, l.130
15	1640s–1660s	I				GIM, Muz.sobr., no.3822, l.142
16	Second half of 17th century		I			GBL, f.98, no. 1186, l.102v
17	Late 16th to early 17th century			I		GBL, f.304, no. 8663, l.152v

Date	Ristle Ploughs	Sokhas	Sokha-harrows	Harrows	Source
18 End of 17th century		I			BAN 34.3.4, l.187v
19 c. 1669		I			GBL, f.299, no. 236, l.108v
20 17th century	I				GIM, Muz.sobr., no.108, l.181v
21 End of 17th century		I		I	GBL, f.310, no. 446, l.47
22 End of 17th century		I			GIM, sobr. Uvarova, no.867, l.7
23 End of 17th century		I			GIM, Muz.sobr., no.265, l.15
24 End of 17th century		I			GIM, sobr. Uvarova, no.867, l.171v
TOTAL	2	19	5	3	

* f = fond; l = list.

From the available archaeological evidence, insofar as we can interpret it on the basis of published materials, the commonest implement part found has been the sokha iron. Often such parts seem to belong to a two-tined implement which probably had the tines set at a fairly steep angle to the ground. The part often measured between 24 and 35 cm in length, about a third of this being taken up by the socket. These are the approximate measurements of several finds from Moscow and Novgorod. There is some evidence that, at least in fifteenth–sixteenth century Moscow, fairly standard irons were being produced, and this may be linked with the development of three-field rotations at this period.[1] These irons did not have their tips bent forward; this suggests they entered the ground at a less steep angle than earlier implements. Pre-Mongol sokha irons seem to have been between 14 and 26 cm in length, between 5 and 7.5 cm wide and weighed about 650 gm.[2] The earliest irons were small, with lengthy sockets at least as wide as the blade; later they became larger, with shorter sockets narrower than the working part and were often asymmetrical. The weight of later irons is said to have been several

[1] Gorskii in *Ocherki russkoi kul'tury*, ch. I, 68–9. Unfortunately, Chernetsov, *Tools and tillage*, II.1.47 misdates the Moscow irons to the fourteenth century and fails to give the correct reference for the clearly asymmetric sokha iron said to be from sixteenth-century Kursk.

[2] Kolchin, *MIA*, 32.87. Cp. Kir'yanov, *MIA*, 65.348.

times that of pre-Mongol examples.[1] It seems virtually impossible on the basis of available data to say anything about regional variations.

There seems general agreement between archaeological and miniature evidence, then, that the sokha was the commonest implement in Russia after the Mongol invasions of the thirteenth century. As Kochin, on the basis of an immense knowledge of the documentary sources, expresses it: 'written

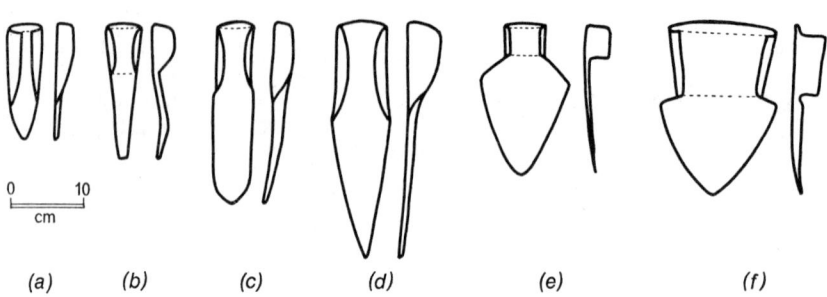

| (a) | (b) | (c) | (d) | (e) | (f) |

Fig. 1. Irons:

(a) Alekseev, *Polotskaya zemlya*, 112, 114; published without indication of size, but this eleventh–thirteenth-century part from Polotsk is said to be from a two-tined implement and similar to finds in Novogrudok, Smolensk and Pskov. The profile has been added and is based on conjecture only.

(b) Novgorod, twelfth–thirteenth centuries; (c) Novgorod, sixteenth century; (d) Moscow, fifteenth century; Kir'yanov, *MIA*, 65.316, 319–20.

(e), (f) Kolchin, *MIA*, 32.87–8; Raiki fortified settlement, thirteenth century.

Notes: In (a) the socket is wider than the working part, round in section and massive relative to the implement as a whole. Type (b) has a socket which is also wider than the working part, but is much shorter. Types (c) and (d) are more massive than type (b). Types (e) and (f) have sockets narrower than the blade and as such are sometimes called winged types.

documents of the thirteenth–fifteenth and sixteenth centuries offer numerous proofs of two-tined sokhas, but the three-tined sokha is completely unknown to them'.[2] The two-tined sokha was the basic tillage implement. 'Written sources of the fourteenth–fifteenth centuries give us sufficient material to talk of the universal use of the sokha in north-east and north-west Rus' and, moreover, as the basic implement for working the soil.'[3] This may perhaps

---

[1] Gorskii in *Ocherki russkoi kul'tury*, ch. 1, 67; he gives no weights.

[2] *Sel'skoe khozyaistvo*, 62. Kir'yanov seems virtually alone in recognising a three-tined sokha as a widespread early implement.

[3] *Ibid.* 72.

not be true in detail, as we have seen, but the general impression is accurate enough. Difficulties come when we try to be more precise and to locate the origin of the sokha and its distribution as the dominant implement in time and space.

Taking the formula 'wherever the axe, scythe and sokha have gone', as indicating the use of a particular implement Rozhkov distinguished areas in which in the sixteenth century the plough was not used at all, where the plough was sometimes used, usually with the sokha, where the two were used equally and finally where the plough was the predominant implement.[1] The objection to this procedure is that the formula was a cliché and should not be taken as evidence for the real situation.[2] It has been pointed out, moreover, that 'for certain typical agricultural uezds, for instance, Vereya, Ryazan' and Yaroslavl', we find not a single instance of either one or the other. It would, of course, be wrong on this basis to assume that the sokha and the plough were not present in these uezds.'[3] Gorskii, therefore, much more cautiously comes to the conclusion that 'it is most probable that the two-tined sokha was the predominant type in the sixteenth century'.[4] Of this there seems indeed to be little doubt.

Doubts arise when we turn to the problem of when and where the sokha originated and how it spread over the east European plain. Technically speaking, it could have originated at any time from the neolithic culture of Tripol'e onwards.[5] There is little evidence to suggest a southern origin, however; no forked implements are known in the Balkan peninsula.[6] Moreover, the distribution of finds suggests that the forest zone was the main area for the sokha.[7] Indeed, Tret'yakov argued that it was mainly a northern implement; he stressed that slash and burn tillage was carried out only with the sokha in the north, especially in Karelia.[8] Naidich-Moskalenko alone implies that it was an implement of large fields and argues that 'the sokha, which appeared with the opening up of large areas cleared of forest, displaced the ard, and on these tilled soils the more productive two-tined sokha replaced it'.[9] Moreover, Tret'yakov maintained that the sokha could scarcely have existed before the advent of the broad bladed iron axe which he

[1] Rozhkov, Sel'skoe khozyaistvo, 111–16.

[2] This error continues to exist: Kochin, Sel'skoe khozyaistvo, 72–3. Kochin also takes terms for implements when used as tax units to represent real implement types: ibid. 43. Naidich-Moskalenko actually cites the Rozhkov passage in Kushner, Russkie, I. 59.

[3] Gorskii, MISKh, III.23. It seems curious that Suzdal' has been omitted from consideration here.

[4] Ibid. 24. See also his Ocherki, 46.

[5] Smith, Origins, 57–9; Map II shows distribution of pre-fourteenth-century implement parts found.

[6] According to Professor Bratanić, in conversation.

[7] See, for example, the very schematic map given by Novikov, MISKh, v.462.

[8] 'Podsechnoe zemledelie', 7, 25, 27, 32.   [9] In Kushner, Russkie, I.58.

dated from the tenth century in eastern Europe.[1] Again, the sokha was 'basically an implement of the feudal countryside, growing up with it and surviving with it almost to our day thanks to the relics of serf relations, long preserved in the Russian village'.[2] Thus, one might conclude that the implications of Tret'yakov's views are that the sokha might have arisen in the northern coniferous forests which the East Slavs were colonising from about 1000 A.D. In an early postwar work, Tret'yakov argued the same case, pointing out that 'the earliest implement irons (soshniki) of sokhas and ploughs in the forest regions are those known in the Volga-Kama district. They are of the period when the Bolgar kingdom was formed. The earliest implement iron known in the Slavonic north comes from the settlement of Staraya Ladoga and is evidently eighth century. Somewhat later, in the eleventh–twelfth centuries, implement irons were widespread. In Slavonic monuments of this period they are found quite frequently.'[3] The evidence available to me relating to finds does not support the latter part of Tret'yakov's statement; relatively few finds of implement irons seem to have been published. Nevertheless the general picture he gives seems plausible, though it must be borne in mind that the term used (soshnik), and indeed other terms for implement parts, continue to be carelessly employed by many Russian authors; this sometimes means that objects are not always distinguished which are in fact dissimilar.[4]

In his first survey of what is known of the appearance of the sokha, Gorskii hinted, but not very explicitly, at its having been in existence before the thirteenth century. He has stressed that a reference to the sokha either as an implement or as a tax unit in 1275 coincided fairly closely in date with two Novgorod letters on birch bark which also refer to the sokha.[5] 'What has been said, of course, does not mean that two-tined tillage implements of the sokha type in general appeared precisely in the thirteenth century.' In his later work he has been more definite: 'both archaeological materials and information from documents lead to a fairly precise time, the second half of the thirteenth to the early fourteenth century. Evidently, approximately from this time it became necessary to separate the two-tined sokha from other types of tillage implements by a special name.'[6]

---

[1] 'Podsechnoe zemledelie', 21.   [2] Ibid. 26.

[3] IKDR, I.56. Kochin regards the Staraya Ladoga find as tenth–twelfth century; Sel'skoe khozyaistvo, 70. Chernetsov, Tools and tillage, II.1.41–3, similarly ascribes the origin of the sokha to the ninth–tenth centuries.

[4] See a similar complaint in Kochin, Sel'skoe khozyaistvo, 52; he regards the lack of clarity as of recent origin.

[5] Ocherki, 45–6; and in Ocherki russkoi kul'tury, ch. I, 59. Tatishchev, Istoriya rossiiskaya, v.51. One birch bark letter, no. 142 (not 78, as Gorskii says), is translated in Smith, Enserfment, 32–3; tree-ring dating makes it after 1299 or 1306. The other, no. 68, is dated 1268–81.

[6] Gorskii in Ocherki russkoi kul'tury, ch. I, 71.

More usefully, Levasheva writes that the evidence 'points to the existence in the thirteenth century of two-share sokhas with asymmetrical, closely set irons, both in the west and in the east of the forest zone'.[1] The finds from Grodno in the west and Pan'zha in the east, leaving out the chance finds from the Moscow oblast (which are somewhat loosely dated as 'earlier than the fifteenth century', or fifteenth century, prior to 1468, or even 'up to the eighteenth century'), indicate that the sokha had had time to spread over a considerable area of the forest zone by the fourteenth century.[2] The two-tined sokha is said to have been common in thirteenth century Lithuania.[3] There is nothing surprising, therefore, in Kir'yanov finding that the two-tined sokha with board was established in the Novgorod area by about 1500.[4] In fact, it seems that this date is much too late.

The area of distribution of the sokha is generally agreed to cover the Slav and Baltic areas very roughly north of the fiftieth parallel in Europe. There appear to be discrepancies in detail, but Moszyński's statement that it is a compact area seems to be valid.[5] The area includes eastern Poland, the north-west Ukraine, Belorussia and Russia in Europe, with an extension into Siberia, as well as east Prussia, Lithuania, Latvia, Estonia, Finland and one small part of Sweden.[6] Within this vast area, there are a number of sub-areas in which particular forms of sokha may be distinguished, for example the south-west, where an implement with a double ground-wrest is character-istic, or where the sokha is found in conjunction with another implement, the saban in the south-east, for example.[7] Moszyński's theory on the sokha is that a single original forked implement led to both one-way and two-way types of sokha, and that therefore the centre of origin is likely to be located in an area where these forks have survived; for example in the western core of Poland, northern Russia, including Poles'e, Belorussia or the Ukraine.[8] It might be objected that the dung-fork, a characteristic implement of sandy soils, is unlikely to have provided the origin for the sokha, which is the characteristic implement of slash and burn cultivation, a system which is un-suited to sands. Perhaps, therefore, an area has to be found where sands met or were intermingled with other soil types suitable for slash and burn in con-ditions favouring the extension of cultivation to new areas.

---

[1] *TrGIM*, 32.37.

[2] Rabinovich, *KS*, 57.89; Rabinovich, in *Arkheologicheskie pamyatniki Moskvy i Podmo-skov'ya*, 76; Rikman, *KS*, 57.89.

[3] *TrIE*, XXXII.5–6; *SE* (1960), 3.94–115.   [4] *MIA*, 65.350, 362.

[5] *Kultura ludowa Słowian*, cz. I, 163.   [6] *Ibid*.

[7] Moszyński's map may be compared with that provided by Novikov in *MISKh*, v.462. Unfortunately both maps are on such small scales that it is hard to make detailed comparison; moreover, neither author provides evidence for his map. Another dis-tribution map of the western one-sided sokha is provided by Professor Bratanić in *Selected Papers*, 226. See also Hensel, *Słowiańszczyzna wczesnośredniowieczna*, 41.

[8] *Kultura ludowa Słowian*, cz. I, 163–5, 144–5.

By careful consideration of the linguistic evidence, especially of that relating to implement parts in the border zones it might be possible to locate a probable origin somewhat more precisely. In fact, some work of this sort has already been done. By means of an analysis of the terms Kustaa Vilkuna has shown that two types of sokha reached Finland in two distinct waves.[1] The first he associates with a borrowing from the Novgorod–Volkov region and may be associated with a new technique of slash and burn tillage based on the new forked implement; this he dates to the twelfth century.[2] Second, the western Finnish type of sokha was found south of the Gulf of Finland and spread into Häme, Satakunta, and eastern Bothnia in the twelfth–thirteenth centuries.[3] This implement had handles raised above the level of the shafts, the rear ends of the shafts being made of roots sticking up like a crane's neck (hence *kurki, kurjet, kured*). This appears to be the type of sokha in which in Russia the shafts are called *kryuchki* (hooks).[4]

The present state of the evidence, then, suggests that the sokha developed in the forest zone of European Russia, though not necessarily in the coniferous part of that zone, even though the approximate coincidence in dates may indicate that it spread rapidly in the northern part of the forest zone and extended into Karelia and across the Gulf of Finland in association with the colonisation of the coniferous forest by the Slavs.[5] Moreover, the fact that by the twelfth–thirteenth centuries at least two types of sokha had become localised and distinct implies that the origin of the implement had occurred some time earlier. Thus, views such as those of Zelenin, who argued that the implement was 'a Great Russian invention related, moreover, to that distant period when Belorussians and Great Russians had not finally separated out', i.e. to the fourteenth or fifteenth century, are seen to be mistaken in ascribing so late an origin.[6] Zelenin's case seems, at least in part, to have been based on the mistaken belief that the word *sokha* only occurs in the sources from the fourteenth century. Similarly, Levasheva's argument that the two-tined sokha was still in process of establishing itself in the Yur'ev-Pol'skii area in the sixteenth century is, as Kochin has pointed out, based on a misinterpretation of a text.[7] In another passage, she points out that a sokha with asym-

---

[1] *Die Pfluggeräte Finnlands* (Vortrag auf dem II. Finnougristenkongress in Helsinki am 24 VIII 1965), 17; *Die Pfluggeräte Finnlands, Sonderdruck: SF*, 16 is a more extended discussion of his subject matter.

[2] Vilkuna, *Die Pfluggeräte Finnlands*, 18–20. See also his map on p. 35.

[3] *Ibid.* 17–18, 21.

[4] Zelenin, *Russkaya sokha*, 36–7.

[5] This view finds some support in the fact that the two pairs of asymmetrical sokha irons dated to the eleventh–thirteenth centuries found in the Kostroma and Vologda regions are both within the coniferous zone. Levasheva, *TrGIM*, 32.31, 35; Kochin, *Sel'skoe khozyaistvo*, 70.

[6] *Russkaya sokha*, 121; see also 181.

[7] Levasheva, *TrGIM*, 32.26–7; Kochin, *Sel'skoe khozyaistvo*, 61.

metric shares and a spade existed in the thirteenth century.[1] It seems, therefore, that the sokha was a widespread, and most likely the dominant, implement, certainly on peasant farms, throughout the area here considered from the time of the Mongol invasions of the thirteenth century. By the fifteenth–sixteenth centuries irons, when used, had become much larger than in pre-Mongol times, probably entered the soil at a less steep angle and, at least in Moscow, seem to have been produced to a fairly uniform pattern. These changes are probably to be associated with the spread of three-course rotations on fields rather than closes (*nivy*), at this period.

The other tillage implements which were in use, as has been seen from the evidence of the miniatures, included the ristle plough, a knife or coulter mounted on a beam; it would presumably have been used in conjunction with some other implement, either sokha, plough or an intermediate type capable of turning a slice. Little can be said of this implement; its infrequent occurrence in the miniatures tallies with what would be expected in a forested area. The known examples are from the seventeenth century, a time when forest clearance had progressed sufficiently for there to be areas where natural glades and man-made clearings and wastes were extended or incorporated into the regular arable fields, and at the same time turfy *chernozems* were being cultivated in addition to the forest *podzols*.[2] Such grassy cover would call for the use of a ristle plough or a coulter to open up the soil beneath it.

The illustrations show, first, a curious curving prong, broad at the top and narrowing to a point, without any iron indicated.[3] Gorskii interprets it as a single-tined ard or sokha, a *cherkusha,* or a simplified drawing of a *kosulya;* he argues that the 'comparatively great height of the implement' suggests it may be a ristle plough, but according to my estimate, this implement is the lowest of all, being only about 55 cm high.[4] In fact it is this lowness of the implement which suggests it may indeed be something like a coulter mounted on a sokha frame; it has been estimated that early coulters gave a clearance of only 35–50 cm between soil and beam.[5] The second illustration of a single pronged implement Gorskii interprets as a *kosulya.*[6] The original is very small, 2 × 3 cm and insufficiently clear to be interpreted reliably. Gorskii claims that the tip entered the ground almost horizontally; the reproduction, however, shows something which looks as if it curves upward. It may be that the dark line Gorskii takes as the forward curve of the tip is something else and that we have here a slightly forward raked coulter-like object, apparently saw-edged. Again, the height of the implement appears to be low.

The question of the single-clawed implements like the ristle plough and of their relationship to the *kosulya,* a fixed board soleless plough, seems to have

---

[1] Levasheva, *TrGIM*, 32.37.   [2] See p. 130 below.
[3] Gorskii, *MISKh*, vi.30; this refers to table 1, no. 15.
[4] Cp. *Ocherki russkoi kul'tury*, ch. 1, 77.   [5] Smith, *Origins*, 100.
[6] *MISKh*, vi.31; this refers to table 1, no. 20.

received little attention. Zelenin lists the following terms for such implements: *chertezh, chertets, razrez, rezets, otrez,* but mentions only one of these, *otrez,* as old.[1] In the early seventeenth century, the tsar Aleksei Mikhailovich had 1130 *otrezy* in a store at Izmailovo, but only 400 sokhas.[2] This suggests that at least in some circumstances in the seventeenth century such implements were of importance. Possibly they were of particular importance in estates which were opening up new, especially turfy ground. A number of implements or implement parts of this type were mentioned in monastic estate books.[3] They included *otrez, chertets, rezets, chertezh* and a term not found elsewhere, *ponada.* Thus, all the terms Zelenin listed, save *razrez,* are found in seventeenth century sources. It seems possible, however, that some at least of this group of terms (*otrez, chertets, chertezh,* for instance) refer not to whole implements but only to a part, a coulter or a knife of some sort. This would mean reinterpreting the figures given for implements in Aleksei Mikhailovich's Izmailovo store. That such terms were used of implement *parts* is shown at least for the Northern Dvina area by a will of 1570 and a monastic account book of the mid sixteenth century. The testator, evidently a man of substance, left 'a plough with coulter and share [*s chertsom i s lemekhom*], a sokha with ard irons and another plough' to a monastery. The monastery's blacksmith 'forged two shares and two coulters [*dva chertsa*] and ard irons'.[4] Clearly in this context coulters, and not implements, are indicated.

Zaozerskii, in his impressive study of the tsar's estate, lists numbers of implements sent out in 1663 as follows.[5]

Sent to:	True ploughs	Ploughs without soles
Dedilov	100	500
Skopin and Romanov	250	50
Domodedovo	300	300
Izmailovo	100	200
Chashnikovo	100	100

He concluded that the true plough tended to be more used in the southern uezds while towards Moscow the ploughs without soles were able to compete with it. Nevertheless there is the curious report of John Tradescant the Elder, who visited the mouth of the Northern Dvina in 1618 and was a good

[1] *Russkaya sokha,* 63. In addition, in the nineteenth century, the term *rezak* was used; Naidich-Moskalenko in Kushner, *Russkie,* I. 57. All these terms derive from roots indicating scratching or cutting.

[2] Zelenin, *Russkaya sokha,* 64, citing *Zemledel'cheskii zhurnal Imp. Mosk. Obshch. sel'sk. khoz.* (1821), no. II, 152. I have been unable to find any reference to this report in Zaozerskii, *Tsarskaya votchina XVII v.*

[3] Kochin, *Sel'skoe khozyaistvo,* 76.

[4] *SGKE,* 1.181, col. 185; AOIA, f. 60, op. 1, d. 48, l.4 etc.; both cited by Reznikov, *MISKh,* IV.83.   [5] *Tsarskaya votchina XVII v.,* 103, based on *AAE,* VI, no. 138.

observer. He stated that 'ther land, so much as I have scene, is for the earable fine gentill land of light mould, like Norfolke land, without stons; ther maner of plowes like oure, but not so neat, muche like to Essex ploughes, with wheels, but the wheels very evill made'.[1] Probably in the coastal plain and similar areas, the plough was well established by the seventeenth century.

The origin of the plough without sole (*kosulya*) is of some importance since it has been argued that there was a connection between the appearance of this implement and the rise of Moscow.[2] Gorskii has quite correctly refuted the argument, which was without any evidence to support it.[3] The term *kosulya* seems to appear early in the seventeenth century. The Saviour of Priluka monastery's account book for 1608–9 refers to the purchase of sokha-kosulyas and kosulya-like sokhas.[4] What is not known, however, is whether the implement itself appeared at this time or whether it had already been in use for some while. Moreover, it seems most likely that the seventeenth-century *kosulya* was a form of fixed-spade sokha which cut and turned a slice, but not the single-share implement; the latter seems only to have emerged around the middle of the eighteenth century.[5] This means that archaeological evidence is unlikely to help us to a solution. As Zelenin put it 'there is decidedly no evidence on the past of the *kosulya*. We have here great scope for personal guesses.'[6] It may have originated around 1600 when, as we have seen, it is first mentioned in the available sources, but it seems to have been mainly an implement of the great estates, not a peasant implement, and possibly it should be associated with the ristle ploughs and clearance and the expansion into new, grassy areas.

The ard (*ralo*), a hook or stave scratch plough with a single prong, is little mentioned in the sources with which we are concerned; but the word often occurs in a term for the basic land tax (*poral'e*) in the Novgorod territories, especially the Onega and Northern Dvina areas.[7] It is found mainly in clerical sources and in translations. A fourteenth–fifteenth-century Pskov charter, however, uses the term ard in a way reminiscent of the triple formula when it lays down that intruders are not to enter the land 'with either ard or scythe'.[8] Indeed, it seems probable that simple scratch ploughs of this type continued in use among peasants, though evidence is hard to come by since such implements might well be entirely of wood and therefore unlikely to leave anything to be found by the archaeologists. On the other hand, it may be that in this area the term *ralo* referred to the two-pronged implement usually known as the sokha; this is suggested by the fact that there are Finnish dialects where the sokha is called *atra*.[9]

---

[1] Cited in von Hamel, *England and Russia*, 276.   [2] Smirnov, *VI* (1946), no. 2–3.
[3] *Ocherki*, 40–1.   [4] Kochin, *Sel'skoe khozyaistvo*, 72.
[5] Zelenin, *Russkaya sokha*, 77.   [6] *Ibid.* 73.
[7] Danilova, *Ocherki*, 57, 65; Gorskii in *Ocherki russkoi kul'tury*, ch. 1, 81–2.
[8] *NPG*, no. 11.   [9] Vilkuna, *Die fluggeräte Finnlands*, 16.

There is also relatively little evidence about the plough. This is understandable; the true plough cuts and turns a slice and was unsuited for use in many areas because of the forested nature of the land. It was found in exceptional areas such as the Dvina mouth.[1] Where grass cover had to be dealt with, the ristle plough or soleless sokhas or ploughs were able to cope and were less demanding of metal and draught power and less costly.[2] The documentary evidence shows that in terms of tax units the plough (*plug*) was estimated as the equivalent of six *obzha* units.[3] The *obzha* was 'one man tilling with one horse' and this was evidently the usual peasant holding in the Novgorod territories in the late fifteenth century.[4] Kochin points out that not a single peasant tenement was valued at a plough among all the thousands listed in the Novgorod registers of inquisition.[5]

Tenements and obzhas in the Novgorod lands*

Pyatina	Tenements	Obzhas	Obzhas per tenement
Derevskaya	17,348	18,610	1.07
Vot'	11,069	14,701	1.33
Shelon'	5,597	8,104	1.45

* Based on Kochin, *Sel'skoe khozyaistvo*, 319.

In fact, if we consider the evidence for ploughs we can see they were probably restricted mainly to richer estates, especially monasteries in the central areas of the Moscow state. On more open lands, such as Ryazan' and the area of Opol'e in the Rostov–Suzdal' principality, there have been a few twelfth–thirteenth-century finds (one coulter from the Suzdal' kremlin and two winged shares and two coulters, possibly from ploughs).[6] Documentary evidence, however, may call in question whether the plough, even on richer estates, became noteworthy much before about 1400. It is true that a *yarlyk* or authorisation from the Khan Mengu-Temir and dated 1267 refers to 'any tribute or tax per plough' (*po pluzhnoe*) and to the collectors of such tax (*popluzhniki*).[7] These references, however, are from a late-fourteenth- or early-fifteenth-century Russian translation from the Uigur original.[8] In some supplementary articles to *Russkaya Pravda*, usually dated to the end of the thirteenth or early fourteenth century there is a reference to rye sown 'on two ploughs' (evidently a land measure).[9] Again, however, it should be borne in

---

[1] See above.
[2] Cp. Zelenin's rejection of Efimenko's and Rozhkov's views on the importance of the plough in the 16th century in *Russkaya sokha*, 127–8.
[3] *PSRL*, xxv.319–20. *GVN*, no. 21 (1448–61), takes two tax-unit sokhas as equal to a plough, but this sokha has two horses and a third available.
[4] Gnevushev, *Ocherki*, 102; Andriyashev, *Materialy*, i.LXIV; Cp. Danilova, *Ocherki*, 24 and appendix II.   [5] *Sel'skoe khozyaistvo*, 74; also 58.
[6] Levasheva, *TrGIM*, 32.32, 34; Mongait, *Ryazanskaya zemlya*, 258–9.
[7] *PRP*, iii.467–8.   [8] *Ibid.* 464.   [9] *PRP*, i.207, 217.

mind that this section of the supplementary articles is from a fifteenth- or sixteenth-century manuscript.[1]

The fifteenth-century evidence seems to give a clear picture. A will of the 1440s mentions ploughland (*pluzhishche*) worked by a slave family on a heritable estate.[2] At about the same time a gift from an educated landowner, who also had slaves, included 'five oxen and two pairs of plough *topeki*'.[3] The term 'plough' was used as a tax unit around the middle of the fifteenth century both in Novgorod, in its relations with Moscow, and on the lands of the rich Trinity monastery of St Sergius.[4] The Novgorod reference has been taken as evidence of the use of the plough on the Novgorod peasant farm, even though 'it is true that the plough was, evidently, a comparative rarity in the mid fifteenth century'.[5] The text does not seem to justify this interpretation. In fact, Gorskii is probably right to stress that there is no evidence for the plough in the basic Novgorod territory at this time, though it seems to have been found in the Dvina area.[6] Finally, in a lawsuit about 1495-7 monastic officials referred to 'tilled plough land', but the peasants in the case significantly omitted any such mention and the justice tried to establish whether there was any 'tilled plough *or sokha* lands' there (my stress).[7] This all suggests that wealthier, better organised estates, often using slave labour, might have ploughs, but also implies that such implements would not be characteristically used by peasants.

Sokhas with more than two tines (here called sokha-harrows) and other harrow-like implements were probably of greater importance than ploughs at peasant level. Such multi-tined sokhas should be treated as harrows since, unlike the two-tined variants, they could not be tilted in use so as to clear a furrow; they could only be used to scratch and scuffle the soil.[8] The miniatures considered earlier show four three-tined sokhas and three harrows (table 1, nos. 10, 11, 12, 13 and 3, 14, 21) which thus formed the largest group other than the sokhas. The curious massive three-tined sokha from the Life of Sergii of Radonezh has been discounted from these calculations.[9] The other three-tined forms are difficult to understand. The implements are such that they would have to be held in an upright position while working, yet the horse is

---

[1] Tikhomirov, *Posobie*, 17-18.   [2] *ASEI*, I, no. 228.

[3] *ASEI*, III, no. 490. This gift appears to be the same as that recorded in a late seventeenth-century inventory of the monastery receiving it, even though the placename has been read differently (Borzhovskoe, instead of Borisovskoe); *ASEI*, III, p. 481. It is not known what *topeki* were.

[4] *GVN*, no. 21 (1448-61); *ASEI*, I, nos. 260, 261.   [5] Danilova, *Ocherki*, 32.

[6] Gorskii, in *Ocherki russkoi kul'tury*, ch. 1, 78-9; *LZAK*, xxxv, no. 57, 159 (1495-6).

[7] *ASEI*, III, no. 209.

[8] Gorskii in *Ocherki russkoi kul'tury*, ch. 1, does not make this distinction.

[9] No. 17. Gorskii, *Ocherki*, 42-3 is sceptical about it; Kochin, *Sel'skoe khozyaistvo*, 56-8, regards it as unreal and conventional. It is taken seriously by Chernetsov, *Tools and tillage*, II.1.46, 48, as regards harness.

saddled as if to be ridden, which was usual for harrowing. These implements, then, might have had two workers. Possibly this is a conventional indication of harrowing, though the implements show enough detail to seem convincing.

The true harrows are of two types. The first consists of a number of split fir logs taken from the trunk with projecting branches left on.[1] The traces are attached to the sides of the implement, which has its planks bound together by an interwoven withy. The second type consists of a frame made up from sticks which are bound together into a grid pattern; at the intersections spikes are inserted. The one example of this type we have in the miniatures (table 1, no. 14; plate 6) has a bow attachment at the front from which the trace runs to the horse. This implement has thirty-six prongs, like more modern examples which are usually squares with thirty-six or twenty-five prongs.[2]

The three miniature illustrations showing harrows in use depict them working in seed; or at least the harrowing is taking place after seeding. All the scenes also show the sokha in use. Of the three-tined implements, two at least (possibly four) are drawn by saddled horses. The well-known massive three-tined implement shown in the Life of Sergii of Radonezh does not have this probable indication that the implement was of the harrow type. It is possible that some of these implements were used in slash and burn. The vegetable shown above the tillage scene (plate 2; table 1, no. 10) might be the turnip (*Brassica rapa* or *Brassica narus*) which was often cultivated in the north on burnt patches.[3]

Little work seems to have been done on field systems and layout in medieval Russia. We know virtually nothing of the work units on which the implements described above were used. No doubt at some time the work will be done and there are already a few indications that Russian historians are following archaeologists and ethnographers in investigating problems *in situ*. For the moment, however, it is difficult to establish any clear picture of the village or other settlement as an economic unit. We are forced to try to make good this gap in our knowledge by examining documents, not places. This is not a satisfactory procedure. Evidence derived from the terminology of one document may not apply in other localities or at other times and should, in any event, be checked in the field. The picture elicited from the examination of a single document, the register of Toropets of 1540 which happens to include a fair number of field and settlement terms is not one which stresses a traditional three-field system.[4]

Even though the sixteenth-century officers of the inquisition make their esti-

[1] Table 1, nos. 3 and 21 seem to be of this type.
[2] Eighteenth-century evidence for 25- and 36-toothed implements is given by Reznikov, *MISKh*, IV.84; for nineteenth–twentieth-century evidence, see *Khozyaistvo i byt*, 24–6 and ills. 4 and 6; *Russkie*, I.54–6.
[3] See below p. 39. Kir'yanov, *MIA*, 65.357.
[4] 'Toropetskaya kniga' published in *AE za 1963*, 277–357.

mates of estates held by service in terms of a three-field distribution of land (so much in one field, 'and in two at the same rate'), in reality it is various individual fields, patches and clearances in the forest which predominate. A normal lot of peasant land may have been about 30 chets (about 16 ha), but closes seem to have been about 2 or 3 chets (1 or 1½ ha) in size and a strip 1 chet (½ ha).[1] It is probable that this held good even on large estates; for the peasant holding with which we are concerned a three-field distribution appears still more unlikely in most cases.

The term 'three-field system' in Britain seems to have little precise significance; even if a settlement, or manor, had three arable fields, within this framework there could still be wide variations in cropping and fallowing practice; in particular, the number of units in a crop rotation need not correspond with the number of fields.[2] The Russian term *trekhpol'e* as used by Russian historians suffers from at least as great a degree of imprecision. In his extremely interesting book on the economic situation of the peasants in fourteenth–fifteenth-century north-eastern Rus', Gorskii avoids this error; he regards three-field as 'a fallow grain system with a three-field rotation', i.e. what we would term a three-course system.[3] He also has the virtue of stating that, 'taken by themselves, mentions of spring- and of winter-sown grains, and of fallow, cannot be taken as unquestionable proof of a fallow-grain system in the form of three-field'.[4] He concludes that 'indications in the sources of tilled land being divided into three fields are the weightiest evidence'.[5] Such evidence can be adduced from documents of the second half of the fifteenth century, but Gorskii has argued that the content of some documents suggests the presence of a three-course system even at the beginning of the fifteenth century, though he points out that it was not absolutely dominant, nor the sole cultivation system.[6]

In his important book on farming in Rus' from the end of the thirteenth to the early sixteenth centuries Kochin makes much more exaggerated claims. He fails to define his use of 'three-field'; in fact, he appears to accept evidence for spring- and winter-sown grains, and especially for fallow, as enough to prove the existence of a three-course system. More than this, though, he claims, on the basis of two documents, that 'this is the date (mid-fourteenth century) we should give to the evidence of three-field in these villages, moreover, a three-field which had become usual by that time'.[7] Neither document, in fact, gives proof of the existence of a three-field system. The first, a will dated 1392–1427, includes the following: 'As to my grain, whether rye or spring-sown, it is to go to them according to their lots. As to the rye they are

---

1 Smith, *Forschungen zur osteuropäischen Geschichte*, 18.125–37.
2 Butlin, *AHR*, IX.II.101–2.
3 *Ocherki*, 31. Cp. his section in *Ocherki russkoi kul'tury*, ch. 1.
4 *Ocherki*, 32.   5 *Ibid.*   6 *Ocherki*, 33, 37.
7 Kochin, *Sel'skoe khozyaistvo*, 158.

now to sow in the earth, that is to go to them half each.'[1] The second document deals with an exchange of lands 'with the grain which is in the earth and on the fields, rye and spring sown, and stocks of hay...'[2] Kochin regards this as 'a completely clear picture of an economy with three fields: on the first field is the sown winter crop, rye (grain in the earth), on the second and third the already harvested spring-sown crop and (winter) rye'.[3] Moreover this picture should be generalised for the central part of north-eastern Rus', at least by the early and middle years of the fifteenth century.[4] It is extended to peasant lands by the simple device of taking references to winter-sown and spring-sown grain stocks on estates as evidence for peasant production, which may in itself, of course, be true.[5] Moreover, the rural settlements of hamlet type had 'large plots of field, of tilled land'.[6]

The picture which emerges from an examination of evidence relating to three-field layout in the late fifteenth to early sixteenth centuries is one of problems arising largely as a result of continuing internal colonisation in a forest area with small, widely dispersed settlements and low population densities.[7] Monasteries and some landed proprietors, metropolitans, princes, lords and, in one case, a group of peasants, held land organised in three fields in the second half of the fifteenth century, according to the documents considered. In three cases witnesses claimed to remember up to seventy years back; if this testimony was accepted there would be some evidence for a three-field layout in the early years of the fifteenth century. The bulk of evidence, however, suggests such a layout from about the 1460s, increasing greatly from the 1480s. This dating fits well with evidence of other changes, such as the development of the sokha already dealt with. Differences of opinion both about the administrative subordination of particular areas of land and often about the nature of the worked areas are largely resolvable in terms of a conflict of systems. The officials of the black peasants, who were sometimes organised, possibly in communes, locally disposed of the Grand Prince's lands among themselves; these they worked in units known as hamlets. Cultivation was often intermittent and recalls outfield, rather than anything akin to a three-course rotation on one set of three fields, to judge from the lengthy periods before disputes were taken up.[8] Except for one case (appendix 1, no. 9), we have no mention of peasants having three fields. The mid-fifteenth-century picture of the ploughman (*ratai*) 'not painted on the wall, but tilling on the close' remained characteristic at peasant level into the

---

[1] *ASEI*, I, no. 11.   [2] *ASEI*, I, no. 58.   [3] *Sel'skoe khozyaistvo*, 157.   [4] *Ibid*. 173.

[5] *Ibid*. 167f. Kochin, however, makes this very criticism of Kaufman's views on pp. 182, 184.

[6] *Ibid*. 239. Cp. 187–8 for Novgorod.   [7] See appendix 1.

[8] The 1497 Law Code prescribed a limitation on cases of disputed lands, six years for land of the Grand Prince, three years (the period of a rotation) for others. See translation in Dewey, *Muscovite Judicial Texts*, 20.

sixteenth century.[1] There seems, then, to be little explicit evidence for Kochin's hamlets with 'large plots of field, of tilled land'. The larger proprietors, servitors, whether noble or not, and especially the monasteries, were organising and rationalising, as well as extending, their lands at this period. We have evidence of such lands being reorganised and made into distinct and compact blocks (appendix 1, nos. 2, 16 and 19). It is on such lands that we find that the 'third fields' characteristically occur.

But what were these third fields? The evidence we have suggests they might vary in estimated size from one tenement to thirty or so, or from six to fifty chets; they might be lands, wastes, sites, clearances, hamlets or ryefields; they might include forest and they could be relocated. In fact, the general impression derived from these documents is that by no means in every case does reference to three fields mean that there was a three-course rotation regularly organized on fields of similar size, nor that such fields were always under frequent surveillance. It was only on certain lands of the metropolitan and the Grand Prince that the land seems to have been measured at this time (nos. 6, 11, 12, 13). Moreover, some of these 'measurements', perhaps most, were estimations, in terms of seed grain, rather than precise surveys made with a line. In most cases part of an estate was regarded as the area of the third field; this might have on it genuine fields, open tilled arable, but more often it indicated anything from new clearances, wastes or sites often used for hay, to long-established hamlets.

It may be that such third fields were a transitional stage in the emergence of more regularly organised estates of which monasteries give us many examples.[2] The field in the transitional stage would then perhaps have indicated no more than the area from which exactions of a particular crop were to be made in any one year. The crop itself would be grown by dependent peasants on their traditional clearances and patches of every sort in the forest or on more regular fields where that was possible. But of this internal life we know little.

Another hypothesis is that, as in Lithuania before the mid sixteenth century, even three-field layout, when it existed, might be used with three courses (winter-sown, spring-sown and fallow) but in a two-field system, i.e. one of the three fields was sown with both the winter- and spring-sown crops.[3] Indeed, the Lithuanian and Polish evidence suggests that the three-field system with a regular sequence of winter-sown, spring-sown and fallow on fields of approximately the same size is unlikely to be much earlier than the sixteenth century in Russia, even on the estates of lords.[4] Earlier evidence of three fields,

---

[1] 'Nakazanie otsa k synu' in *Issledovaniya istochnikov po istorii russkogo yazyka i pis'-mennosti*, 290.

[2] A similar view of three-field layout in Lithuania is taken by Yurginis in *EzhAI* (1962), 95–100.

[3] *Ibid.* 99.

[4] *Ibid.* 95–100; Łowmianski, *Studja nad początkami społeczeństwa i państwa Litewskiego*, I.214–19; Gieysztor, *Prace z dziejów Polski feodalnej*, 71–9.

or of two crops and fallow, refers to what Łowmianski called archaic three-field, a three-course sequence, not necessarily in strict rotation.[1] The spread of a three-field layout in the last two decades of the fifteenth century would, therefore, be transitional between what might be called close farming – the use of a great variety of natural and man-made forest clearances – and a regular sequence of three courses on each of three fields of approximately the same size. The latter involved a degree of forest clearance and control of resources more often found on larger units, as we have seen.

The peasants in a volost or a village sometimes tilled or mowed over another peasant's boundary, to judge by the 1497 Law Code (article 62), and this might indicate intermingled strips in some cases; perhaps that is why there is no mention in this article of the more usual, though smaller, hamlets. Our records in the main reflect conflicts over disputed boundaries between holdings or estates and it is not till the seventeenth–eighteenth centuries that, in the north, we have detailed accounts of village arable from the field registers (*verevnye knigi*) which notes courses (*smeny, peremeny*) divided into fields (*polya, kony*), each tenement having a portion in every field of every course and in each meadow, fallow field and areas beyond the fields (*zakraina*). Distant clearings and new patches in the forest, not included in the rotation, were held similarly.[2] Such registers survived in the north together with the free peasantry there; but it seems likely that similar arrangements existed in other areas, wherever settlements were of sufficient size, prior to the absorption of the black peasant lands into service and heritable estates. In order to distribute land in this manner it had to be measured; this was done by volost officials with a rope of 64, 32 or 16 sazhens (136, 68 or 34 m).[3] References to equalising in the sixteenth–seventeenth-century documents are concerned with the allocation of tax burdens in terms of land held, not with land re-distribution.[4]

For the central areas we know that the recording of land by the officers of the Moscow prince was only gradually achieved. In the fourteenth- and fifteenth-century deeds peasants had to pay tribute and certain other main taxes 'according to their strength' (*po sile*), whereas the sixteenth-century formula was 'according to the registers' or 'according to what the officer of the inquisition records'.[5] Veselovskii stressed that 'both in the fifteenth and sixteenth centuries, as well as in the seventeenth, the basic principle of tax in sokha-units and by vyts was according to strength';[6] but this vagueness

---

[1] Cp. Veselovskii, *IGAIMK*, 139.32.

[2] Efimenko, 'Krest'yanskoe zemlevladenie', IV.214; Dovnar-Zapol'skii, *LZAK*, xv.

[3] Efimenko, 'Krest'yanskoe zemlevladenie', 247. In the seventeenth century the Service Tenure Department usually issued ropes of 80 and 30 sazhens; Sedashev, *Ocherki i materialy*, 42. The standard desyatina in the sixteenth–seventeenth centuries was 30 × 80 sazhens; the chet was half this area; Veselovskii, *Soshnoe pis'mo*, II.363.

[4] Efimenko, 'Krest'yanskoe zemlevladenie', 245; Veselovskii, *Soshnoe pis'mo*, I.286.

[5] Kashtanov; unpublished communication.   [6] *Soshnoe pis'mo*, II.79.

reflects the fact that officers of the inquisition, even in the early sixteenth century, had no right to enter old-established lands unless a court case arose. This distinction between old lands and the new ones liable to inquisition emerged in the second half of the fifteenth century. The Grand Prince was only able to establish the right to carry out inquisitions on the great boyar and monastic estates after the internecine wars of the second quarter of the fifteenth century. The function of the first officers of the inquisition was evidently largely limited to establishing boundaries and arriving at decisions about land use.[1] All this was part of the development of the administrative system as Moscow's power grew. Although the taking over of black peasant lands and enserfment of the peasants had gone far in the intervening century, it was the reforms of Ivan IV in the middle of the sixteenth century which led to a fully developed system of land allocations to servitors. These were always in terms of a three-field notation.

The evidence of the 1589 Law Code shows that at peasant level, on the one hand, hamlets might have fields with intermingled arable strips allocated among their inhabitants, as well as hayfields and other appurtenances, all shared and liable to adjustment in terms of some standard allocation (§§159, 160); such arable may have been organised in a three-field rotation. The villages and hamlets sometimes had arable contiguous with that of other settlements (§ 172) and the bounds were sometimes marked, presumably by a furrow or balk, sometimes by a fence (§§ 172, 168, 169).[2] Abandoned individual arable overgrown with young trees was at the disposal of the settlement (§ 174). On the other hand, the Code also shows that distant plots were tilled, both those within walking distance (§ 171) and those more distant (§ 161); the latter were permitted to be tilled in this way for only three years and then had to be sold or purchased and fenced. Presumably this was intended as a reform to avoid land disputes between settlements and to bring additional areas into the taxable category. Thus, we have a picture of increasing regularisation of peasant arable, especially within the settlement or commune, but also of a mixed economy: regular tilled fields by the settlement being combined with the continued existence of some tilled areas in the forest.

It was not until the 1620s, however, that detailed instructions on surveys carried out in order to distribute estates to servitors seem to have been issued to the officers of the inquisition.[3] These, inaugurated probably as a result of dissatisfaction with the early arrangements, combined with the need for extensive rearrangements in the aftermath of the Time of Troubles, remained in use until the 1670s. Sedashev has calculated that even in the early seventeenth century the areas surveyed were too great to be measured in the time taken.[4] So areas were probably determined, as also in the previous century, by measuring one field, usually the fallow field, using local testimony for the size

[1] See *AGR*, I.70; Shumakov, *Obzor*, II.80–116. [2] *Sudebniki XV–XVI vekov*, 402–4.
[3] Sedashev, *Ocherki i materialy*, 19. [4] *Ibid.* 16–17.

of the other two and checking against the officer's own impression and estimation by eye.[1]

Thus, the three-field formula (seen in appendix 1, no. 10) which came into use in the late fifteenth century and became the regular notation for the sixteenth-century inquisitions should not of itself be taken as an indication of a precise, fully developed three-field system with compulsory rotations and repartitions even at the level of the estate or manor. There could be, and no doubt were, many variations of farming practice within that notational framework.

It is certain that even in the 16th century, when the Moscow state had not yet acquired the rich black earth of the south and the east, there was a great variety of methods [of cultivation]; in the 17th century this would increase, but none of this contradicts the view that the accounting of land in three fields arose on the basis of a three-course rotation [*trekhpol'nago sevooborota*]. This method for the measuring and notation of land arose and became established in the central parts of the state, where three-course rotation was the dominant system; it was then extended by the Moscow officers of inquisition to areas of the state where the agricultural set-up deviated more or less from typical three-field forms.[2]

It should be added that our limited sources suggest that even the notational three-field only gradually extended from the great estates, especially those which were well organised, to the smaller ones. The spread of a three-field system at peasant level was even slower where most settlements were small and land, including forest, abundant.

There were, of course, variations by area. In places where the land was free of forest, and settlement was relatively dense, a three-field layout may have developed among some peasants by the sixteenth century. For much of European Russia, however, especially the central and northern parts, settlements were small and not nucleated. The pattern is reflected in some of the earliest estate maps known for Russia, which probably date from the second half of the seventeenth century. They show a very schematic three-field layout around the central settlement, itself often little more than a manor house and church with outbuildings, while the surrounding forest is pitted with wastes, each of which bears the name of its initiator or holder.[3]

In very general terms, the major zonal types of rural settlement now distinguished have their roots both in natural conditions which mediate most agricultural activity, and in the history of colonisation going back many centuries.[4] An extreme view is that 'the majority of collective farm settle-

---

[1] *Ibid.* 28.   [2] Veselovskii, *Soshnoe pis'mo*, II.365.

[3] See Goldenberg in *PI*, VII.296–347. I am grateful for the help given me by V. N. Shumilov, former Director of the Central State Archive for Ancient Documents, and especially for his permission to make use of the estate maps reproduced here (plates 10, 11).

[4] Kovalev, *Sel'skoe rasselenie*, 154–75, discusses current zonal types of settlement for the areas which are dealt with here.

ments in the country's long-settled regions exist in the same places where they once arose and were noted in the fifteenth–seventeenth-century registers of inquisition and enumeration, largely preserving the former picture of settlement'.[1] This may be doubted, but the northern area remains an area of small, dispersed settlement. In the central areas the consolidation of the Moscow state, based on grants of land and peasants to servitors, led to immense changes. Free peasant organisation virtually disappeared. The peasant tenements, whether paying labour rent or rent in kind or in money, became part of a larger administrative or management unit which was organised in terms of a three-field notational system and actually had three fields at the central settlement, often combined with a variety of other forms in the rest of the manor or estate. Such small out-settlements were liable to shift their location from time to time.

The main crop grown on field or patch was rye, usually winter sown. Rye in general was known as *rozh'* (with variants *rezh'*, *rzhina*, *rzhitsa*, *rzhichka*); spring-sown rye, however, was called *yaritsa*.[2] It is almost always listed first when grains are mentioned in the documentary sources. It was found throughout European Russia, though in the more northerly areas it was replaced by barley. Throughout most of the area, however, rye was the basic grain and has remained the staple into modern times.[3] It was clearly a peasant crop at all times. In 1229, for example, 'rye did not grow throughout the whole of our land and grain was dear'.[4] In 1435 in Pskov 'frost killed the rye and there was a loss of peasant grain'.[5] An entry for 1468 states that 'there was heavy rain, because of our sins, when the peasants reaped the rye in July'; the rain continued for four months and was so disruptive that 'not much rye was sown in the villages... and the peasants were in much need'.[6] The numerous references to rye in other regions make it most likely that these quotations relating to Pskov are not exceptional.[7]

Barley, due to its increased frost resistance, was grown further north. In 1466, for example, 'the spring grain was all frozen, except for the barley'.[8] In the northern parts of the Novgorod territories, around Lake Vyg to the north of Lake Onega, along the Northern Dvina, and in the White Sea area barley seems to have been the main, sometimes the only, grain.[9] It was also found in more southerly areas such as Radonezh, Moscow, Tver', Maloyaroslavets, Yaroslavl', Kashin, Kostroma, Vladimir, Suzdal', Klin, Nizhnii Novgorod, Toropets.[10] Evidence for barley taken from documentary mention

[1] *Ibid.* 107. [2] Antropovy, *Rozh'*, 5. [3] Antropovy, *Rozh'*, 179.
[4] Sreznevskii, *Materialy*, III.204. [5] *PL* vyp. 1 and 3. [6] *Ibid.*
[7] See Gorskii, *Ocherki*, 23–4 and Kochin, *Sel'skoe khozyaistvo*, 218; Rozhkov, *Sel'skoe khozyaistvo*, 118–19 for a guide to documentary references.
[8] *ULS*, 86.
[9] Kochin, *Sel'skoe khozyaistvo*, 218; Bakhteev, *MIZ*, II.228; Rozhkov, *Sel'skoe khozyaistvo*, 118–19.
[10] Gorskii, *Ocherki*, 26; Rozhkov, *Sel'skoe khozyaistvo*, 119.

of beer should be discounted. Malt may be produced from other grains and also might be obtained by trade rather than grown on the spot.

Wheat was highly esteemed as a grain; it was used for the bread consumed at communion, as well as for certain other types of bread and rolls, but as it is more demanding as regards both soil and climate, it was not at all as widely grown as rye, barley or oats.[1] Moreover, even when its presence is attested in an area it is probable that the quantity grown was small.[2] For the area around present day Kuibyshev, however, it appears that wheat may have been the dominant grain in the sixteenth century, perhaps because the forest clearance had already made great progress.[3]

Emmer was also cultivated in Russia, but does not appear to have been of great importance.[4]

Oats was probably the second most important grain crop and was eaten by man and horse. Kochin has suggested that for the latter purpose it was more important on lords' estates, for animals for the forces and for those in transport than on peasant farms.[5] It seems at any rate to have been widespread throughout central Russia as well as in Novgorod.[6] As it was used directly both for man and beast and also to make malt and for other purposes it was clearly an important grain crop.

Buckwheat (*Polygonum fagopyrum*) seems to have appeared in Russia about the time of the thirteenth-century Mongol invasions; earlier evidence relates almost entirely to the steppe. At Grodno, for example, buckwheat was only found in the twelfth–thirteenth-century layers as an admixture to rye. In the thirteenth–fourteenth century layers, it was found as an independent grain and in greater quantity.[7] The linguistic evidence points in the same direction. The Russian, Ukrainian and Belorussian terms (*grecha, grechikha, grechka*) as well as the Lithuanian and Latvian (*grikal, griki*) derive from the term for a Greek, since the grain evidently reached Russia by means of merchants trading with Greece.[8] Estonians and Finns, however, have terms derived from the name Tatar (*tatar, tatri, tattari*). This problem does not seem to have been seriously investigated and needs further work, but it looks as if buckwheat was regarded as an import when it was first encountered in European Russia. From the early fifteenth century, documentary evidence is available and helps to establish its distribution. In some parts it was evidently a common grain; in sixteenth-century Pskov a row of stalls in the market was devoted to its sale.[9] There is

---

[1] Gorskii, *Ocherki*, 25; Kochin, *Sel'skoe khozyaistvo*, 219, *ASEI*, I, no. 71; Yakubtsiner, *MIZ*, II.141–2. [2] Rozhkov, *Sel'skoe khozyaistvo*, 119.
[3] Yakubtsiner, *MIZ*, II.76–7; Fat'yanov, 'Opyt'.
[4] Lyubomirov, 'O kul'ture polby v Rossii', *Tr. po prikladnoi bot., gen. i sel.* (1928), XVIII.1.
[5] Kochin, *Sel'skoe khozyaistvo*, 220.
[6] Gorskii, *Ocherki*, 25; Mordvinkina, *MISKh*, IV.321–2.
[7] Voronin, *MIA*, 41.56, 168; Kir'yanov, *MIA*, 41.205–10.
[8] Krotov, *MISKh*, IV.419–21; Fasmer, *Etymologicheskii slovar'*, I.457.
[9] Krotov, *MISKh*, IV.423.

also a reference to the export of buckwheat at this time.[1] The crop seems, however, to have been somewhat more widespread in southern areas than in the north.[2]

Millet is also known from the fourteenth–fifteenth centuries, but finds have been restricted to Moscow and Radonezh, and even these may be imports.[3]

Harvesting of grain seems always to have been done with a sickle; the scythe was only used for hay. The harvested grain, bound in sheaves, was often dried artificially; a sixteenth-century visitor pointed out that 'by this method the grain is hardened by the smoke and it can be moved without fear of rotting and kept in granaries'.[4] Small two-storey structures, called *oviny*, built away from the house because of fire risks, were used, especially in the north; both surface and semi-dugout types are known.[5] There is early evidence for the fire in the lower chamber of such drying kilns being addressed in prayer, a fact which underlines the importance of this activity.[6] These kilns were usually located by the threshing floor (*gumno*, *ladon'*, *dolon'*, *tok*) which was sometimes itself within a structure.[7] On isolated peasant farms, however, and on small units generally, the threshing floor is likely to have been of the type described by a late-sixteenth-century ambassador to Russia. 'The principal nobles have structural threshing floors, but ordinary people flood some level space in front of those huts we spoke of [i.e. drying kilns] and when the ice is hard they knock out the grain.'[8]

Threshing would be done either by knocking the sheaf against a horizont-ally mounted pole or by beating the sheaf with sticks or flails. The first method, which at least in recent times was sometimes used by women, survived in places till the twentieth century. It appears to be ancient, even though we have no direct evidence for its early existence.[9] It was probably used particu-larly to obtain seed grain and also the long, undamaged, straw needed for thatching, weaving and plaiting; rye straw is particularly long and soft, so that it was well suited even for rope making. Such straw would also provide many of the smaller containers needed about the house and farm. Perhaps it was because of the weaving and plaiting that women might undertake this form of threshing. The second method was traditionally men's work. It might be carried out with the help of simple wooden sticks with a natural thickening, if possible a spade-like root, at the end; such an implement (called *kichiga*, *chap*, or *kolotilka*) survived in northern Russia into the nineteenth century, being

---

1 *Vrem.MOIDR*, kn. 8, otd. II, 1.

2 Krotov, *MISKh*, IV.427; Kochin, *Sel'skoe khozyaistvo*, 222.

3 Gorskii, *Ocherki*, 27.   4 Daniel Printz, *SRL*, 2.728.

5 Blomkvist, *TrIE*, n.s. XXXI.295–8; Bezhkovich, *Khozyaistvo i byt*, 158–9. The term *ovin* is known from the 14th century in the Novgorod territory. Another term for the drying kiln (*oset'* or *aset'*) evidently existed in Toropets, see p. 179 below.

6 Sreznevskii, *Materialy*, II.592; *PRP*, I.245 (late thirteenth century).

7 Olearii, *Opisanie*, illustrates a covered threshing floor and drying kiln (on map between pp. 360 and 361).   8 Daniel Printz, *SRL*, 2.728.   9 Kushner, *Russkie*, I.88.

then used mainly for flax.[1] Flails seem to have been the usual implements for threshing grain, certainly where larger quantities were involved. Threshing with flails was often carried out by equal numbers of men steadily working their way along two rows of sheaves laid out with the ears next to one another in the middle. A nineteenth–twentieth-century norm was 60 sheaves a day, but this tells us little of earlier times when sheaf size may have been different.[2]

Winnowing was usually carried out with wooden shovels, the husks being separated by the wind; but simple sieves were also sometimes used.[3]

The sixteenth-century household management manual, 'Domostroi', gives a vivid picture of storage of grain and other produce in a rich household. The grains, and grain products such as malt, had to be sound, not mildewed, contaminated by mice or otherwise endangered; the various items were stored in barrels (*bochki*), tubs (*kadi*, *kadushki*), chests (*sunduki*), bins (*zakromy*, *suseki*), boxes (*korobi*), and other firmly covered containers in the grain-store (*zhitnitsa*).[4] Peasant grain-stores were often raised on piles, sometimes on inverted pine stumps with a frill of lopped off branches intended to bar the way to rodents.[5] Barns (*ambary*) seem usually to have been somewhat larger structures. The word itself is said to have come from Persian via the Tatars; it was probably not used at peasant level until the seventeenth century or later; it ultimately displaced the term *zhitnitsa*.[6] The new term presumably reflects changes in the agricultural system. Cellars and cold stores (*ledniki*) were also used to store bread.[7] Damp and mice were evidently seen as the main dangers.

Prior to the seventeenth century windmills were rare among the Russians and almost all mills are likely to have been water-driven.[8] A first reaction might be to regard this as odd. The frost-free period in the forest zone of European Russia was limited and this might be thought to restrict the use of water-mills. To some extent this is so; water-mills frequently worked only in spring and autumn, but this fitted well with the cycle of farm work. Moreover, windmills were more complex structures than many water-mills, and hence more costly. They begin to work when wind speed reaches three to five metres per second.[9] Average annual wind speeds in European Russia appear to satisfy this minimum.[10] The dense forest of early Russia, however, reduced surface wind speeds and this probably accounts for the enormous log bases on which many windmills in Russia were raised. It seems that surface

[1] Kushner, *Russkie*, I.87–8; Bezhkovich, *Khozyaistvo i byt*, 34.
[2] Bezhkovich, *Khozyaistvo i byt*, 35.   [3] *Russkie*, I.94–5.
[4] 'Domostroi', §50, 52, 63.
[5] These were the 'huts on chicken's feet' of Russian fairy tales.
[6] Blomkvist, *TrIE*, n.s. XXXI.308; Bezhkovich, *Khozyaistvo i byt*, 156–7.
[7] Domostroi, §54.
[8] Ponomarev, *Vozniknovenie i razvitie vetryanoi mel'nitsy*, 57.
[9] Vinter & Fateev, *Ispol'zovanie energii vetra v sel'skom khozyaistve*. This agrees well with the ten feet per second given in McConnell's *Notebook of agricultural facts and figures*.
[10] Krzhizhanovskii, *Energeticheskie resursy SSSR*, II.339–46.

speeds suitable for windmills are now observed at meteorological stations in Moscow and Kazan' for up to a third of the year.[1] In earlier times, when forest was more extensive, it is likely that throughout the zone with which we are concerned the minimum surface speeds required occurred even less frequently. The raised wooden structures on which Russian wind-mills with their four, six or eight sails were built are shown in what early illustrations are known and may be compared with similar structures for nineteenth- and twentieth-century mills.[2]

Unfortunately, the historical evidence on mills does not seem to have been examined in much detail. In dealing with thirteenth–fifteenth-century crafts and trades in the countryside, for example, Rybakov only distinguished hand-mills from water-mills, but attempted no discussion of what types of water-mill existed.[3] Various types, however, were distinguished by contemporaries and it seems possible, chiefly on the basis of later, mainly nineteenth-century, data, to indicate certain main types.[4]

An important distinction may be made between those water-mills which had horizontal and those which had vertical wheels. The former type (*mutovka, kolotovka*) were generally quite small mills with the wheel mounted at the base of a vertical shaft with the set of stones immediately above it; the shaft passed through the lower stone and supported the upper stone. The wheel was thus internal to the building; as there was no gearing the stone rotated at the same rate as the wheel (about 30 revolutions a minute). The output of such mills was small, in sixteenth-century Kazan half a chet a day, and they were most suited for use by a family, not for any purpose requiring large quantities of flour.[5] Their intermittent use means too, that full-time millers were not required. These easily and cheaply constructed mills could be built virtually anywhere where there was a stream and so the need for windmills was reduced. In the mid sixteenth century the stones for a simple mill of this type

---

[1] Material supplied by Meteorological Office, M.O.13c: my thanks to H. H. Lamb for this information. See also Lebedev, *Klimat SSSR*. 154, 161.

[2] See plate 7, *Dioptra ili zertsalo mirozritel'noe* (GIM, Muz. sobr. no. 2709, l. 53), undated marginal drawing; Ponomarev, *Vozniknovenie*, 58; Petrov, in *Sbornik arkheologicheskago instituta*, kn.3, 118 f. (1694, but recorded in register of 1622); *Kniga ob izbranii na tsarstvo Velikago gosudarya, tsarya i velikago knyazya Mikhaila Fedorovicha* shows seven mills near the Presnya; the illustration of the scene in 1619 was done between 1672 and 1678; all mills are raised on pyramid bases. It seems unlikely that windmills were built in Russia much before this; see Ponomarev, *Vozniknovenie*, 57. For modern examples see: *The Studio*, 'Peasant art in Russia', 1912, no. 320; *Russkie*, I.306; Bekzhovich, *Khozyaistvo i byt*, ill. 125; Makovetskii, *Pamyatniki narodnogo zodchestva russkogo severa* and his *Pamyatniki narodnogo zodchestva Verkhnego Povolzh'ya*, passim.

[3] *Remeslo drevnei Rusi*, 565–70.

[4] Gorskaya, *MISKh*, VI.36–7 has made similar criticisms of the lack of work on grain processing.

[5] TsGADA, f. 1209, kn. 152, l. 178 v. This is identical with the minimum given by modern ethnographic evidence, Blomkvist, *TrIE, n.s.* XXXI.327.

cost little more than half a ruble, those for a large mill almost six rubles.[1] Many west European works point out that such mills are found in Hither Asia, through the Balkans and as far west as the Pyrenees and as far north as the Shetlands and Norway.[2] This simple type of mill, requiring only a small fall of water, was also widespread in eastern Europe. They occurred both in Toropets in the west of Russia on peasant lands and in Kazan' on the east; the term *mutovka* indicates a churn staff, so I call these churn mills.[3] They require a running stream, but not necessarily a millpond, and this is why examples are sometimes noted as working in spring and autumn. The second term for such horizontal mills (*kolotovka*) is also used of mills which had pestles instead of, or in addition to, millstones; these produced oatmeal and hulled products.[4]

The evidence we have, however, does not enable much more to be said with certainty, There appears to be no special term for mills with vertical wheels (later known as *kolesukhi*), though these may be indicated when 'big', or 'foreign', wheels are mentioned, occasionally both elements are combined.[5] Such wheels might be undershot, overshot or breast-mills, but the sources dealt with here say nothing of this.[6] Gorskaya, however, claims that mills with big wheels were 'not undershot but the more productive overshot wheels which are always associated with an artificial dam'.[7] Clearly, 'stone' in the terminology used in the Moscow area stands for sets of stones. It is not clear whether 'wheel' is used in the same sense; but it seems likely that two-wheeled mills had at least two sets of stones. The items which refer to two-wheeled mills seem to associate them with foreignness.[8] The same to be true of the mills with two sets of stones.[9] It may be, therefore, at about this time that foreigners, probably from the west, were introducing mills working with two sets of stones into the Moscow State; because of the gearing required these would be much more complicated machines than the churn mills.[10] It is also noteworthy that

---

[1] *AGR*, I, no. 57 cited in Gorskaya, *MISKh*, VI.45.

[2] Bloch, *Annales d'histoire économique et sociale*, VII.538–63; Singer, *History of technology*, II.593–5; Bennett & Elton, *History of corn-milling*, 2.12; Curwen, *Antiquity*, 18.136–46. For an interesting account of such mills in Ireland and Scotland see MacAdam, *Ulster Archaeology*, 4.6–15.

[3] See pp. 178, 209 below. Bezhkovich, *Khozyaistvo i byt*, 165; *Russkie*, I.122–3, illus. on p. 308; Blomkvist, *TrIE, n.s.*, XXXI.326–7. They are sometimes known in English as Norse mills, though in seventeenth-century Co. Down they were known as ladle mills, Bennett & Elton, *History of corn-milling*, 2.16.

[4] Baburova in *Russkie*, I.124. The mill mentioned in *PKMG*, I.I.7, appears to be of this type although it is not designated as *kolotovka*.

[5] *PKMG*, I.I.54, 55, 149; *AFZ*, I, no. 28; Gorskaya, *MISKh*, VI.41.

[6] For a seventeenth-century reference to an overshot mill, see Blomkvist, *TrIE, n.s.* XXXI.324.

[7] *MISKh*, VI.45.   [8] *PKMG*, I.I.55, 73, 54, 149, 199.

[9] *PKMG*, I.I.55, 223; *AFZ*, I, no. 28.

[10] In the west, however, mills with double races and hence separate wheels with no cog gearing did occur; Bennett & Elton, *History of corn-milling*, 2.17.

two of the millers' names recorded were foreign. Such larger and more complicated mills were evidently only to be found in peasant hands when leased by a group of peasants; they were evidently beyond the resources of the individual peasant farm, unlike the churn mills. Moreover, they might involve the work of a professional miller, even though, due to climatic conditions, the mill might not operate the year round. Gorskaya, however, believes that, judging by the money income from mills noted in monastic account books, 'mills worked evenly the whole year, with a very small natural rise in output during the autumn months. In other words, we see no attempt to mill a considerable quantity of flour and transport for sale by sledge.'[1]

Apart from grains, peas and lentils are attested by finds and documentary evidence, but they were probably not of much importance.[2] They seem to have been items of diet for monks and officials, rather than for peasants.

One crop of importance for peasant diet in some parts of Russia was the turnip (*Brassica rapa*). Turnip plots or fields (*repishcha*) are mentioned in the documents for several northern areas such as Bezhets Verkh, Kostroma, Beloozero and the northern territories of Novgorod, especially in the Dvina area.[3] This rapid growing crop was particularly suited to northern conditions. It was able to benefit from the extended hours of daylight in the northern summer and it was also frequently sown on burnt patches. Like barley it was found quite far north, reaching beyond the sixty-ninth parallel in places.[4] Moreover, the methods of cultivation were suited to either slash and burn or field tillage with the sokha. The seed was broadcast on the stubble of the harvested winter-sown grain or on a burnt patch and then simply harrowed in.[5] It might also be sown in June for harvesting from August.[6] The existence of a special term for the turnip sown on burnt patches (*lyadinnaya* or *lyadnaya repa*) suggests that this was a common crop.[7] In the early thirteenth century, the turnip is mentioned as being sold in Novgorod by the load.[8] It is possible, if we allow for artistic licence, that the plant shown above a harrow-like sokha in one of the miniatures, is a turnip, though no precise identification can be made (see p. 26).

Hemp seems to have been fairly widely distributed in the central and northern areas.[9] In the Novgorod, Pskov and Smolensk areas, it is particularly noticeable. It may have been sown in special patches (*konoplyaniki*) rather than in the general course, and was probably not grown on every peasant holding.

[1] *MISKh*, VI.54.
[2] Gorskii, *Ocherki*, 28; Kochin, *Sel'skoe khozyaistvo*, 222.
[3] Kochin, *Sel'skoe khozyaistvo*, 223; Rozhkov, *Sel'skoe khozyaistvo*, 119.
[4] Efron Brokgauz, *Entsiklopedicheskii slovar'*, 53.480.
[5] *Ibid*; Kochin, *Sel'skoe khozyaistvo*, 88; Blomkvist, *TrIE, n.s.* XXXI.315.
[6] Kochin, *Sel'skoe khozyaistvo*, 223.
[7] Efron Brokgauz, *Entsiklopedicheskii slovar'*, 53.480; Burnashev, *Opyt*, I.375.
[8] *NIL*, 54 (1215).
[9] Gorskii, *Ocherki*, 30; Rozhkov, *Sel'skoe khozyaistvo*, 122.

Hemp is sometimes mentioned in the sources jointly with peas; it seems that these crops may have been grown together. Peas and hemp, for instance, are included in a sixteenth-century list of 'flour and every sort of provision' amid other items all of which are grains.[1] The register of inquisition for Toropets in 1540 lists peas and hemp jointly when they are mentioned as items of a lord's income.[2] Certainly hemp seed was an item of diet and its high oil content made it especially attractive to Russian taste. Hemp fibre was less used for rope-making as straw was frequently substituted. In 1536, Pskov was ordered to send supplies to a military expedition and these included 360 chetverts of peas and the same quantity of hemp seed.[3]

Flax was grown more widely than hemp and was the basic vegetable fibre plant. The Pskov area was an important centre for fibre flax growing and in the fourteenth to fifteenth centuries seed was exported to neighbouring regions, to the north, the north-east and into Belorussia as well as abroad.[4] Between the thirteenth and the sixteenth centuries, flax growing in the Pskov and Novgorod areas was considerably developed.[5] Flax was grown on almost every peasant holding in the Derevskaya and Vot pyatinas of Novgorod.[6] It may, indeed, have been grown by every peasant in the area.[7] In 1472 in Pskov a charter relating to flax and kept in the town chest was torn up by the town head 'and all the peasants rejoiced greatly for it had been in the chest for 8 years and there had been much oppression and loss caused the peasants in that time'.[8] The flax was evidently sown in relatively small areas within the general field course. The units were called *zagony*; from nineteenth-century evidence, we know that these units were then reckoned to be half a desyatina, or just over an acre. The same term, however, means a selion, the work unit of the arable field, and this is probably its meaning in the sixteenth-century texts in which it occurs.[9]

These, then, seem to be the main crops grown on peasant holdings, though there is also evidence for some others, such as poppy and mustard. The mass crops, however, were undoubtedly rye, the main foodstuff, and flax, the main fibre.

---

[1] 'Domostroi', §52.   [2] *AE za 1963 g.*, 277f.   [3] *PL*, 2.228.
[4] Sizov, *Problemy botaniki*, II.120, 163.   [5] Sizov, *MIZ*, II.418.
[6] *Ibid.* citing Gnevushev, *Ocherki*, I.202–7.   [7] Kochin, *Sel'skoe khozyaistvo*, 223.
[8] *PL*, 2.186.   [9] Rozhkov, *Sel'skoe khozyaistvo*, 121; Burnashev, *Opyt*, I.215.

2

The hayfields: livestock

The scythe, the second element of the three-fold formula, was not used for harvesting grain, but only for taking hay in Russia.[1] The scythe referred to was initially of the short-handled type (*kosa-gorbusha*): the long-handled type, similar to the scythe in use in modern times, probably came into use in Russia in the fifteenth or sixteenth century; illustrations in a sixteenth-century manuscript psalter are the first evidence for it.[2] The former type seems, in any event, to have been dominant throughout the period with which we are concerned. On the basis of finds mainly from the Kievan period a number of subtypes have been identified.[3] These are shown in table 2. In general, unlike the smaller scythes of the first millennium A.D., they are somewhat more like the short-handled scythes which have continued in use in northern and forest areas till modern times. The blade, however, tends to be shorter and broader. No handles of ancient implements appear to have been re-covered but, to judge from modern examples, the handle was probably about the same length as the blade and might be either straight or with a curve.[4]

Judging by modern ethnographic evidence, implements such as these were held in both hands and cuts were made to both right and left. Their advantage was that they were well suited to cutting the relatively soft forest grasses and, since they could be used in a confined space, they were well adapted to work around trees or boulders.[5] It is for these reasons that they survived so long in the north.

The cut hay was dried, raked and stacked, or carted and stored in haylofts (*senniki*). Stacks sometimes had to be fenced as other peasants evidently had pasture rights, or at least rights of way, over the meadows.[6] The stacks or ricks (*stogi, kopny*) were evidently used from the top and this suggests, as does the use of these terms as images in literature, that they were cone-shaped and fairly small.[7] The inquisitions of the sixteenth and seventeenth centuries record hayfields in terms of ricks (*kopny*). A seventeenth century source, not a register of inquisition, defines a rick as '2 sazhens and an arshin up and

[1] Levasheva, *TrGIM*, 32.60.
[2] Levasheva, *TrGIM*, 32.91; the illustrations depict not only the figure of Death with a
[3] scythe, but also scenes of daily life which include scythes.
   Ibid. 87–91.   [4] Toren in Kushner, *Russkie*, 1.64–5; Bezhkovich, *Khozyaistvo i byt*, 32.
[5] Levasheva, *TrGIM*, 32.91; Toren in Kushner, *Russkie*, 1.64.
[6] *ASEI*, II, no. 154 (1450–86).   [7] 'Pskovskaya sudnaya gramota' in *IZ*, 6.237.

TABLE 2. *Short-handled scythe types*
(eleventh to fourteenth centuries)
(measurements in cm)

	Blade			Curve		Extension of blade + lug
	length	breadth	breadth/length	angles	depth	
	A–B	D–E	D–E/A–B		C–D/A–B	A–F
Novgorod	*45–46*	*3–4*	1/11–1/15	$B° > A°$	1/5–1/8	10–11.5
Central Russian	*33–50*	*4–5*	1/7–1/13	$A° > B°$	1/8–1/9	9–10
Lower Volga	*36–45*	*3–4*	1/9–1/15	$A° > B°$	1/6–1/23	
Baltic	*42–45*	*2–5*	1/8–1/22	$A° > B°$	1/8–1/6	
South Russian	*35–37*	*4–5*	1/7–1/9	$A° > B°$	1/28–1/10	6–8

Source: Levasheva, *TrGIM*, 32.88–90; Levasheva's illustrations have no scale; this has been added from information in her text and measurements (shown in italic) then estimated; not a satisfactory procedure.

See also Kolchin, *MIA*, 32.95 on similar pre-Mongol types.

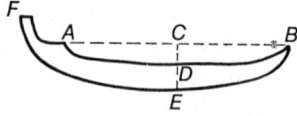

over, and 3 sazhens round', i.e. about 5 m and 6½ m.[1] It has been calculated that the weight of such a rick of hay, called a measured rick (*mernaya kopna*), was about 240 kg (15 puds). There was also a smaller rick, called a drag rick (*volokovaya kopna*), two-thirds the amount of the 2-sazhen one, the measured rick was presumably too heavy to be dragged.[2] Abramovich argues that neither of these was used by the officers of the inquisiton when they recorded agricultural holdings, but a third rick (*melkaya volokovaya kopna*) of 80 kg (5 puds), taken as the equivalent of a tenth of a desyatina, a little more than a tenth of a hectare.[3] This implies that a yield of about 730 kg per hectare was regarded as normal. Average present day hay production for the non-Black Earth areas of European Russia for dry-valley and water meadow lands in the forest are about 880 to 1,000 kg per hectare.[4] In forest, particularly dense coniferous forest, yields are much reduced.[5]

[1] Ustyugov, *IZ*, 19.324–5.   [2] Abramovich, *PI*, XI.366.   [3] *Ibid.* 368.
[4] *Sel'skokhozyaistvennaya entsiklopediya*, IV.456; Efron Brokgauz, *Entsiklopedicheskii slovar'*, 32.357.   [5] Yurre & Anikin, *Ispol'zovanie lesnykh bogatstv*, 75.

Hay, of course, was taken from widely differing types of land. Water meadows were particularly favoured, as were any flat areas between rivers. The term *navolok* meant both 'a portage' and also 'a pasture or meadow by the shore'.[1] The need for fodder, however, was so great, largely owing to the long period for which livestock had to be stalled, that every possible patch of grass had to be made use of.[2] Thus, we find numerous arguments over pasture and hay resulting in court cases; princes and monasteries were concerned to secure their hayfields, and even remote patches in the forest, which had to be mown when a trip could be made to them (*naezdom kosyat*), were brought into use.

The proportion of hayfields to arable varied widely. One hamlet with 33 ha of arable had 20 ricks, while another place had 106 ha of peasant arable and 600 ricks of hay.[3] Thus, while the main feed for animals was undoubtedly hay, other sources of feedstuffs were probably important in certain cases. Leafy fodder was doubtless a useful supplement where suitable forest trees were available. According to evidence from the Vologda area, aspen, birch and lime have been used in recent times.[4] Additives such as caraway may also have been used.[5] Oats were given to horses, but it is unknown how widespread this was; the evidence available refers to oats fed to officials' horses.[6] As Kochin writes 'justly or unjustly the monasteries multiplied their places for hay; the horses were well supplied with grain as its sown area was expanded and dues in kind were increased. The peasant was unable to give his horse what the monastic authorities gave their riding horses'.[7] After field work had been completed livestock were probably set free to feed themselves in the forest until autumn.[8] Metal and wooden bells (*botala*) for hanging around animals' necks are known from early times.[9]

The horse was the main work animal on a holding. The evidence considered above on implements suggests this conclusion and it seems to be supported by the opinion of most of those who have examined the question.[10] Only Rozhkov seems to have tried to make a case for oxen.[11] One man tilling with one horse was the usual picture for Novgorod; the *obzha*, a 'primary,

---

[1] Podvysotskii, *Slovar'*, 96; 'Toropetskaya kniga', f. 170v.
[2] Traditionally animals were stalled from the feast of the Protecting Veil of the Mother of God (*Pokrov Bogoroditsy*) to St George's Day, i.e. 14 October to 6 May by our calendar, a period of 204 days. A sixteenth-century account book of the Joseph of Volokolamsk monastery recorded expenditure of hay from 1 October to 1 May; Kochin, *Sel'skoe khozyaistvo*, 259; Shchepetov, *IZ*, 18.112.
[3] *AFZ*, I, no. 52, 65; no. 166, 152; both documents are dated 1498–9.
[4] *Kolkhoznoe proizvodstvo* (1952), 5.40.   [5] *Rasteniya primenyaemye v bytu*, 125.
[6] Gorskii, *Ocherki*, 25.   [7] Kochin, *Sel'skoe khozyaistvo*, 259.
[8] Rozhkov, *Sel'skoe khozyaistvo*, 126; Aristov, *Promyshlennost' drevnei Rusi*, 47; Veselovskii, *IGAIMK*, 139.33–6, 131.
[9] Levasheva, *TrGIM*, 32.85, 92; Kochin, *Sel'skoe khozyaistvo*, 273.
[10] Kochin, *Sel'skoe khozyaistvo*, 255. Gorskii, *Ocherki*, 48, 65, 70.
[11] Rozhkov, *Sel'skoe khozyaistvo*, 123f.

real tax unit, not to be divided further' was so defined and this picture seems likely to represent the situation for the whole area.[1] The horse in the forest area was small to medium in size; steppe horses were somewhat larger and wider boned than the animals represented by the finds from Novgorod, Pskov, Ladoga, Moscow and Grodno. The vast majority of Russian horses are estimated to have stood between 12 and 14 hands.[2] Oxen as work animals were only found in a few places in the north, but were somewhat more widespread in the central areas; they occurred, however, only on monastic estates in most cases.[3] Only in the south were oxen commonly used for tillage; but even there the horse was probably the main tillage animal.

The horse and the ox were not only used for tillage; they were also eaten, and most of the finds of animal bones in archaeological excavations are from kitchen waste. Evidence on numbers of the main domestic animals, based on available archaeological evidence, is summarised in table 3. From this we can see how much less significant the horse was as an item of consumption than as a work animal. Nevertheless, despite clerical prohibitions (themselves suggesting the presence of the act being forbidden), the horse was evidently of continuing importance as an item of diet, though only to a limited extent.

TABLE 3. *Finds of domestic animals*
(numbers)

Site	Century	Cattle	Pigs	Sheep and goats	Horses	TOTALS
Novgorod						
Yaroslav's Palace	13th–15th	309	34	48	5	396
	15th–17th	47	14	18	2	81
Slavna	13th–14th	109	61	15	13	198
Nerevskii konets	13th	68	40	14	16	138
	14th	140	86	28	23	277
	15th	228	103	28	26	385
Moscow						
Zaryad'e	14th–15th	360	433	207	44	1,044
	16th–17th	170	222	149	37	578
Staritsa	14th–15th	4	7	6	2	19
Tushkov	10th–15th	53	126	45	15	239
Izheslavskoe ⎫ fortified sites		12	8	3	2	25
Zhokinskoe ⎬ (*gorodishcha*)	12th–13th	8	5	8	2	23
Lyubyanskoe ⎭		1	1	–	1	3
Lipinskoe rural settlement (*selishche*)	10th–13th	6	18	5	2	31

Source: Tsalkin, *MIA*, 51.175, 177, 180.

[1] Gorskii, *Ocherki*, 49; *PSRL*, xxv.319–20; p. 24 above.

[2] Tsalkin, *MIA*, 51.55, 81, 90–3.

[3] Rozhkov, *Sel'skoe khozyaistvo*, 124–6; Gorskii, *Ocherki*, 50; Kochin, *Sel'skoe khozyaistvo*, 255–6.

The proportion of finds shows that it was consumed not only in times of famine as seems to be implied by twelfth- and thirteenth-century chronicle entries.[1] Tsalkin has pointed out that the less damaged horse bones, sometimes made the basis of the argument that the horse was not eaten, are those of the lower limbs on which there was little meat.

In general, however, it is not possible to make any accurate estimate of the importance of the various animals in the diet. We know something of their numbers, but virtually nothing of their live and dead weights. As regards numbers of domestic animals, Novgorod, especially the area of Yaroslav's Palace, seems to have been somewhat exceptional, with a higher proportion of cattle than elsewhere; Nerevskii konets shows a lower proportion. Elsewhere the pig was numerically the dominant domestic animal, except in the group of twelfth–thirteenth-century fortified sites. The size of pigs varied a good deal from site to site; all were small by modern standards, and may even have been smaller than the pigs found in neolithic Swiss lake dwellings. Most of those for which it has been possible to establish age were less than eighteen months and a considerable proportion were very young, including some suckling pigs.[2] This suggests that the implication of some additional thirteenth–fourteenth-century articles to the old Russian Law, *Russkaya Pravda,* that half the pig herd was overwintered, may not be too far out.[3]

Sheep seem to have been similar to the Romanov breed and the Ryazan' long lean-tailed sheep.[4] They were thus small in size and were of a hornless type. Goats were all of the *Capra Prisca* type, and were much smaller than modern ones.[5] Both sheep and goats were eaten when young, many at less than a year. Old animals are extremely rare among the finds.[6]

As regards cattle, Tsalkin considers there was only one type involved, although there seems to have been considerable variations between localities. Most cattle stood from 95 to 105 cm in height; they are compared by Tsalkin with the small cattle investigated in European Russia in the 1870s, and with the Meshchera and Oka cattle.[7] The first were kept in primitive conditions, mainly for the manure they gave; they reached a dead weight of 3–8 puds (roughly 50–130 kg). The Meshchera cattle seem to have been of similar size, the live weight of an adult cow sometimes not reaching 12 puds (less than 200 kg). The Oka cattle on the water meadows along the middle Oka were somewhat larger. They averaged 119 cm in height and a live weight of 406 kg. Thus, Tsalkin concludes that poor feeding resulted in the small cattle which are found at old Russian sites and which are similar in size, and probably in dead weight, to the animals later known as *taskanka* or *goremychka* which, as adults, might weigh as little as 75 kg and have to be carried from the stall at the end of the winter.[8]

---

[1] Tsalkin, *MIA*, 51.145–6; Smith, *Origins*, 116–17.   [2] Tsalkin, *MIA*, 51.98, 103–4.
[3] *PRP*, I.208, 217.   [4] Tsalkin, *MIA*, 51.113–14.   [5] *Ibid.* 106, 108.   [6] *Ibid.* 105.
[7] *Ibid.* 47–51.   [8] Gol'msten, *PIMK* (1933), 3–4.62–4.

The scythe, then, aptly symbolises the problem of livestock in Russia. Since animals had to be stalled for the long winter, the demand for fodder was great, yet at the same time yields of hay in Russia are extremely low. The short-handled scythe can be used to take hay from every scrap of open land in the forest, around every tree stump and boulder. The animals kept were, primarily, the horse which was the most important tillage animal, and was also eaten. In the main, however, meat consumption, certainly that from domestic animals, was small; what there was came mainly from young animals, probably those slaughtered as winter came on. Meat also came, as we shall see, from wild animals. Cattle probably provided most of the meat from domestic animals, even though the animals were very small in size; pigs were so small that their greater numbers probably failed to give pig meat anything like its proportional weight in meat consumption. At the end of the fifteenth century a Beloozero customs charter equated ten sides of meat and thirty piglets.[1]

[1] *ASEI*, III, no. 23, §6. These items were also taken as equal to twenty geese, thirty ducks, twenty hares or ten sheep for toll purposes.

3
The forest: gathering and extractive industry

The third element of the three-fold formula indicating the limits of the holding was the axe. This implement signified the forest, a constant element of Russian culture in the centuries following the Mongol invasions. Russia was a wood civilisation, as has often been said. The East Slavs occupied the forests of eastern Europe; this environment was the core of the Russian state in Europe; it was the source of supplies for many of man's needs and should not be thought of merely as a temporary and unsatisfying refuge from the alleged abundance of the steppes.[1] The excavations at Novgorod well illustrate the great range of objects which were made of wood, though these objects far from exhaust the use to which timber was put in medieval Russian society.[2] Attitudes to the forest were probably ambivalent. On the one hand, the forest had to be cleared in order to establish cultivation; on the other, it afforded shelter from enemies and much that could be used to supplement income from agriculture. Indeed, where gathering grew into extractive industry it might offer an alternative to agriculture. We are mainly concerned, however, with gathering as the constant companion of agriculture.

MATERIALS

Building materials

We start with the axe itself. Trees were felled by axe only; houses and other buildings were built without the help of almost any other implement. 'The ability to wield the axe was a necessity of life for a man born among the forests. This helped to develop the carpenter's craft among the broadest mass of the people from which talented master builders emerged to create the most important urban constructions – fortress walls, cathedral churches, princes' palaces.'[3] Our interest, however, is concentrated on materials for rural, rather than urban structures and for these the axe was still more important.[4] Even in the eighteenth century a foreign observer could note that 'the whole

[1] Smith, *Origins*, 32, 75–80.
[2] For a good survey of wooden objects from the Novgorod excavations, see Kolchin, *Novgorodskie drevnosti*.
[3] Makovetskii, *Arkhitektura russkogo narodnogo zhilishcha*, 14.   [4] Kerblay, *L'isba*, 32.

47

Architecture wants no other Tool but a Hatchet, which their Carpenters understand to handle with more Skill than those of any Nation whatsoever'.[1] The saw was not used in village constructions until the seventeenth century when it seems to some extent to have replaced the adze.[2] The axe used was broad bladed, with the cutting edge extended towards the butt and a straight handle; there appear to have been only marginal variations by regions.[3]

Most houses were built of timber; even in the south, where daub was used, houses were timber framed. Unfortunately, there is little archaeological evidence about peasant houses in the post-Mongol period, since most excavations have taken place in towns.[4] The reports of the Novgorod archaeological expedition have provided much relevant material, though we should not assume, as Zasurtsev rightly points out, that information from this populous town necessarily applied to rural settlements. Nevertheless, it seems reasonable to believe that the materials used did not differ much, if at all. The life of a house seems to have averaged twenty years, though sometimes it might last twice that.[5] The length of life was extended by using oak timbers, more resistant to damage from moisture, for the bottom layers of the log constructions; in the main, however, buildings were of pine. Spruce was rarer and larch was only occasionally found. In Moscow, on the other hand, spruce, together with oak, seems to have been the main building material, especially for dwellings.[6] In dwelling houses, the length of beam varied between 4 and 5 metres in Moscow and between 3 and 13 in Novgorod.[7] Thus, in the latter area variation in size was proportionately greater than in Kholmogory and Ostrov, other northern towns for which we have details, where length was about 4 to 8 metres in the sixteenth century.[8] Lime was used for the window surrounds. Often windows were of wicker type, an opening covered by a sliding piece of wood; this long continued to be usual in peasant dwellings, but sometimes mica or fish bladders were used as lights. Oak was used for carved columns.[9] The timber was evidently floated and brought on drags to the site where it was to be used. Moss, usually from the group *Sphagna cymbifolia,* was used to seal gaps between beams.[10] Sheds caulked with moss (*mokh*) were known as *mshaniki.*[11]

Timber was not only used for houses (*khoromy*) with their heated (*izby*) and unheated rooms (*seni, kleti, povalushi*), their storerooms (*pogreby*) and ice-

[1] Weber, *The Present State of Russia,* I.337.   [2] Rybakov, *Remeslo drevnei Rusi,* 183, 558.

[3] Artsikhovski, *Drevnerusskie miniatyury,* 82–3; Levasheva, *TrGIM,* 32.42–3, 47–8; Kolchin, *MIA,* 65.26–30.

[4] See, for example, the complaint by Shennikov in *SA* (1970), 4.244.

[5] Zasurtsev, *MIA,* 123.164, 17   [6] Rabinovich, *SE* (1952), 3.72.

[7] Zasurtsev, *MIA,* 123. 19.

[8] Makovetskii, *Arkhitektura russkogo narodnogo zhilishcha,* 20.   [9] *Ibid.* 15.

[10]Petrov, *KS,* 11.47; Blomkvist, *TrIE,* XXXI.78. Cp. the contract of 1622 quoted by Makovetskii, *Arkhitektura russkogo narodnogo zhilishcha,* 22.

[11]*PKMG,* 2.326.

TABLE 4. Woods used for various objects excavated in Novgorod

	Number	Pine	Spruce	Juniper	Oak	Ash	Maple	Birch	Lime	Alder	Willow	Aspen	Elm	Wych elm	Hazel	Rowan	Apple	Pear	Bird cherry	Spindle tree	Chestnut	Beech	Walnut	Larch	Yew	Silver-fir	Cedar	Box
Buildings	8	4	3		1																							
Cornices etc.	7	3	2		1																			1				
Pavements	2	1	1																									
Conduits	2	2																										
Oars	12	7	1		1		2																	1				
Rowlocks	6	2	1						1		1													1				
Turned vessels	139	5	7	2	1	44	45	8	2	25		3			8	1												
Cut vessels	49	4					12	8	4	3	2	1		2	3	1			1									
Spoons	129	7		9		4	84	9		1	2												1			1	1	
Cut dippers	46	3	4		3	4	13	9	1	2	2								1	1							1	
Buckets etc.	54	35	9						1	2				2	1	3	3	2		1				3	3	1	1	
Barrels	7				7																							
Cooper's hoops	6		6																									
Spades	21	12	6	2	3																							
Crooks, sticks	6						3									1	1											
Churn staffs	1																											
Ribs	8	1	4																					3				
Sledge parts	6				5	1																						
Machine parts	58	22		9	12	1	3	4		1																2		
Rollers, pulleys, bearings	9		9		2		1																			2		
Shoe lasts	6	2						3					1													1		
Ornamental knobs	27	4	4						14						1		1		1					1				
Miscellaneous knobs & spheres	14	1	2		3		2	2		2		1									1							
Combs	85						1	1			1																	82
Turned boxes	1																											1
Knife handles	6						1						1		1				3							1		
Children's toys	12	2		1	1	1	1																					
Tops	10	6			2				1																			
Miscellaneous	163	46	21	4	15	7	11	13		6	4	2	7	2	2	2	1	2			1	1		14	3	3	1	1

49

Source: Vikhrov & Kolchin *TrIL*, LI.143. Cp. Vikhrov, *TrIL*, XXXVII.273–8; Kolchin, *Novgorodskie drevnosti*, 12.
I am indebted to R. G. Richens, Director of the Commonwealth Bureau of Plant Breeding & Genetics, for advice in the distinction between *vyaz* and *il'm*.

houses (*ledniki*), but also for outbuildings such as bath-houses (*bani, myl'ni*), stalls (*khlevy, volovni*), drying barns (*oviny*) and other structures essential to the running of the holding. A large village in the Tver uezd, in the sixteenth century, for example, had '54 heated houses [*izby*], 63 halls [*seni*], 9 small outhouses [*povalushki*], 20 storerooms and 13 bath-houses, 34 stalls, 13 drying barns, 10 ox-stalls'.[1]

Frequently, too, the holding was surrounded by a fence (*ogorod, tyn*) and the yard with its buildings was entered through a solid, timber gate. The Law Code of 1550 laid down that the dwelling-payment, due when a peasant moved from one lord to another, was to be charged per gate.[2] Clearly, one peasant unit was normally found within the fenced yard and its gate at this time. Cut branches were used for fences (called *osek*) built to protect remote fields from animal predators; these were to have 7 decent timbers.[3] Such fences had to be substantial to keep out such powerful animals as bears. The 1589 Law Code laid down that fences round threshing floors were to have 9 solid timbers.

The weirs set across rivers to enable fish to be taken used considerable quantities of timber. The following details have been taken from an inquisition relating to Beloozero uezd in 1585 and no doubt, since this was the Tsar's weir and is described in such detail, it was exceptionally large, but peasant settlements or communes may have had similar though smaller structures. 'There are 18 piles in that weir; and 60 timbers of 6 or 7 sazhens [13 or 15 metres] from big trees go into that weir and for the cross timbers and beams 108 whole 12-sazhen [26 metre] trees; of medium trees for weights, for supports and for forks 60 4-sazhen [8½ metre] trees; and of small trees, 900 6-sazhen [13 metre] poles as a stock; and in the ferry-raft, of big trees 6 14-sazhen [30 metre] stakes go into it, and for the cross-ties 25 7-sazhen [15 metre] boards and pieces.'[4] Bridges, pavements and corduroy roads were also built of timber.[5] But it is only the first of such structures which were likely to be found in the countryside.

Other structures which demanded considerable quantities of timber were mills, especially windmills. As we have seen, water-mills existed in Russia, but their use was restricted by the fact that European Russia was flat and it was not always easy to obtain an adequate head of water. Ice, too, was a great danger, particularly when it broke and was carried downstream in the spring. Many water-mills only operate in spring and autumn.[6] We might, therefore, expect to find more windmills than water-driven ones. The evidence we have does not enable us to say which type predominated; it seems likely that conditions varied in different areas. What is certain is that

---

[1] *PKMG*, II.309.   [2] *Sudebnik 1550g.*, §88; in Smith, *Enserfment*, 92–3.
[3] *Sudebnik 1550g.*, §169. See, for example, *NPG*, no. 27; *ASEI*, I, nos. 408, 457.
[4] *PKMG*, I.ii.415.   [5] Artsikhovsky, *Drevnerusskie miniatyury*, 69–70.
[6] See chapter 8 for examples.

the cover of forest was so dense that ground-level wind speeds were reduced.[1] It was therefore necessary to build mills with high bases, sometimes of considerable size which used great quantities of timber.[2] One example recently reported from northern Russia had more than 120 tree trunks incorporated in the supporting base, quite apart from the timber used in the mill proper. A drawing in the margin of a seventeenth century(?) manuscript shows a very similar type of construction (plate 7a).

Sand, clay, lime and stone were also dug and used as building materials in areas where they were found, and they were occasionally brought from considerable distances for important works. The white stone used in building some of the Kievan period churches in Vladimir was floated from the Kama area, a distance of roughly 700 kilometres.[3] Usually, however, the distances involved were much less than this. The attractive, rosy, shell-encrusted stone used in some Novgorod churches, such as the Saviour on Nereditsa Hill (built in 1198), or the beautifully restored Peter and Paul in the Tanners (1406), was floated down from Korosten' near Staraya Rusa. Mica was used in some windows.[4] Such building, however, was almost entirely undertaken by princes or the church. Even for them it was far from common. Timber, not stone, was the usual material. The sixteenth-century English merchants, for instance, reported that 'their churches are built of timber and the towers of their churches for the most part are covered with shingle boordes'.[5] Excellent shingles, made of aspen and modelled on examples recovered from excavations, may be seen on the church of Peter and Paul just mentioned. Stone or brick private buildings were very few and belonged only to powerful nobles, such as the Novgorod town head, Yuri Ontsiforovich.[6] In seventeenth century Pskov, however, stone buildings became somewhat more widespread.[7] There is no evidence to suggest any peasant stone buildings.

Fuels

Touchwood or tinder to catch the spark struck from flint or, probably more commonly, the heat generated by rubbing was provided by the dry spores of shelf fungi. The Russian term for the Polyporaceae (*trutovye griby*) shows the 'rub' root (*trut*) and *trut* or *trud* itself meant a shelf fungus growing on birch trees. *Trut* also meant tinder and *berezovyi trut* (lit. birch tinder) meant touchwood. Probably the fungi used particularly for this purpose were

[1] See p. 36 above.
[2] Makovetskii, *Arkhitektura russkogo narodnogo zhilishcha*, 215–16; 311–21 illustrates many types of wooden windmills. See also his *Pamyatniki narodnogo zodchestva russkogo severa*, 172–7, and his *Pamyatniki narodnogo zodchestva verkhnego Povolzh'ya*, 110–13.
[3] Solov'ev, *Istoriya Rossii*, II.47.   [4] Zasurtsev, *MIA*, 123.43.
[5] Hakluyt, *Principal navigations*, I.291.   [6] Zasurtsev, *MIA*, 123.68.
[7] Spegal'skii, *MIA*, no. 119.

Fomes fomentarius and *F. igniarius*.[1] It is said that in some northern districts this method of getting fire was in use till recent times.[2] Such dried fungi would also be used for maintaining a smouldering fire.

The main fuel was timber. It was evidently stacked in open-sided out-buildings or outside the buildings, as it often is today. The Pskov chronicle records that on 13 November 1478, there was a flood and, among other things, stakes and firewood were carried downstream past the town.[3] Wood was burnt in the stove found in every peasant hut. Usually such stoves had no stone or brick chimney; the stove aperture was blocked up with a board. Frequently this board or the accumulated soot was liable to catch fire in these wood houses with their straw thatching.

It seems possible that some Slavs may have known crude petroleum from the south, but much more important for the period we are dealing with was the oil derived from sea mammals found in northern waters, such as whales and seals. Whale oil seems to have been traded southwards through Russia and beyond from early times; it is said to have been used for illumination in a lighthouse on the Bosporus.[4]

The English merchants of the Muscovy Company were much interested in the production of trane-oil for import to Britain. In the seventeenth century the white whale (*Delphinapterus leucas*), 'a greate white fish in the baye of St Nicholas which tumbles up like the porpis', and the seal *Phoca barbata* were used to provide oil.[5] This raw material, however, was restricted to the White Sea area. Fletcher noted that 'their greater men use much wax for their lights – the poorer and meaner sort, birch dried in their stoves and cut into long shivers which they call *luchiny*'.[6] Illumination was usually by wood spills or torches of pine or birch which remained the main method in peasant homes into the first half of the nineteenth century.[7] 'Firewood and spills, the first concern' runs the proverb.[8] Lanthorns of mica, which must have used candles, were much admired by Fletcher in the sixteenth century but were probably only for the rich.[9] Candlesticks appear in ordinary town houses in Moscow from the sixteenth century.[10] Candles, however, long remained restricted to cleric and noble.

Implements

It has been claimed that even up to the mid nineteenth century, ninety per cent of peasant implements and objects of every day use were of wood.[11]

---

[1] Wasson and Wasson, *Mushrooms, Russia and History*, 1.112–18.

[2] Shul'ga, *Griby nashikh lesov*, 70–1.   [3] *PL*, 2.218.

[4] Sreznevskii, *Materialy*, 1.1210 (s.v. kitov) (*c.* 1250); Evliya Efendi, *Narrative of travels*, 1.2.73.   [5] James, 31a:6 and 7.   [6] *Of the Russe Commonwealth*, ch. 3.

[7] Bezhkovich, *Khozyaistvo i byt*, 178; Artsikhovskii, *Drevnerusskie miniatyury*, 190.

[8] *Drova da luchina – pervaya kruchina.*   [9] *Of the Russe Commonwealth*, ch. 3.

[10] Rabinovich, *SE* (1952), 3.74.   [11] Levasheva, *TrGIM*, 33.61.

It is difficult to see what common unit could be used to make such an estimation, but there can be little doubt that wood was a very common material for peasant implements in the period dealt with.

The implements of tillage, sokhas, ards, and ploughs, were basically of wood, though sometimes equipped with iron tips or shares. Harrows, too, though sometimes evidently iron shod, could be entirely of wood, whether simple drags or of the frame type; judging from 16th-century evidence, the frame type was sometimes woven, presumably from withies.[1] There were also harrows formed from fir planks with the branches left on to act as teeth.[2]

Spades seem to have been usually entirely of wood or of wood rimmed with iron, judging from certain illuminated manuscript evidence in one collection; of 50 illustrations of spades 32 were entirely of wood, 14 had iron bindings and 4 were entirely of iron.[3] Other hand tools such as rakes and forks were also usually entirely of wood. Threshing sticks and flails were of wood joined by leather strips.

Most of the implements for working flax and hemp, from retting and scutching to the final stage of weaving, were also of wood.[4]

Means of transport

Skis, illustrated for the fifteenth century, snow shoes, sledges, drags and four-wheeled carts with solid or spoked wheels, shafts and sometimes the bow harness, as well as boats of various types, were all produced from timber gathered from the forests.[5] These were the basic means of transport throughout the period.

Furniture and minor equipment

The indoor equipment of shelter for man and animal was limited. Even in the nineteenth century the interiors of peasant homes had little more than a variety of benches, shelves and chests, all of wood.[6] Most of this equipment was made in the process of building the house and formed an integral part of it; it was built in, only a table, a bench and perhaps a chest or box being movable. Such, for example, was the 'maydan the planke in the Cobacke ("a Rus. taverne") on which they plaie at dice and cardes'.[7] Not all chests and boxes were iron-bound, even in the eighteenth and nineteenth centuries.

Smaller, but essential items to be found in all peasant homes would include

[1] See plate 6.
[2] See Smith, *Origins*, ills. 30–2; Bezhkovich, *Khozyaistvo i byt*, ills. 4 and 6.
[3] Artsikhovsky, *Drevnerusskie miniatyury*, 23, 71, 83.
[4] For a reconstruction of a tenth–thirteenth-century loom see Levinson-Nechaeva, *TrGIM*, 33.18–19.
[5] Artsikhovskii, *Drevnerusskie miniatyury*, 68, 92; Levasheva, *TrGIM*, 33.69–74.
[6] Blomkvist, *TrIE, n.s.* XXXI.419f.; Bezhkovich, *Khozyaistvo i byt*, 172f.
[7] James, 50a:20; 10a:17.

wooden buckets or barrels, yokes for carrying loads, flour bins and dough troughs; a hearth broom and baking shovel; a milk churn, scoops and bowls, platters and spoons, a pestle and mortar, baskets and punnets, a trough and washing dolly, seed lips, perhaps a rattle for driving stock from the forest and pipes. All these were made in the home either from wood or bark, or else woven from bast of lime or of birch bark, pine roots or straw.[1] Even buckets were sometimes of woven material, and there is a fourteenth-century reference to horse harness of bast.[2] In 1592 a monastic account book mentioned 'bast for shafts and for tugs' (na podvoi i na guzhi).[3] Mats were woven of lime bast.[4] Pottery was used, but seems customarily to have been kept for best wherever possible; potting seems to have been weakly developed in villages in the post-Mongol period.[5] Wood and bast provided many substitutes, and bast surrounds were sometimes fitted to protect pottery.[6]

Ropes were made of bast or from roots (as well as from cultivated hemp and from rye straws).

Writing materials etc.

Parchment and, from the end of the fourteenth century, paper were known and used in Russia. At peasant level, however, birch bark seems likely to have been a more important writing material. It was not the thin, white, outer layer, but the full thickness of the bark which was used and the characters were inscribed with a stylus, usually of bone. Finds of such documents, some evidently coming from minor officials on estates or from peasants, are known from Novgorod (over 400 documents have been published), Smolensk, Vitebsk and Staraya Rusa.[7]

On important documents seals were of wax mostly gathered from tree-hives, but it seems likely that it was only in the north, where there was a tradition of some peasant literacy reinforced by the presence of free, un-enserfed, peasants, that such documents would be found in peasant hands.

Tally sticks (birki, plashki) were evidently quite common. Many have been found in seventeenth-century layers at Moscow, and later village examples are known from ethnographic material.[8] Wooden calendars were evidently used, perhaps from pre-Mongol times.[9]

---

1 Rabinovich, SE (1955), 4.51, shows that spoons were home-made in fifteenth–sixteenth-century Moscow; Bezhkovich, Khozyaistvo i byt, 180f.
2 RIB, VI.175.   3 Shchepetov, IZ, 18.109.   4 James, 67.10–12.
5 Rybakov, Remeslo drevnei Rusi, 561–2.   6 Rabinovich, SE (1955), 4.45–7.
7 See Novgorodskie gramoty na bereste, edited by Artsikhovskii and others; Avdusin, SA (1957), 1.248–9; Ist. SSSR (1967), 2.214.
8 Rabinovich in Arkheologicheskie pamyatniki Moskvy, 66–7; Bezhkovich, Khozyaistvo i byt, 40.
9 Maisterov & Prosvirkina, 'Narodnye derevyannye Kalendari'; Konstantinov, 'Narodnye reznye Kalendari', 91.

Clothing and textiles

Fur and leather clothing were worn in Russia, but for the post-Mongol period there is little evidence that such materials obtained from wild animals were used for peasant clothing. While in Kiev Rus' bear pelts were considered the poor man's winter clothing, the later documents only mention bear hunting as the preserve of the lords.[1] Probably sheepskin winter clothing was more usual from early Mongol times. The fur coat (*shuba*) frequently worn by peasants was defined in the seventeenth century by Richard James as 'made of Nagai sheepe skins which they weare inward the inwart the winter and turned in summer'.[2] It is possible, however, that some peasants were able to draw on the resources of the forest for pelts. Richard James also mentioned 'leasnic, camushnic [i.e. foresters] men that kill beares and losshes [elk] and all wilde beastes to which end they goe in the wilde woods abought Micholes day in sommer and return not untill Christmas or after'.[3] It seems most unlikely that hunting expeditions of five months or so were lordly pastimes; these men were professional hunters, but, of course, we do not know whether their gains went to peasant or lord.

Footwear was frequently of woven bast ('basket shewes' as James called them), though in some areas, such as Novgorod, where livestock husbandry was well developed, leather footwear seems to have been common, but finds of bast shoes have been made in other towns, such as Moscow.[4] Ethnographic material shows that three or four young limes would be cut to produce one pair of such footwear, which lasted only about a month.[5]

Miscellaneous

Giles Fletcher, who was ambassador to Russia in 1588, noted that 'the women...use to paint their faces with white and red colours, so visibly, that every man may perceive it'.[6] The white used was apparently white lead, the rouge may have been saffron, but this use of cosmetics did not spread into the countryside until the eighteenth century.[7] Coloured earths and ores were also used, mainly in monasteries, for the painting of frescoes, icons and illuminated manuscripts.[8] Colours associated with certain areas included Kaluga ochre, Pskov vermilion, probably derived from the parasite *Porphyrophera polonica* found on certain roots, Moscow and Yaroslavl' white lead, Yaroslavl' red lead, Kashin and Rzhev lake, Kopor'e green.[9] Saffron, produced

[1] Artsikhovskii in *IKDR*, I.236; *AFZ*, II, no. 87; *ASEI*, III, nos. 92, 119, etc.
[2] James, 65:16.   [3] James, 48a:7–10.
[4] Rabinovich, *SE* (1952), 3.61 (fifteenth–sixteenth century).
[5] Levasheva, *TrGIM*, 33.90.   [6] Hakluyt, *Principal Navigations*, II, 336.
[7] Luk'yanov, *Istoriya khimicheskikh promyslov*, IV.36–7.   [8] *Ibid.* IV.66–77.
[9] *Ibid.* IV.59; Levasheva, *TrGIM*, 33.102.

from the stigmas and tops of the styles of *Crocus sativus autumnalis,* was used to give yellow and red colourants, while *Carthamus tinctorius* gave an inferior but similar product.[1] In the main, however, all these colours are unlikely to have been of much importance to the peasant as consumer, though it was probably peasant labour which gathered the raw materials. Levasheva has, however, pointed out that there were numerous wild plants available which could be used for vegetable dyes.[2] She lists 20 yellows, 10 reds, 10 blacks, 9 blues and 7 greens. Most of these plants are widespread throughout the area of European Russia, but some are more restricted. Woad, for instance, is fairly rare and occurs along the Don and Sosna in the south-east and along the Oka; it is also found in the Baltic area.[3] These colourants were probably used to dye wollen and linen cloth.

Potash was produced in Russia at least from the fifteenth century.[4] It was extremely cheap; half a ton cost 12 kopeks about the middle of the seventeenth century. At this period the demand for potash increased greatly as glass manufacturing developed; potash was also in demand for soap making.

Much more important at peasant level were vegetable substances used for tanning and dyeing; often the same substance did both. Tanning was a peasant domestic process from Kievan times up to the nineteenth century.[5] Vegetable tannins seem to have been obtained mainly from the bark of oak, willow and spruce, but birch, alder, larch and elm were also used.[6] Other plants which may have been used include *Potentilla* and *Poligonum* sp., sorrel, iris (*Iris pseudacorus*), heath (*Calluna vulgaris*), hazel, ash, spiraea (*Filipendulla ulmaria*) and *Andromeda polifolia*.[7] For the initial processing of the hides, lime and ash was used. Thus all the materials were obtained by extractive activity.

A number of lubricants and preservatives were derived from the oleo-resins of trees. The exudate of pine trees, apparently usually obtained by heating the roots in closed pits, gave resin, tar, pitch and other products.[8] Tar, possibly derived from birch rather than pine, since the term used is *degot'*, was also an item of internal trade in Beloozero in the late fifteenth century,[9] though this may have been being exported. A barrel for birch resin was found in a house yard, destroyed in 1468 in Moscow.[10] The consumption of such materials within any peasant household, however, was likely to be quite small; moreover, peasant needs are likely to have been met from their own winter production.

1 Luk'yanov, *Istoriya khimicheskikh promyslov*, IV.566. 2 Levasheva, *TrGIM*, 33.96–101.
3 Luk'yanov, *Istoriya khimicheskikh promyslov*, IV.113. 4 *Ibid.* II.7.
5 Rybakov, *Remeslo drevnei Rusi*, 182. 6 Levasheva, *TrGIM*, 33.46, 96.
7 *Ibid.* 97–101.
8 Sreznevsky, *Materialy*, II, 893 (thirteenth century) s.v. *pek"l*; Luk'yanov, *Istoriya khimicheskikh promyslov*, III.424f.
9 *PRP*, III.176. There is, however, a good deal of confusion between *degot'* and *smola*; see, for example, Luk'yanov, *Istoriya khimicheskikh promyslov*, III.425, 428, 432.
10 Rabinovich, *SE* (1955), 4.44–5.

FOOD

Berries, nuts, fruit

The simplest form of gathering for food is the collection of plant products which are ready to eat. Wild berries which are still collected in the forest zone of European Russia include the barberry (*Berberis vulgaris*), cowberry (*Vaccinium vitis idaea*), cranberry or bogberry (*V. oxycoccus*), bilberry (*V. myrtillus*), bog whortleberry (*V. uliginosum*), dewberry (*Rubus caesius*), stone bramble (*R. saxatilis*), raspberry (*R. idaeus*), cloudberry (*R. chamaemorus*), wild strawberry (*Fragaria vesca*) and white and black currants.[1] Hips of the dog rose (*Rosa canina*) and the cinnamon rose (*R. cinnamomea*) were also used for food, as were the fruits of the guelder rose (*Viburnum opulus*), sorb-apples (*Sorbus*) and bird cherries (*Padus racemosa*).[2] Yields of cowberry are said to reach 500 kg per ha where there are fields of them in coniferous forest.[3] The bilberry averages 100–200 kg per ha, but may reach as much as 500 in spruce forests.[4] Yields such as this would mean that such fruits could contribute usefully to the diet.[5] There is, however, little direct evidence about the consumption of such fruits in earlier times. In part, this is due to the nature of our sources, which are little concerned with peasant diet; in part, though, it is due to the fact that most of these berries are easily spoiled if transported when fresh and so are unlikely to enter trade, and hence, possibly, our documents, on any scale. Consumption on the spot, then, would involve minimum losses. In fact, such gathering for food would often combine well with extensive forest farming, such as slash and burn.

Cowberry, bog whortleberry, raspberry and cloudberry seeds were among those found in ninth–tenth-century layers at Staraya Ladoga.[6] In the late sixteenth century, Giles Fletcher listed among the native commodities of Russia wild cherries, 'rasps, strawberries and hurtilberries, with many other beries in great quantitie in every wood and hedge'.[7] In the early seventeenth century, an English traveller to Russia listed under berries, in his Russian–English word list, strawberries and 'respires'; then he gave the Russian names of eight more plants for which he gave no equivalents. These were: cowberry, guelder rose, bilberry, crab apple, black currant, stone bramble, bird

---

[1] Rozhko, *Yagody i ikh lechebnye svoistva*. On the etymology of *klyukva* (cranberry) and *moroshka* (cloudberry) see Kleiber, *Scando-Slavica*, VIII.220–3.

[2] Umnikov, *Plody*, 37, 75, 17.

[3] Yurre & Anikin, *Ispol'zovanie lesnykh bogatstv*, 106.   [4] *Ibid*. 109–10.

[5] In 1968–9 the USSR procurement agencies received nearly a quarter of a million tons of wild fruits and berries, including 47,297 tons of cranberry and cowberry; Obozov i dr., *Pobochnye pol'zovaniya*, 4. The same source gives harvests of 80 kg per ha for bilberry and cowberry as practicable and, incidentally, points out that in cash terms such crops are now about seven times more valuable than the timber yield; *ibid*. 9.

[6] Smith, *Origins*, 38.   [7] *On the Russe Commonwealth*, ch. 3.

cherry and cloudberry.[1] Elsewhere, he mentions the Russian words for bog whortleberry, blueberry and bogberry.[2] Of these, the cowberry is considered to have its centre of origin in the pine forests of the north, the bilberry and bogberry are found wild throughout the northern and central Russian forests and the cloudberry is common in north European peat bogs.[3] James described the last as a 'pretty big red soure marras berrie, which the russes eat with honie'; sugar-covered it can now be found on sale in Russian shops. Honey was also used to prepare 'a kinde of Russ. marmalet made of hony and berries';[4] this consisted of thin layers of crushed fruit and other materials dried on boards in the sun.

Apart from their use as food, fresh or preserved, berries were also used for their flavours, particularly in the making of kvass. This Russian beverage, now sold on the streets from trailer-cisterns, is of ancient origin and, together with mead and beer, was probably the main drink in the period we are dealing with. There are frequent references to it in the 'Gardens of health' (*vertogrady*), herbals mainly of the seventeenth century.[5] A present-day book states that, of fruits and berries, 'those mainly used in preparing kvass are wild apple, pear, bogberry, cowberry, cloudberry, wild strawberry, black currant, raspberry and other berries'.[6] The latter include the barberry and sorb-apples.

Thus, it seems that wild berries, throughout the centuries we are dealing with, made an important subsidiary contribution to peasant diet, as well as being included in the provisions of lords.[7] The list of berries gathered nowadays is mostly confirmed by what little early evidence we have and by their continuing use as flavours for the traditional drink, kvass, and, of course, for metheglyn. It also seems likely, even though we have little evidence, that many of the jams, juices and other preparations made from these berries in recent times were known to the inhabitants of Russia many centuries ago.

Hazel nuts were also collected. Documents refer to hazels in trade and when describing the bounds of land held.[8] There are also references to apple trees, presumably wild ones, on boundaries and to 'a pear fenced with a fence' in a swamp in Suzdal' uezd.[9] It seems unlikely that cultivated trees would have been situated on boundaries far from the dwelling; moreover, some of those mentioned bore marks possibly indicating ownership rather than a boundary in the first instance.

Fungi

Fungi have been collected and eaten by Slavs from the earliest times. There may be little documentary evidence for this but the elaborate, and very

---

[1] James, 8a:12–22.   [2] *Ibid.* 11.21, 22; 43.4.   [3] Umnikov, *Plody*, 59, 78, 71, 75.
[4] James, 67:6.   [5] Bogoyavlenskii, *Drevnerusskoe vrachevanie*, 53.
[6] Korolev, *Russkii kvas*, 4.
[7] A charter of 1577 mentions berries and milk-caps (*Lactarius* sp.), *Ist. arkhiv*, III.281.
[8] *AFZ*, II, nos. 7, 74, 123; *ASEI*, III, nos. 23, 56.   [9] *ASEI*, I, no. 485.

expressive, Russian popular terminology for a wide range of edible fungi leaves us in no doubt.[1] On the other hand only two popular terms apply to non-edible fungi; *poganki* are things not worth eating (not only fungi, but, for example, fish too);[2] *mukhomor*, 'fly-killer' is applied to certain *Amanitae* (*A. virosa, A. mappa, A. muscaria* and *A. pantherina*) not all of which are deadly poisonous. Some fungi (*syroezhki*) might be eaten uncooked; there are about 60 species known in the Soviet Union and all are edible if pickled in brine. Those considered best are *Russula vesca, R. virescens, R. claroflava* and *R. paludosa*; but they are in the third of the four categories into which edible fungi are now divided in Russia. The first category includes *Boletus edulis* (most sensibly permitted by the church to be eaten during all fasts save that of Holy Week), *Lactarius deliciosus* and *resimus*. The Russian words for these (*ryzhiki, gruzdi*) are given without translation by Richard James in his seventeenth-century word list.[3] He also gives two Russian words meaning fungi (*guby* and *griby*); the latter is a general term, possibly connected etymologically with our grub, grubbing, but the former is linked with roots meaning sponge, and in the north of Russia, this term indicated any edible fungi. This is shown, for example, in sixteenth-century ecclesiastical literature laying down that 'fish are not to be caught nor berries nor fungi (*guby*) carried'.[4] The Wassons say that in the sixteenth–seventeenth centuries, *griby* indicated boleti, while *gruzdi* meant gilled fungi.[5] A wooden model of a fungus (unidentified) was found during excavations in Novgorod.[6] Fungi, in several dishes, were included among the items served in the Patriarch's household in the late seventeenth century.[7] The truffle was sought with the help of trained bears;[8] but it is not clear when this began. Such luxuries were evidently for the nobles. To the peasant, however, fungi provided a useful supplement to a diet short in proteins and fats. Dried fungi contain up to 73 per cent protein and 2 per cent fat.[9] Moreover, the forest gave a substantial harvest and one which could be easily preserved and stored. Present day estimates are that in good years the harvest is about 5 million tons of fungi in the RSFSR and 1.2–1.8 million are used by the Soviet food industry.[10] It is very hard to assess yields per unit area; on some small areas they may reach the equivalent of 500 kg per ha but the average will be very much less, if we allow for the fact that a harvest may be taken from perhaps only a tenth of any forest area, and for those spoilt.[11] A recent survey shows yields in different types of forest in Karelia of from 6 to 52 kg per ha.[12] Vladimir oblast in the 1960s had yields

---

[1] See the magnificent work by Wasson & Wasson, *Mushrooms, Russia and History*, 2 vols.
[2] James, 43:4. [3] James, 10b:29, 31. [4] Sreznevskii, *Materialy*, I.606.
[5] *Mushrooms*, II.188. [6] Rumyantsev, *KS*, 72.97.
[7] *Raskhodnaya kniga*, entry for 17 March 1699.
[8] Yurre & Anikin, *Ispol'zovanie lesnykh bogatstv*, 105. [9] *Ibid.* 95.
[10] *Ibid.* 93; Obozov *i dr.*, *Pobochnye pol'zovaniya*, 4. In 1968–9 nearly 40,000 tons of wild fungi were pickled or dried.
[11] Vasil'kov, *Griby*, 4; Galakhov, *Izuchaite griby*, 48. [12] Shubin, *Griby*, 74.

around 2.5 kg per ha, Bryansk oblast about 1–3 kg per ha according to the year.[1]

Miscellaneous food plants

In pre-Mongol times sorrel (*Rumex* sp.), marsh marigold (*Caltha palustris*), nettles (*Urtica urens*) and blind nettles (*Lamium album*) were used as food; it seems likely that such use continued into the centuries with which we are concerned, though in modern times these plants have mainly continued in popular use for medicinal purposes.[2] The use of sorrel, and, to some extent, of nettles and blind nettles as food has also continued till the present day. There seems to be no evidence for the continued use of flote-grass (*Glyceria fluitans*), perhaps because of its somewhat restricted habitat. It is clear, however, from the fragmentary information at our disposal that other plants were gathered for food. Richard James mentions goutwort (*Aegopodium podagraria*) 'a herbe, which in the springe they gather out of the woods and eate much of it in pottage and with their meale'.[3] This, together with bear-garlic or ramson (*Allium ursinum*), would probably be used to make something like *botvin'ya*, 'a kinde of porridge or hodgpodg made of boild beets and onions', or the delicious cold soup made with *kvas* and spring onions and called *okroshka*.[4] James records that 'abought Toule [i.e. Tula] it [bear-garlic] growes in abundance in the woodes and marasses and they sell and eate much of it'.[5]

Fish

Fish was a very important item of diet in early Russia, but most historians have paid little attention to the question of what fishes were available to the population. Kochin, for example, wrote that 'the same fish bred in the lakes and rivers in the thirteenth–fifteenth centuries as do now'.[6] He lists about twenty varieties of fish, but does not identify them all. In fact, however, it is clear from a combination of documentary and archaeological evidence that such a list may be expanded and also that there have been certain changes in the distribution of some species since the seventeenth century.

The most valuable noble, as opposed to common (*krasnaya* as opposed to *belaya*) fish were sturgeons and salmon; broadly speaking, the sturgeons were to be found in the south, the salmon in the north. The stellate sturgeon (*Acipenser stellatus*) was still found as far north as Simbirsk on the Volga in the seventeenth century, whereas now it mainly enters the southern Bug and only individuals are to be found in the Dnepr.[7] In the seventeenth century, Richard

---

1 Obozov *i dr.*, *Pobochnye pol'zovaniya*, 73.
2 Smith, *Origins*, 38; *Lekarstvennye rasteniya dikorastuyushchie*, 85–6, 251, 81, 177.
3 James, 65:22.   4 James, 19a:23.   5 James, 58a:24.
6 Kochin, *Sel'skoe khozyaistvo*, 292. Mal'm, *TrGIM*, 32.117 makes the same point as regards the period back to the tenth century.
7 Kirikov, *Promyslovye zhivotnye*, 124, 258.

James considered this fish 'verie delicate and holesomer' than the great sturgeon (*Huso huso*) or the Russian sturgeon (*Acipenser guldenstadti*).[1] At that time the great sturgeon was to be found in the Vyatka and Sheksna; in the eighteenth century it was still to be found in the Pripyat'.[2] This fish was so valued that there are several instructions that the whole catch was to be handed over to the authorities, whereas it was usual to have to hand over only a tenth of the catch of other sturgeons. The Russian sturgeon was also found on the Pripyat' in the sixteenth century; in the seventeenth century it was fished on the Oka and the Vyatka; in the eighteenth century it was still in the Msta, Sheksna and Mologa, but between the seventeenth century and the end of the eighteenth the Russian sturgeon had disappeared from the Samara.[3] In the fourteenth–fifteenth-century treaties between Novgorod and its princes there was regular mention of the prince's sturgeon fisher and he presumably operated in the lower Volkhov.[4] A Novgorod birch bark letter of about 1400 refers to the despatch of a small bucket of sturgeon.[5]

The sterlet (*Acipenser ruthenus*), like the Russian sturgeon, was found as late as the eighteenth century in the Msta, Sheksna, Mologa and upper reaches of the Volga; it had been abundant in the Oka in the seventeenth century and still survives there and in general is more widespread than other sturgeons in the waters of European Russia.[6] The sea sturgeon (*Acipenser sturio*) was fished in the Volkhov, Svir' and other rivers prior to the eighteenth century; in that century it was caught in Lake Ladoga, but its distribution and stocks have been drastically reduced in modern times.[7] There are few references to the ship sturgeon (*Acipenser nudiventris*) in the period with which we are concerned and they are insufficient to form any coherent pattern.

The next most valuable group of fishes were the salmons, but there is often difficulty in distinguishing the different species, a difficulty reflected by confusion in the terms used. It therefore seems best to treat the salmons – the true salmon (*Salmo salar*) (*semga*), salmon trout (*S. trutta*) (*lokh* in old texts, later *losos'*), taimen (*Huso taimen*) (*taimen'*) and sea trout (*S. trutta trutta*) (*losos'-taimen'*, *taimen'* or *kumzha*) – as a single group. The salmon was the noble fish of the north and there are several references to salmon in the Novgorod birch bark letters. One probably from the second half of the fourteenth century includes the earliest known use of the word *taimen'*.[8] Another from a level dated to 1340–60 appears to be a trader's note of salmon collected from eleven persons, of whom at least four were related to one another.[9] Four supplied one salmon each, but there was one who contributed thirteen. Another from the 1370s instructs someone to 'take ten salmon from

[1] James, 26a:20–2.   [2] Kirikov, *Promyslovye zhivotnye*, 35, 44, 58, 257.
[3] *Ibid.* 28, 35, 45, 58, 258.   [4] *GVN*, nos. 10–12, 15, 17, 20, 22, 27, 29, 35, 40, 47.
[5] *NGB* (1956–7), no. 259.
[6] Kirikov, *Promyslovye zhivotnye*, 35, 44, 45, 50, 54, 125, 258.
[7] *Ibid.* 28, 35, 38.   [8] *NGB* (1956–7), no. 280.   [9] *NGB* (1953–4), no. 92.

the Kanunikovs and take another ten from Sanilka Beshkov'.[1] These documentary references to salmon are not confirmed by archaeological evidence. In an important article on fish finds from tenth–fourteenth-century Novgorod, E. K. Sychevskaya has pointed out that sturgeon appear to be underrepresented and salmon are totally absent.[2] In the eighteenth century, salmon were still being caught in the Volkhov, Syaz', Pasha and Svir', as well as the Neva and far up the Western Dvina and in the areas of Polotsk, Vitebsk and Toropets.[3] In the nineteenth century, salmon became extinct in certain northern rivers due to increasing pollution from activities such as the rafting of timber.[4]

Two other valuable fish were also evidently caught and consumed at least till the early seventeenth century, though, again, there are no finds from tenth–fourteenth-century Novgorod. These were the Caspian inconnu (*Stenodus leucichthys Guldenstadti*), called by Richard James 'the white sammon of Volga', and the nelma (*S. leucichthys nelma*), described by him as 'a fish of the Dwina with a taile forked like a sammon in propert also not unlike, but tasted like a bream. In the Dwina they are not big, but from Pechora they have them of exceeding length.'[5] The distribution of the inconnu has been very sharply restricted in European Russia over the last three centuries and its stocks reduced.[6]

It may well be that the underrepresentation of such noble fish in the Novgorod finds analysed by Sychevskaya is due to the fact that her material came from a residential area of the city, Nerevskii konets; this kitchen waste probably reflects the fish diet of the general populace, insofar as various factors influencing survival and recognition do not entirely distort the picture. Noble fish were probably consumed almost exclusively by restricted categories of the population, mainly princes and their noble retainers, as well as the higher levels of the clergy and the monasteries.

A number of other fishes were also consumed by the general populace. These included certain whitefish other than the Volkhov whitefish (*Coregonus lavaretus baeri*): the European whitefish (*C. l. lavaretus*), called the 'Neva whitefish' (*nevskii sig*) in Russian, the omul (*C. autumnalis*), the vendace (*C. albula*), and probably the peled (reading R. James' *circa* as *syrok*, not, as Larin does, as *chirok*).[7] The importance of the whitefish at a later stage of Novgorod's history is shown by its inclusion in the town's crest. It was caught and marketed not only in Novgorod and Pskov in the sixteenth century, but also in Ostashkovo in the central area of the Moscow State.[8] Among the cyprinids for which there is documentary evidence, though they are not all attested in

[1] *NGB* (1955), no. 186.   [2] Sychevskaya, *SA* (1965), I.236–56.
[3] Kirikov, *Promyslovye zhivotnye*, 36, 259.   [4] *Ibid.* 19.   [5] James, 9b:52; 18b:33.
[6] Kirikov, *Promyslovye zhivotnye*, 270.
[7] James, 36a:13 (translated by him as 'a lamprey', but Larin, 234, interprets it as *Coregonus nasutus*).   [8] Man'kov, *Tseny*, 54.

Sychevskaya's finds, there are wild and pond carp (*Cyprinus carpio*), tench (*Tinca tinca*) and bream (*Abramis brama*); there is also evidence for the cut-tooth (*Rutilus frisii*) and the Siberian roach (*R. rutilus lacustris*) from the early seventeenth century as well as the rudd (*Scardinius erythrophthalmus*) and burbot (*Lota vulgaris*). The delicious stint or Baltic lake smelt (*Osmerus e. eperlanus – forma spirinchus*), now restricted to the Pskov area, was probably more widespread than nowadays in the north-west of Russia. The smelt (*O. eperlanus*) and the White Sea or Arctic smelt (*O. dentex*) were also known.

TABLE 5. *Fishes in Novgorod, tenth–fourteenth centuries*
(percentages)

	scales	bones		scales	bones
Russian sturgeon	–	2.6	Bream	30.1	31.5
Volkhov whitefish	0.14	1.82	Blue bream	9.6	1.7
Pike	7.5	15.2	Vimba	–	0.24
Roach	1.1	0.24	Chekhon	2.7	0.24
Chub	0.07	0.24	Crucian carp	0.05	–
Ide	0.86	0.24	Sheatfish	–	3.47
Asp	0.07	0.24	Pike-perch	21.5	40.6
Bleak	0.21	–	Perch	0.6	6.7
White bream	22.8	1.7	Ruffe	0.33	–

Source: Sychevskaya, *SA* (1965), 1.252.
Note: totals do not sum to 100, presumably because some items were not identified.

All the fishes so far mentioned were caught in rivers or ponds. Of sea fish, probably the most important was the herring. The first evidence for the import of herring into Novgorod comes from the early fifteenth century.[1]

Other sea fish were mainly obtained from the White Sea. These included flounders, turbot, cod, capelin and navaga (*Eleginus navaga*) which, though very sweet and delicate, was considered by many Russians in the seventeenth century to be unfit to eat.[2]

This list is clearly incomplete. Eels and crayfish were certainly consumed as were some marine mammals in those parts where they were found.[3]

The general question of fish in early Russian diet appears to have been somewhat neglected, and a considerable amount of research will be required in order to achieve a fuller picture. In particular, it seems necessary to pay greater heed to the question of the consumption of the poorer quality fishes, rather than of the noble ones which so frequently attract our attention. We

[1] *GVN*, no. 56; Khoroshkevich, *Torgovlya*, 333.   [2] James, 43:2.
[3] In addition, I have been unable to find equivalents for a number of terms met with. These are: *vykhlokh* (from *lokh*?), *vykholka*, *kil'chevshchina* (from *kil'ki*?), *kostogolov*, *loduga*, *loduzhina*, *mol'*, *ostrets*, *sabel'shchina*, *sushch*, *tarabora*, *khokhli*, *chabak*.

know many details of the fish dishes served at the tables of princes lay and clerical, but very little of those eaten by the peasants.

The means used to take fishes varied a good deal. Hooks similar to those from Kievan times continued to be used in later centuries.[1] Spoon-baits are found more rarely. In the Novgorod area in the nineteenth century, hooks were used to take such fishes as pike-perch, perch, ide and ruffe, bream and burbot; these fishes had, in earlier times, been among the cheaper sort, and this suggests that, though relatively few hooks have been found in excavations, their use was widespread.[2] Numerous finds of weights for use on lines also suggest that line fishing was widespread.[3] Fish-spears were used, probably mainly in the shallow waters of the spring floods and for night fishing with lights during the spawning season; in the nineteenth century dried pine roots were used for illumination as they burnt without crackling.[4] Fish-spears were also used in winter; a hole would be made in the ice and the fish, short of oxygen, would hasten to it and be speared.[5] Once the ice was thick enough, simple traps could be cut in it, but this ingenious method would, of course, leave no trace and we cannot be certain when it was first used.[6]

Traps of wicker or nets were also used to take fish, but it is difficult to derive any clear picture from the terms used for the various types. 'The names of covered traps are exceedingly numerous and varied. Frequently one and the same implement has different names in different places and, on the contrary, implements varying in construction are called by one and the same name. Sometimes one and the same name is given to both net and wicker traps.'[7]

The simplest traps were open labyrinths (*koty, kottsy, kot'tsi*); some had a throat and were called *mordy*. One form of the latter type is called *vanda,* a word which recalls a term for smelt (*vandysh*) used in Beloozero in the mid sixteenth century.[8] Other traps woven of withies and apparently widespread in north and central European Russia, at least in the nineteenth century, though also known from some parts several centuries earlier, were called *naraty, naroty, neroty, noroty,* cognate with the modern *nereda.*[9] *V'r'zhi, vershi,* known from the fourteenth century, were apparently basket traps, though these terms seem also to be used of open, cone-shaped traps 2–2.5 metres long which are now used to take pike and downstream migrant fish.[10] *Kuritsy* also seem

[1] Mal'm, *TrGIM*, 32.117–19.   [2] Bezhkovich, *Khozyaistvo i byt*, 65, 66.
[3] Rabinovich, *SE* (1955), 4.37–9.
[4] Kolchin, *MIA*, 65.76–8; Bezhkovich, *Khozyaistvo i byt*, 67.
[5] Mal'm, *TrGIM*, 32.122.   [6] Mukhomediyarov, *Prosteishie rybolovnye orudiya*, 30–1.
[7] Baranov & Bessonova, *Derevyannye rybolovnye lovushki*, 31.
[8] Sreznevskii, *Materialy*, I.226; cp. the Ustyuga customs charter of 1599; *Khrestomatiya,* XVI–XVII vv., 34.
[9] 'Toropetskaya kniga', f. 20v; *OOS*, 123; Baranov & Bessonova, *Derevyannye rybolovnye lovushki.*
[10] *NGB* (1956–7), no. 248; Baranov & Bessonova, *Derevyannye rybolovnye lovushki;* 'Toropetskaya kniga', f. 20v; *OOS*, 96.

to have been used to take pike.[1] Some of these traps were not constructed of withies, but with nets. *Merezha*, a term known from Kievan times, indicated one such closed net trap; *versha, venter', fitil'* were also terms with a similar meaning; in the sixteenth century peasants were making such nets for sale.[2] In the main, nets were of two broad types: *seti* were nets (or snares) in a broad sense, while *nevody* were trap nets or seines, some apparently of considerable size. In the 1560s in Kazan' the Archbishop's seines had to be carried by horses.[3] In the fifteenth century a seine was taken as a sokha unit for tax purposes.[4] *Keregody* appear to have been a special sort of fairly widespread seine.[5]

Seines, of course, required several men and boats to work them and it is for this reason, as well as the considerable cost of such large nets, that we find peasants fishing 'with nets and traps, but there is no seine fishing'.[6] Trapping with nets may have been the most widespread means used by peasants to take fish.

The fish taken were treated in a number of ways, if not consumed while still fresh. Fish were sometimes preserved by freezing. They were also sun-dried, cured and prepared in brine.[7] Caviare and other roe products were consumed in considerable quantities throughout the period here dealt with. Sales of caviare are mentioned in the Beloozero customs charter of 1497, for instance.[8] Richard James mentioned perch and whitefish caviare as well as pressed caviare 'which is pict clean from the guts and skinne and dried more in the sunne and kneaded into rownde lumbes, which they so put into great bladders', and it cuts as smooth as any bread and is in most esteem with them'.[9] He also referred to 'a rod of sturgeon, which they drie and eate rawe'.[10] This probably refers to what is known as *vyaziga*, the dried gelatinous spine of the fish. The sun drying of the south was probably replaced in central areas by smoke curing; a sixteenth-century curing chamber has been uncovered in Moscow.[11] Unfortunately, the report does not tell us what fish were cured. Man'kov has suggested that the populace of towns bought the cheaper varieties of fish: bream, pike-perch, stint, vendace and perch.[12] This seems most probable. Moreover, a very similar list is given in a sixteenth-century inquisition, apparently as a definition of the non-noble fish found in the numerous lakes around Toropets.[13] This lists: bream, pike, roach (?) (*plotitsa*), perch and asp. It is likely that these fishes, widespread throughout the waters

---

[1] On *kuritsa*, see Podvysotskii, *Slovar'*, 126.
[2] Mal'm, *TrGIM*, 32.124; Sreznevskii, *Materialy*, II.128; Jablonskis, *Lietuviški žodžiai*, I; Bezhkovich, *Khozyaistvo i byt*, 66.
[3] TsGADA, fond 1209, kn. 152, f. 241v.   [4] *GVN*, no. 21.
[5] 'Toropetskaya kniga', *passim*; Sreznevskii, *Materialy*, I.1206; *OOSD*, 79.
[6] 'Toropetskaya kniga', f.143v (f.55v).   [7] Mal'm, *TrGIM*, 32.128–9.
[8] *ASEI*, III, no. 23.   [9] James, 54a:18, 19; 51a:19.   [10] *Ibid.* 69:5.
[11] Rabinovich, *SE* (1955), 4.39.   [12] Man'kov, *Tseny*, 54.
[13] 'Toropetskaya kniga', f. 224v.

of European Russia and taken by net, hook or spear, provided the greater part of the protein intake of the majority of the population. It has been calculated that at the end of the nineteenth century, fish protein amounted to about 43 per cent of the meat protein consumed in Russia.[1] It seems likely that in earlier centuries the proportion was much higher than this, because of the abundance of fish and the relative simplicity of taking them. Finally, the more widespread distribution of certain species in earlier times seems to have applied throughout the period dealt with here. It was mainly from the eighteenth century onwards, and especially from the nineteenth century, that overfishing and pollution of the waters resulted in appreciable changes in the nature of the fish catch.

Game

In general, historians seem to have paid little attention to the part played by hunting in the provision of meat, especially at peasant level, in the period with which we are concerned. For example, Kochin has written that hunting helped in the provision of food for the peasant, but only mentions that hare was eaten, although forbidden by the church.[2] Gorskii does not mention meat consumption, though his section on hunting is in many ways superior to that of Kochin.[3] Mal'm has a little to say on the subject, pointing out that hunting was carried out 'to get furs and in part meat. In the excavations of towns and settlements of rural type the bones of wild animals and birds, mainly those whose flesh was used for food, are always encountered.'[4]

Certainly available information is inadequate and does not enable us to make an analysis in any depth. Nevertheless, there is enough material for the beginnings of a pattern to be discerned; it must be stressed, however, that there seems to be no information on rural sites. The nearest data we have is that from Lebedka and Lipinskoe, both rural sites (*selishcha*), but only one of which lasted into the thirteenth century. Most of our information (table 6) comes from town sites; the data from Kiev comes from three separate sites, two of which are now within the city limits, the third being twelve kilometres to the north.

The first point which stands out is the low percentages of wild animals found at most town sites. The exception is Grodno. This site from the western border of the Kievan state has been included simply for the sake of comparison. The information here relates to Lithuanian, not Russian, occupation and is to be explained partly by the town's situation both in a frontier position, where local supplies would be important and livestock husbandry might be exposed to danger, and in an area of dense forest which has continued into modern times to be famed for abundance of game. Kiev is the only other town site which shows more than ten per cent of wild animals

[1] E. A. Grimm, cited in Mal'm, *TrGIM*, 32.117.   [2] Kochin, *Sel'skoe khozyaistvo*, 288–9.
[3] Gorskii, *Ocherki*, 87–92.   [4] Mal'm, *TrGIM*, 32.107.

TABLE 6. *Relative proportions of wild animals to total number of animals*

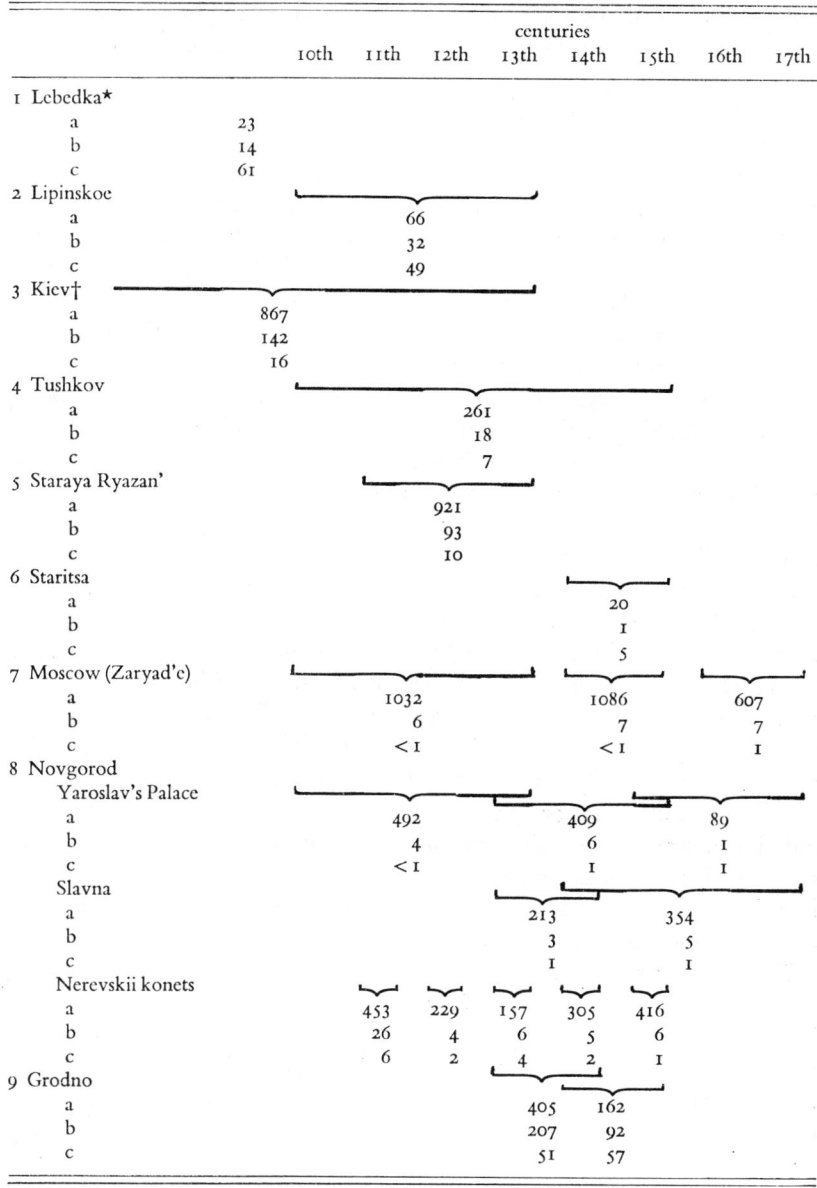

	centuries							
	10th	11th	12th	13th	14th	15th	16th	17th
1 Lebedka*								
a	23							
b	14							
c	61							
2 Lipinskoe								
a			66					
b			32					
c			49					
3 Kiev†								
a	867							
b	142							
c	16							
4 Tushkov								
a				261				
b				18				
c				7				
5 Staraya Ryazan'								
a			921					
b			93					
c			10					
6 Staritsa								
a					20			
b					1			
c					5			
7 Moscow (Zaryad'e)								
a			1032		1086		607	
b			6		7		7	
c			<1		<1		1	
8 Novgorod								
Yaroslav's Palace								
a			492		409		89	
b			4		6		1	
c			<1		1		1	
Slavna								
a				213		354		
b				3		5		
c				1		1		
Nerevskii konets								
a		453	229	157	305	416		
b		26	4	6	5	6		
c		6	2	4	2	1		
9 Grodno								
a				405	162			
b				207	92			
c				51	57			

Source: Tsalkin, *MIA*, 51.131–3, 137, 175–8.
a. Total number of animals.
b. Number of wild animals.
c. Wild animals as percentage of total (to nearest whole number).
* Early second millennium A.D.
† 6th–13th centuries.

present; this may be due to the fact that data from three sites have been combined. A breakdown into the individual elements might help to explain this rather high proportion, though the location of both Kiev and Staraya Ryazan' in the forest steppe zone might also help to account for the wealth of game. The picture from Moscow and Novgorod, both very large and important towns in this period, seems quite consistent. The proportion of wild animals is very small. The data for Nerevskii konets gives a somewhat higher percentage, probably, according to Tsalkin, because there were large landowners living there.[1] It would seem necessary, however, to confirm the location of the finds more precisely before accepting this explanation, since the area may not be unique in this respect, and there were certainly many artisan houses in the area.[2] There is written evidence that the prince in Novgorod has fowling (*gogol'nye*, i.e. golden-eye) runs and hare-hunts in the late thirteenth century.[3] The two rural sites show the much greater importance of wild animals; the proportion is considerable higher than in towns (apart from Grodno), even those in the forest-steppe zone, so the explanation would not seem to lie merely in their location. It appears that hunting was of considerable importance in the countryside, but more evidence is required before a fuller picture can be given.

Details of the wild mammals represented in these finds are shown in table 7. The totals shown in the table suggest that, in terms of individuals, wild animals hunted for meat accounted for about two-thirds of the total number taken. The actual figure given, 78 per cent, is distorted by the exceptional figures for Grodno. If we exclude the Grodno data from the calculation, the result is 65 per cent. Even if we further modify the calculation by also excluding the figures for Staritsa and Moscow, where only meat animals were found, the result is not greatly altered – 62 per cent of the total number of individuals were taken mainly for meat. The number of individuals is small so the significance of the statistical calculation may be doubtful. It should also be borne in mind that these calculations almost certainly underestimate the contribution made to the meat diet by hunting and trapping.

First, the hazards of survival have favoured the larger animals, so that the smaller ones may well be under-represented; and these, such as the hare, are likely to have been of most importance in peasant diet. It seems probable, for example, that the number of hares is not fully reflected in the finds; this may be due partly to their bones breaking down in the soil, partly to dogs being given the scraps from the table, so that the bones entered the soil already broken. Other small animals, not represented in the finds, were also eaten. Even in the twelfth century, a cleric asked 'And of those who live in the villages do some eat squirrel or anything else? This is evil, but more evil than

[1] Tsalkin, *MIA*, 51.131.   [2] See Zasurtsev in *MIA*, 123.5–165.

[3] Tatishchev, *Istoriya rossiiskaya*, v.47; a variant reading refers also to many dogs, *ibid.* 279; Solov'ev, *Istoriya Rossii*, II.163.

TABLE 7. *Meat and fur-bearing animals*
(individuals)

	Meat								Fur-bearing							all wild	Totals			
	elk	red deer	roe deer	Steppe antelope	bison	aurochs	boar	hare	bear	beaver	fox	wolf	badger	marten	otter		meat no.	meat %	fur-bearing no.	fur-bearing %
Lebedka	4		3				2		2	2					1	14	9	64	5	36
Lipinskoe	2	1	9					7	2	5	2			3	1	32	19	59	13	41
Kiev																142★				
Tushkov	7						2	3		4				2		18	12	67	6	33
Staraya Ryazan'	37		4	1			9	5	12	23		2				93	56	60	37	40
Staritsa								1								1	1	100	0	0
Moscow	9							5								14	14	100	0	0
Novgorod Yaroslav's Palace	2								1		1					7	5	71	2	29
Slavna	4					3	2			1		1				8	6	75	2	25
Nerevskii konets	10							1	1	1	2	1	1			17	11	65	6	35
Grodno	40	81	62		34		39	3	1	21	1	2	5	7	4	300	259	86	41	14
Total	115	82	78	1	34	3	54	25	19	57	6	6	6	12	6	504	392	78	112	22

Source: Tsalkin, *MIA*, 51.133, 175, 177–8, 180.
Dates as in table 6.
★ of which 77 per cent ungulates, mainly elk and red deer: not included in totals.

that is the flesh of a strangled animal; it would be better to eat squirrel and beaver, rather than the flesh of a strangled animal.'[1] In the steppe areas various small rodents, such as hamsters and steppe marmots, were eaten in very early times, but we do not know whether this habit of diet continued into later centuries. Secondly, some of the animals here classed as mainly fur-bearing were also eaten. We find, for example, clerical prohibitions about the consumption of bear meat (which can still be bought in Moscow); the eating of beaver has already been mentioned, the badger may also have been eaten.[2] The third factor contributing to an underestimation of the contribution of gathering to the meat diet is the omission of birds from the table given. Grouse, partridges, ducks and geese, as well as larks, cranes, herons and the greatest delicacy, swans, were also eaten, at least in the seventeenth century.[3]

Nevertheless, the proportion of fur-bearing animals is probably also underrepresented in the finds of animal bones since it is likely that in many cases only the pelt would be brought back to the settlement. Thus we can agree with Tsalkin's conclusion that the evidence derived from finds of bones in general underrepresents the part played by hunting in the economy.[4]

It is more difficult to assess the importance of gathering for the meat diet. Kirikov, for example, makes no attempt to do so in a chapter on the former significance of hunting, but limits himself to the statement that 'fur-bearing animals, ungulates, water fowl and grouse were of greatest significance'.[5] Tsalkin himself has pointed out that 'the red deer (not to mention the bison and the elk) gave more meat than cattle, and the boar undoubtedly considerably more than the very small domestic pig'.[6] We still have insufficient evidence to be able to answer the question satisfactorily. Meat, moreover, was not a major element in the diet. Today the lack of many native Russian names for meat dishes may reflect the lack of meat in the national diet in earlier times. The only term I have been able to find is *zharkoe* (roast meat). Richard James mentions only boiled and roast meat in his word list.[7] Today, most meat dishes have names of German (e.g. Schnitzel) or French (e.g. Languette) origin. The only Russian meat dish widely known outside Russia is Boeuf Stroganov, in many Russian restaurants as bastard a dish as its name.[8]

The distribution of wild animals is closely linked with the physical environment and the greater extent, and somewhat different nature, of the forest in the past, together with changes in the densities of human populations, have

---

[1] Sreznevskii, *Materialy*, 1.486. [2] Aristov, *Promyshlennost'*, 76; Tsalkin, *MIA*, 51.134.
[3] 'Domostroi', §64; Richard James, however, remarks that the Russians did not eat the heron, 53:17. Ducks and geese are mentioned in trade in fifteenth-century Beloozero: *ASEI*, III, no. 23, §6.
[4] *MIA*, 51.136. [5] Kirikov, *Izmeneniya (lesnaya zona i lesotundra)*, 29.
[6] *MIA*, 51.138. [7] James, 15a:9 and 10.
[8] The only Russian dishes Brillat-Savarin mentioned as parts of a connoisseur's dinner in 1825 were dried meat, smoked eels and caviare (*The Philosopher in the kitchen*, 275). The first item was not further specified; the other two were both products of gathering.

been of importance in determining the areas where each species might be found. The elk (*Alces alces*), which required dense forest for shelter, was formerly abundant in the western parts of the forest zone of European Russia, especially towards the north, in the Volga–Kama area and it was able to penetrate quite far south onto the steppes by making use of the trees and shrubs which bordered the rivers such as the Don and the Southern Bug.[1] In the main, it was somewhat localised in distribution; this may account for distant hunting expeditions being mounted to obtain it and that, when it was taken, alive or dead, it was sent to the court from quite far away.

The red deer (*Cervus elaphus*) and the reindeer (*Rangifer tarandus*) are not always distinguished in the sources. Red and roe deer (*Capreolus capreolus*) were widespread throughout the forest zone, though their numbers were noticeably small by the seventeenth century in the west and centre.[2] Their distribution was affected by the depth of the snow cover. When the snow reaches above the carpal joint, the animals have to push against the weight of snow in their efforts to forage and they rapidly become exhausted. It has been suggested that a snow depth of 30–40 cm is critical for roe deer and 50–60 cm for red deer in Europe.[3] Moreover, the roe deer favoured the forest steppe, which provided it with excellent natural conditions, and it remained numerous there at least till the eighteenth century.[4] Large numbers were taken where herds crossed rivers.

The steppe antelope (*Saiga tatarica*) was particularly numerous among the steppe ungulates. Its numbers fell as the steppes were taken over by agriculturalists and ploughed up, but it remained abundant into the eighteenth century in the Ukraine.[5] The fact that it was found at Staraya Ryazan' may be due to a migration from dry to moister conditions during the summer drought.

The bison (*Bison bonasus*) was rare on the steppes even in past centuries, though large herds were to be found on the steppes of Podolia in the sixteenth century.[6] In the forest-steppe zone it disappeared soon after settlement and agriculture had led to changes in the physical environment, by the early eighteenth century; prior to that it had been found from Belorussia to the Caucasus.[7] The aurochs (*Bos taurus primigenius*) seems only to have been found in the north-west of European Russia in post-Mongol times.[8] Finds have also been reported from the Kiev area in the period up to the thirteenth century.[9]

1 Kirikov, *Promyslovye zhivotnye*, 24, 27, 43, 58, 80, 134.
2 Kirikov, *Promyslovye zhivotnye*, 27, 43.
3 Formozov in *Devyatnadtsatyi Mezhdunarodnyi geog. kongr.*, 228.
4 Kirikov, *Izmeneniya (stepnaya zona i lesostep')*, 97.
5 Kirikov, *Izmeneniya (stepnaya zona i lesostep')*, 48; *Promyslovye zhivotnye*, 302.
6 Kirikov, *Izmeneniya (stepnaya zona i lesostep')*, 95.
7 Kirikov, *Promyslovye zhivotnye*, 313.
8 Kirikov, *Promyslovye zhivotnye*, 24; Tsalkin, *MIA*, 51.175.
9 Tsalkin, *MIA*, 51.133.

There seem to be no finds or reports later than the fifteenth century, though Kirikov now considers that there may still have been aurochs on the Desna in the sixteenth century.[1]

The boar (*Sus scrofa*) was also formerly widespread, but in the last two or three centuries its distribution, especially in the steppe and forest-steppe, has been sharply curtailed.[2] Until the end of the seventeenth century, however, there was no shortage of this animal in the Ukraine and the forest-steppe area along the Dnepr. It also seems to have been plentiful in the Novgorod and Smolensk areas.[3]

The white hare (*Lepus timidus*) formerly had a very wide distribution, being found even in the forest-steppe in the mid seventeenth century.[4] It has now, however, been restricted to a much more limited area and the European hare (*Lepus europaeus*) has penetrated into the forests and become much more numerous.[5]

The weapons used to take animals included the axe, spear and, probably the most widespread, the bow and arrow.[6] The bow was of the compound type and underwent little change throughout the period. It is believed that the few examples of simple bows found may have been toys for children.[7] Usually the long bow was similar to a late-twelfth-century find from Novgorod. Complete, this would have been about 190 cm in length; it was built of birch and juniper wood, cemented with fish glue, reinforced with animal sinews and protected by an outer covering of birch bark.[8] Other woods such as bird-cherry, pine, elm, aspen, maple, apple and pear, were also used, while the sinews came from deer, elk or oxen. Bows for use when on horseback were somewhat shorter and rarely exceeded 150–160 cm. A set of equipment for mounted archery was known as *sagadak* or *saadak*, a term which seems to have come into use in the fifteenth century from the east.[9] It was in the fifteenth or sixteenth century, too, that the Russian bow came to be strengthened by the addition of horn vibrators.[10] By this time, however, at least as far as the armed forces were concerned, the bow was giving way to firearms. At peasant level, of course, the bow remained important throughout the period we are dealing with.

Most arrow heads from the tenth century and later were of iron. Tenth–fourteenth century finds of bone arrow heads are rare and imitate their iron counterparts.[11] This suggests that iron heads were sometimes too costly. In general, heads seem to have been specialised mainly according to their

---

[1] Kirikov, *Promyslovye zhivotnye*, 69; but see also 79.

[2] Kirikov, *Izmeneniya (stepnaya zona i lesostep')*, 109.

[3] Kirikov, *Promyslovye zhivotnye*, 24, 43, 79, 94.

[4] Kirikov, *Izmeneniya (stepnaya zona i lesostep')*, 73.

[5] Kirikov, *Izmeneniya (lesnaya zona i lesotundra)*, 96–7; *Promyslovye zhivotnye*, 308–9.

[6] Mal'm, *TrGIM*, 32.109.   [7] Medvedev, *Arkh. SSSR, SAI*, E1–36. 8, 10.

[8] *Ibid.* 10–11, 117.   [9] Savvaitov, *Opisanie*, 120.

[10] Medvedev, *Arkh. SSSR*, E1–36. 12.   [11] *Ibid.* 53.

military functions, but one group, the blunt *tomary*, were used to take fur-bearing animals without damaging the pelt.[1] The use of the bow to kill squirrel is shown in a fourteenth-century wood carving in a Stralsund church (plate 9).[2]

Axes, spears and arrows were the main weapons used in the hunt, but traps and snares were probably even more important. The simplest form of trap was the concealed pit into which large animals might fall. Other simple traps which have continued in use until the present consist of a heavy beam held up by a slight support which the animal moves when it takes the bait. Such traps (*kolodki* or *pokolodvy*) were used for beaver.[3] Nooses (*siltsi*, *klyaptsi*) were used for a number of small animals and birds, though, as we have already seen, it was prohibited to eat the meat of strangled animals. Richard James mentions *plenitsa* 'a slidinge knot to ketch ducks'.[4] It seems likely, however, that this, along with the use of other types of snares and nets, was one of the main means by which the peasant got much of the little meat he ate. Nets (*prugli*) were used to take small creatures like hares, as well as birds. Larger nets (*teneta*) were used for larger animals, usually by means of the battue, so that these are likely to have been either co-operative activities, possibly by a commune, as has been suggested by Mal'm, or, more likely, they were restricted to use by the lords.[5] Wicker-work traps (*koshi*) were used to take beaver, and there were also self-release bows (*samostreli*), but these were probably not very widespread.

One form of net requires special mention. This, called *pereves*, was still used in Siberia at the end of the last century.[6] There a clearance of about ten metres wide was made in dense forest, running in the direction of the flight of the duck. On each side a tall pole (called *verèi*) was fixed as high as or higher than the trees; it was attached to the trees by a loop allowing it to be raised or lowered freely. At a man's height from the ground a thick rope was stretched between the poles and held the lower end of the net; this latter was raised by means of reins running through the blocks attached to the tops of the poles. The whole was held by side ropes and a coarser mesh net fitted to the cross rope. The fowler sat by the side of one of the poles with the reins attached to a stick in his hands and as soon as a flight of birds touched the net he would let it drop.[7] It is said that up to a hundred birds could be taken at once by this means.[8] Both Kutepov and Mal'm suggest that ancient examples

---

1 *Ibid.* 55. 2 Reproduced in *Hanse, Downing Street und Deutschlands Lebensraum*, 68–9.
3 Kutepov, *Velikoknyazheskaya i tsarskaya okhota*, I.134–5; Mal'm, *TrGIM*, 32.111.
4 James, 63:22. 5 Mal'm, *TrGIM*, 32.114.
6 Strutosov, however, regarded *pereves* as a place set aside for hunting gear on the spot 'but by no means the net with which they used to, and even now do, trap birds flying constantly in one direction'. *Priroda i okhota*, 1881, t. II, mai: 4.
7 Kutepov, *Velikoknyazheskaya i tsarskaya okhota*, I.131–2; based on an article in *Priroda i okhota* (1892) VI.1–32.
8 Mal'm, *TrGIM*, 32.113.

of this device may have been somewhat different and the net may not have been controlled by the fowler. It seems very unlikely, however, that this was the most widespread means of fowling, as Mal'm claims.[1] It seems much more likely that the more modest snares, nets and 'slidinge knots' were commoner, though it is impossible to document them. Such devices could easily be set up by any peasant in the forest which formed an essential part of any holding, whereas it was only in certain times and places that fowling, like other forms of hunting and trapping, became a specialised activity and supplanted the essentially complex activities of the usual peasant family.

CONDIMENTS AND FLAVOURINGS

Salt

Salt was a normal part of human diet as well as being an essential requirement for the successful keeping of livestock; it has been estimated that daily needs were 12–15 gm for a human adult, 50 for a horse and nearly 100 for a cow.[2] It was also used for preserves and in the preparation of hides. Rock salt called *kolomyya*, was obtained near Galich on the Dnestr and was traded throughout the Russian area even before the Mongol invasions.[3] The princes in this area seem to have distributed salt or the right to mine it to their retinue as part of their salary or grants.[4] But this source of origin was beyond the frontier and, throughout the period dealt with here, salt obtained from the brine of lakes or underground sources, as well as sea salt, was much more important. References to salt being obtained from brine become more frequent from the mid fourteenth century onwards and the locations extend across north European Russia. Rukha or Ryukha, near Pskov, is mentioned in 1363;[5] Galich Sol' is mentioned in 1389;[6] other sites include Staraya Rusa, Vychegda, Vologda, Kostroma, Tot'ma, Nerekhta, Pereyaslavl' Zalesskii and Rostov, apart from the important Northern Dvina and the White Sea areas. Most if not all of this extractive industry, however, did not take place within the usual peasant farm and it will, therefore, not be dealt with here. Nevertheless, the fact that the Russian for 'hospitality' is *khleb-sol'* (i.e. bread-salt) reinforces Nenquin's suggestion that the bread–salt relationship 'might be a very vague echo of the parallelism in agriculture – saltwinning'.[7]

Honey

Honey, mainly from bees (presumably *Apis mellifera mellifera*), though wasp and hornet honey were also known, was gathered by the Slavs from the

---

[1] *Ibid.* 112.   [2] Nenquin, *Salt*, 140.   [3] Rybakov, *Remeslo drevnei Rusi*, 571.
[4] *PSRL*, II.789 (1240).   [5] *PL1*: some take this to refer to Staraya Rusa.
[6] *DDG*, no. 12, 34.   [7] Nenquin, *Salt*, 144.

earliest times.[1] The simplest method of keeping bees was to put a mark as a sign of ownership on a forest tree in which bees had nested. Pines, limes and oaks were preferred, but elms, poplars, and willows were also used when men began to cut hollows in which bees could nest.[2] In the Pskov area such nests evidently existed in the first half of the fourteenth century; a purchase deed refers to bee-trees (*bort'*) 'old-put and new-put and pick [*peshnya*] and bee-garden [*pastke*]'.[3] The 'pick' probably refers to a narrow-bladed adze used for hacking out hollows in trees to accommodate bees, perhaps similar to Levasheva's type I; this type has not been found in Russian contexts after the twelfth century, but it continued in use among the Finnish peoples some of whom, the Mordva, for instance, were famed beekeepers.[4] Moreover, the Mordva seem to have gone over to the hiving of bees only towards the sixteenth century.[5] The term *pastke* is not known from other sources, but Marasinova suggests it may have the meaning of modern *paseka* (bee-garden). *Bort'* and *peshnya* appear to be distinguished in a fifteenth-century charter from the same area, presumably in the sense of a natural and artificial tree-hollow for bees.[6] Thus, this purchase deed may be taken to distinguish between hives or skeps hung or fixed in trees, hives made in a tree trunk, and, perhaps, an area where hives or skeps were kept, probably near the house.

A will of 1392–1427 mentions three lots of 'bees' (*troi pchely*), presumably hives, left to the Trinity monastery of St Sergius in Kinela volost.[7] Hives are also mentioned in fifteenth-century Novgorod documents and there is a reference to skeps, probably made of bast, in 1500.[8] Hive tools for taking out the combs, very like their modern counterparts, are known from Kievan times.[9]

Thus it seems likely that starting from the fourteenth century beekeeping in man-made nests or hives gradually spread in north-west Russia, but this change from gathering to farming was evidently very slow. In the seventeenth century Richard James referred to 'a kinde of lesser sorte of beare with a whitish ringe on the necke, verie fierce and so calld [*bortnik*] because he uses to clime trees for honie and robs the boores *borts*'.[10] He added that 'these trees have all the bowes lopt of untill the top, and to take their honie, they have long leather ropes etc.' Crampons were also used. Smoke was used to drive bees from the nest[11]. Peasant beekeeping evidently retained many of the ancient methods so vividly depicted in the Tale of Petr and Fevroniya.[12] Concern for the preservation of trees as valuable property, even though

[1] Bogoyavlenskii, *Drevnerusskoe vrachevanie*, 59; Smith, *Origins*, 45–6.
[2] Mal'm, *TrGIM*, 32. 131. See also the Kazan' evidence, p. 213 below.
[3] *NPG*, no. 7. On *peshnya* see Galton, *Survey*, 13.
[4] Levasheva, *TrGIM*, 32.40–1; Mal'm, *TrGIM*, 32.135.   [5] Mal'm, *TrGIM*, 32.132.
[6] *GVN*, 345.   [7] *ASEI*, I. no. 11.   [8] *NPK*, II.312; III.6.
[9] Mal'm, *TrGIM*, 32.130, 133, 135.   [10] James, 38a:13–14.
[11] Sreznevskii, *Materialy*, I.764.   [12] *TrODRL*, VII.225f; Smith, *Origins*, 46.

temporarily unoccupied by bees, is shown by the careful way in which a note is made in sixteenth-century registers of inquisition of bee-tree marks used by individuals or by groups of bee-men. Such marks are both described in words, using conventional expressions which suggest the usage was well established, and also depicted.[1]

Honey was used for most of the purposes for which sugar is used today. Probably most of it was used for drinks. Mead, metheglin and a sort of very weak toddy (*sbiten'*) were common. The latter, a hot drink made of honey and water flavoured with herbs such as mint, hops, common tormentil (*Potentilla tormentilla*) or bistort (*Polygonum bistorta*), or with more exotic imported spices, was commonly used in old Russia.[2] Honey was also used for various sweetmeats such as the berry preserve already mentioned (p. 58) and for such dishes as *kut'ya*, 'a kinde of meete made of hole corne and pease boild and honied'; this, however was a ritual dish perhaps eaten on only four days in the year.[3] Honey was eaten with pancakes, gruel and wheat.[4]

It is clear, however, that some areas lacked honey and that not every settlement or peasant holding had a bee-tree or hive. The question of internal trade in honey is, therefore, of some interest. Unfortunately, as Khoroshkevich has pointed out with reference to wax, no special research has been carried out.[5]

Miscellaneous

There were a considerable number of plants which could be gathered and used to add flavour to many dishes. These probably included coriander and caraway.[6] Dill continues to be widely used in the preparation of pickles and such preserved food was very important in Russia because of the long winters. Wormwood, presumably in very small quantities, was used as a spice with meat and fatty dishes such as goose. Tansey, horseradish and thyme were also used, the latter apparently being particularly favoured by the Slavs from very early times. St John's wort was used with other herbs as a bouquet garni. All these were to be found wild in European Russia, but no doubt many of them would be cultivated, or at least planted near the home, from early times.

Juniper berries, often found in pine woods, were used to stimulate the

---

[1] These marks are, unfortunately, rarely reproduced when the sources are published. See 'Toropetskaya kniga', ff. 150 v, 152v, 153v, 155v, etc.; Bakhrushin, 'Knyazheskoe khozyaistvo', 22–3. See Kotkova, in *Issledovaniya po lingvisticheskomu istochnikovedeniyu*, 120–33, on some late-sixteenth- and seventeenth-century bee-marks.

[2] Tea does not seem to be known in Russia till the mid seventeenth century and it did not become common till the establishment of Kyakhta in 1725 enabled the Russians to obtain it direct from China; Prozorovskii, 'Chai', *Domashnyaya beseda*.

[3] James, 27a:11; See also the Life of St Sergius in Fedotov, *A Treasury of Russian Spirituality*, 65.

[4] Gorskii, *Ocherki*, 76; Sreznevskii, *Materialy*, II.122.

[5] Khoroshkevich, *Torgovlya*, 124.   [6] *Rasteniya primenyaemye v bytu*, 125.

appetite. Iceland moss was used as the basis of an appetising drink. Perhaps the only loan from the Russians to us in this field has been angelica. Two forms were used as mild stimulants and as flavourers, *Angelica silvestris* and the less common *A. archangelica*. One stimulant which, if we are to believe Richard James, was far from mild was henbane (*Hyoscyamus niger*), 'a kinde of herbe that beine dried and throwne on the stones in a bathstove will make the bathers beate one another even to death'.[1] Fungi, too, were used as stimulants, particularly *Amanita muscaria*, which could be used to induce intoxication either alone or by being steeped in whortleberry juice.

MEDICINES

As in any peasant society, in ancient Russia a very large number of plants were gathered and used for medicinal purposes. Recently it has been reported that considerably more than 200 species are still used in folk as opposed to scientific medicine; in Belorussia people use 290 species, of which only 130 are used in scientific medicine.[2] It seems unlikely that in earlier times the number will have been less than this. Attitudes to herbal medicine seem to have altered over time, however. There was always an element of magic about the herbal art, but fear of this, or at least hostility to it, mainly from the church, seems to have been the dominant attitude till about the sixteenth century. In the thirteenth century Ilya of Novgorod complained of women who took their sick children to wizards 'and not to the priest for prayers'.[3] In the sixteenth century, however, there were a large number of translations, mostly from the west, of herbals and books of health and the first pharmacies were opened in Moscow and a few other towns.[4] Early in the seventeenth century a Pharmacy Department was organised in Moscow with herb gardens and storehouses of raw material. The 'sellar of saves', however, is likely to have preceded the pharmacies by several centuries and to have continued after their establishment.[5] Thus, by the seventeenth century herbal lore seems to have been somewhat more acceptable than in earlier times. It still remained a somewhat hazardous profession, though, and practitioners were burnt at the stake as sorcerers. The term *zelie* continued to mean both medicine and poisoning or the magical use of herbs; the immanent power could evidently be for good or evil.

Fungi, too, were used in folk medicine. These include some, like the honey agaric, which were used internally and others which were only used externally. Of the latter the fly agaric (*Amanita muscaria*) was not only used, as its

---

[1] James, 55:24. [2] *Lekarstvennye rasteniya dikorastuyushchie*, 3.

[3] Quoted by Bogayavlenskii, *Drevnerusskoe vrachevanie*, 14. Cp. the use of Lithuanian *burtnik, burtnikas*, as distinct from the word of foreign origin used for warlock or witch in deeds relating to witch burnings: Jablonskis, *CPHB*, 274.

[4] *Lekarstvennye rasteniya dikorastuyushchie*, 10. [5] James, 14b:55.

name suggests, as an insecticide, but also against rheumatic pains. Fungi or extracts made from them were also used to treat wounds, and, according to late seventeenth-century medical texts, frost bite.[1] Shelf-fungi were apparently used as a sort of poultice. In 1462 'the Grand Prince ordered a shelf-fungus to be burnt on his back because of his illness, against the consumption'.[2] Another source adds 'as is the custom for those ill with consumption'.[3] The patient died soon afterwards.

Products derived from wild animals were also used for medicinal purposes. Some, such as dried toads, seem to be more akin to magic than medicine, but castor, the dried preputial follicles and secretions from the beaver, still appears in the pharmacopoeia. This was perhaps the main such animal product.

It seems unnecessary here to list the wide range of ailments which were treated with what was gathered, but it may be of interest to glance briefly at a few of the health problems which troubled the old Russian countryside.[4] Attempts were made to keep insect pests under control by the use of poisonous plants such as fly agaric, the false hellebore (*Veratrum Lobelianum*) and others. The false hellebore was also used to deal with lice. Tar obtained from spruce was used by mowers 'to stinke away the Moskitos'.[5] Drunkenness could sometimes be cured by adding an infusion of the root of *Asarum europaeum* to the drunkard's liquor; the resulting sickness was sufficient to create an aversion to drink 'in some cases'.[6] The major epidemics of bubonic plague, typhus and other diseases were likely to make greater inroads into the town than the village population. Various dietary measures were very popular, but without effect.[7] The custom of building a church in a day, between matins and evensong – possibly because the timbers gathered from the forest were prepared and probably numbered in advance – seems to have been no more effective.

CONCLUSION

This survey of gathering and extractive industry has not answered the question of how important these activities were relative to agriculture. Nevertheless, it has perhaps shown how important the forest was in supplying a range of materials essential to the economy of the agricultural holding. Buildings and their furnishings, implements, containers and means of transport, mills and weirs, fuel and lighting, writing material and medicines were supplied almost exclusively by what could be got from the forest. Some clothing also came from this source, but the holding itself also made important contributions in

[1] *Lekarstvennye rasteniya dikorastushchie*, 312.
[2] Sreznevskii, *Materialy*, III, 1009.   [3] *Iosafovskaya letopis'*, 53.
[4] *Lekarstvennye rasteniya dikorastushchie*, and Nosal' & Nosal', *Lekarstvennye rasteniya*, give many details.
[5] James, 66:1.   [6] Nosal' & Nosal', *Lekarstvennye rasteniya*, 37.
[7] Vasil'ev & Segal, *Istoriya epidemii v Rossii*, 31.

this respect and so it is much more difficult to ascribe a weight to gathering for this purpose.

A wide range of food items also came from the forest. Some, such as berries and fungi, were probably important as supplements to an otherwise monotonous diet; in relict form such gathering remains as a much livelier tradition among Russians than the annual trip for blackberries with us. Others, such as game and fish, were important as major sources of protein, but, in the period prior to the eighteenth century, fish may have been more important than meat in the diet of the majority of the population. This was not due to any natural or technical limitation; game was still more abundant in general and the peasants had the means and skill to take it. Hunting rights being restricted more and more to the nobility may have stressed the importance of fish in peasant diet, though we cannot be sure of this. It may also have increased the importance of smaller animals in the peasant bag. Gathering as a major source of food, however, seems likely only to have occurred in isolated forest settlements and in a frontier situation. In 1638, for instance, the Don cossacks reporting on the capture of Azov said there were no stocks of grain and that they were feeding themselves on 'fish, game and grass'.[1] It was only as agriculturalists gradually extended their control of the steppe frontier that arable land became important in this area. Even in 1685 it was said that

hitherto along the rivers Khoper and Medveditsa they tilled no arable and sowed no grain, but got grain from the Russian towns and fed themselves with game and fish; but now in those small towns those Cossacks have introduced arable; and hearing that those Cossacks are tilling arable and sowing grain, peasants of the court lands, of those holding by service and holding heritable estates and of monasteries, as well as labourers and boyar villeins, flee to those Cossacks.[2]

A final point to note is that as gathering merges into extractive industry, it not merely supplements agriculture, but may offer an alternative form of economic activity to it. The extraction of salt from brine is the best example of this. In this case, the inputs required were so considerable that they were beyond the means of the average peasant household, or even many groups of such households. Other activities, such as the extraction of oleo-resins from pine and birch, might also become more or less full-time work while undertaken, but could still be accommodated within the round of varied employments usual to the peasant family. They could, therefore, be engaged in without involving any substantive change in the nature of the peasant economic unit.

[1] *Vossoedinenie Ukrainy s Rossiei*, I.203.   [2] *DAI*, no. 17, I, p. 124.

4

The family

The farm unit we take to be indicated by the three-fold formula was based on the labour of the household family. Outside labour was rare, partly no doubt because of the nature of settlement and the dispersion of population resulting from it for much of the area and the period with which we are dealing. There seem also to have been cultural factors involved; the family in Russia had certain characteristic features. We must now ask what the nature of this family was.

Basically, the family seems to have been the nuclear family of the married couple and their young children. The term *dvor* indicated the 'household' in the sense both of a physical complex of land and buildings and also of a group of persons. To a great extent this term probably carried much of the meaning of 'family', at least at peasant level. Terms such as *semiya*, etymologically the closest to the modern Russian for 'family', or *rod* 'kin group', do not seem to have been closely relevant. The former is found in documents issued by princes; the latter has a more extended meaning, something like 'relatives'. Perhaps of greater help to us is the terminology relating to those family members who were not connected by blood or marriage. Such terms are *priimak, vlazen'* and *prikhodets*. All indicated new arrivals, though it seems possible to go a little further than this and to suggest that the first indicated full acceptance into the household, while the other two may not necessarily have done so. *Priimak* (cp. *priimati* to accept), in fact, has survived into the Soviet period; the term for adoption into a collective farm family (*kolkhoznyi dvor*) is *priimachestvo*.[1] At the start of the fifteenth century parents were enjoined to feed an adopted child (*priem'noe ditya*).[2] *Vlazen'* has simply the implication of entry into a new situation or place.[3] *Prikhodets*, literally 'arrival', has a similar meaning.[4] It seems likely then that the household labour force could be maintained by such means. Full adoption, presumably also gave certain property rights. Adoption might be in a more limited form, however, at social rather than family level. The new 'arrival' would be

---

[1] *Yuridicheskii slovar'*, the usual term for adoption is *usynovlenie* (cp. *syn*, 'son').
[2] Sreznevskii, *Materialy*, II.1400.
[3] Sreznevskii, *Materialy*, I.378, gives *v laziny* as meaning 'transfer to a new settlement' in the seventeenth century.
[4] In Toropets such men were of low status, usually they were known by forename only; see p. 182.

entitled, once accepted by the volost or other community, to a share in the commune's land allocation and its tax burden.

Early marriage was usual. In 1410 metropolitan Fotii instructed his subordinates not to marry a girl of less than twelve, 'but marry her when she enters her thirteenth year'.[1] In 1758 Daniel Printz noted that many girls married before puberty, at age ten, and boys at twelve or fifteen.[2] An early seventeenth-century document regarded marriage as normal for female slaves at eighteen, for males at twenty and for a young widow two years after the death of the husband. 'Do not hold the unmarried contrary to the law of God and the rules of the holy fathers,' it declared, 'and fornication and filthy activities shall not multiply among the people'.[3] An Englishman in Russia about a century later than this commented that 'they marry very young in that countrey, sometimes when neither the Bride nor the Bridegroom are thirteen Years of Age'.[4] This may well be an exaggeration for the majority of marriages, but the fact of early marriage seems well established. This indeed appears to be a characteristic pattern for Eastern Europe and has survived into modern times.[5] Possibly ready availability of land encouraged, or at least failed to discourage, early marriage. A further indication of early marriage is to be found in the fact that there are at least three variant forms of a term indicating a man who has an illicit relationship with his daughter-in-law. The terms *snokhar'*, *snokhach* and *snochnik* (cp. *snokha,* daughter-in-law) may indicate differences in such a relationship but also show it was once fairly widespread.[6] The Metropolitan's Justice, a law code dating perhaps from the end of the fifteenth or from the sixteenth century, though based on earlier materials, continued to lay down a fine of 100 grivnas for this relationship, the same as for two brothers sharing one woman; these were the highest fines; bestiality, on the other hand, was cheap at 12 grivnas.[7] One way of attempting to adjust the land–labour ratio given a shortage of hands would be to marry off a young son to an adult but young female; this would account for the custom being fairly widespread. In such cases the wife would live with her husband's family and an adult worker was added to the farm; sexual relations between father and daughter-in-law were, as it were, a bonus which lasted until the young son became sexually mature.[8] The alternative way to increase the

---

[1] *RIB*, VI, no. 33, 275: also 284.   [2] *SRL*, 2.723.

[3] *PRP*, IV.588. For a translation of this statute see Smith, *Enserfment*, 103–7.

[4] Perry, *The State of Russia under the present Czar*, 200. Cp. Weber, *The Present State of Russia*, I.120.

[5] Hajnal, 'European marriage patterns', 101–3; Pisarov, *Narodonaselenie SSSR*, 177–9. An instruction to officers of inquisition in 1676 defines youths (*nedorosli*) as those under 15; cited in Veselovskii, *Soshnoe pis'mo*, II.125. Even in the late nineteenth century the active population was taken to be those between 15 and 50.

[6] Dal', *Tolkovyi slovar'*, IV.249; Cp. Sreznevskii, *Materialy*, III.454.

[7] §33 in *PRP*, III.428.

[8] In Bosnia such a custom continued into the twentieth century (personal communication of Professor Dubić).

labour force would be to bring in a man, a new arrival or an adopted son as mentioned above; but the latter at least might then compete with any natural children for the inheritance. This last point is probably of importance since it seems to suggest that the phenomenon discussed related to peasant families as independent units and was not a means used by lords to adjust the labour force in their estates.

Land normally passed by the male line, but there is some evidence that inheritance by the female line was recognised fairly early. The law code of 1497 laid down that a daughter could inherit all, including land, if a man died without making a will and there was no son; if there was no daughter either, a close relative was to inherit (*vzyati blizhnemu ot ego rodu*).[1] This rule was repeated in the 1550 law code, but was added to in the 1589 version.[2] It was made more explicit by the statement that the close relative should be of the father's or of the mother's line. Commenting on this article, Kopanev came to the conclusion that the possibility of inheritance by the female line 'was close to the manner of inheritance among the peasantry'.[3] Certainly in the north of Russia, where peasant traditional culture survived best, daughters could claim family property including land.[4]

The family, then, was basically the nuclear family of the married couple and their young children, those who, to quote the Code of Laws of 1649, 'live together with their father and mother and not apart from them'.[5] In some circumstances it was the situation as regards the family which determined a man's social situation. For example, the priest's son 'who lives with his father and eats his father's bread' was thereby subject to the administrative authority of the metropolitan, but if he separated and lived apart from his father and ate his own bread he was then subject to the Grand Prince.[6] Brothers, children or nephews might live on the peasant tenement of their fathers or kinsfolk, but they might depart and begin to live on their own holdings.[7] Such holdings, even if held by a bachelor, were treated for certain purposes as though they were a full holding; but in practice a farm seems normally to have required at least the married couple to run it.[8] This was why a priest whose wife died, since he was not allowed to remarry, became known as *rozpop,* a quondam priest; he was allowed to remain in office no more than a year.[9] 'For wich Reason', Perry remarks, 'it is remarkable that the Priests use their Wives better than any Man in the Countrey'.[10] The underlying economic reason, of course, was that the holding could not be worked by the priest alone. Similarly, a widow by no means always continued to

---

[1] *Sudebniki XV–XVI vekov,* I.27.   [2] *Ibid.* 174, 406.   [3] *Ibid.* 542.
[4] *SGKE,* I.74.18.   [5] Chapter XI, §28; translated in Smith, *Enserfment,* 151.
[6] Smith, *Enserfment,* 45–6. This model charter was dated 1404.
[7] *Sobornoe ulozhenie 1649 g.,* XI.24; Smith, *Enserfment,* 149.
[8] *PRP,* IV.588–9 (a statute of 1607); Smith, *Enserfment,* 105.
[9] The term also indicated a defrocked priest.
[10] Perry, *The State of Russia under the present Czar,* 230.

work her deceased husband's holding. Some became host-makers for the church, or lived by begging.

We have virtually no evidence on family size, though the general impression formed from the rate of colonisation and forest clearance is that the birth rate was fairly high, as one would expect with the early marriage pattern, and the survival rate was also high. Three late-fourteenth-century documents give rates of payment due from villeins undertaking a court case without a guarantor; the payments are different for individuals and for families as follows:[1]

Year	Individual rate (dengas)	Family rate (dengas)
1375	20	50
1381	— (12 from horseless peasants)	6
c. 1396	6	18

This suggests that a family was then regarded as containing normally the equivalent of 2.5–3 adults.[2]

Finally, we should remember that adjustments in household size which were important in determining the land–labour ratio could be made by taking in 'new arrivals'; this meant that the household was not always a marriage or blood group, but might include non-relatives, though no slaves or similar dependants. The household was a work group as well as a group related by blood.

There seems to be no evidence for the household of extended kin in Russia in the period dealt with.[3] A widow's gift, dated 1484–1502, to a monastery, was made for the abbot to commemorate thirteen relatives.[4] Given families with three or four children, a widow might have thirteen dead relatives from siblings of herself, her husband and of their parents, even excluding their spouses. The vast majority of peasant households in the registers of inquisition are noted with only one male name. Sometimes another male is noted or a relative mentioned, but this is fairly rare. In one quite exceptional case, seven males were noted in one tenement in Toropets.[5] Veselovskii pointed out that 'relatives, even sons, preferred to go off to free land and establish their own hamlets, and not settle in the old hamlets'.[6] This was a major factor in that creeping colonisation which settled the east European forest zone.

---

[1] *DDG*, nos. 9, 10, 15; Zimin, *PI*, VI.322

[2] This approximates fairly well with a twentieth-century calculation for a peasant family farm of six persons taken as equivalent to 3.5 adults; see p. 93 below.

[3] Unfortunately, the extended kin group is often uncritically accepted in the Western literature, e.g. by Shimkin, 'National forces and ecological adaptations'. See also Elnett, *Historic origin*, 2.

[4] Chaev, 'Severnye gramoty', no. 7, 132.   [5] 'Toropetskaya kniga', f. 184.

[6] *IGAIMK*, 139.30.

5

A production and consumption model

It now seems possible to put forward very tentatively a simplified model of production and consumption in the isolated farm household in the core of European Russia. This household was the basic unit of society for many centuries and throughout the period dealt with here the majority of Russian people lived out their lives within its framework. Yet it is this unit which is almost entirely concealed from us both by the relative paucity of documentary evidence and by the nature of that evidence. This unit is so important for any deep understanding of the long-term history of the Russian peasantry that it seems worthwhile attempting to describe it, however hesitantly.[1]

First, it must be stressed that this attempt at a model starts with the isolated farm. The importance of this is that the resultant picture initially leaves out of account complexities which in all probability affected the majority of Russian medieval farms; here no attention is paid to relations with other farms, with the lord of the land or with the government; my estimates leave out of account dues and taxation, as well as incomings from crafts and trades. This is done for the sake of simplicity. The model would have to be modified to take these complicating factors into account, and be extended by including other elements, in order to approximate to the real situation.

The 'normal' amount of land held by such a farm unit is very hard, if not impossible, to determine. Some very rough idea of size may be suggested, though it must be remembered that precise measurement of land area only developed in modern times. Most land measurement in history has been very flexible, taking account of quality of soil and other factors, and often expressed in terms of the amount of seed required. The *desyatina* (a tenth) was the land unit which occurs in the documents from the late fourteenth century. A deed of 21 October 1391 mentions that peasants have to mow the monastery meadows by desyatinas.[2] It has been suggested that the original unit of which the desyatina was a tenth was the *versta*, a unit of length. A square versta would approximate to 11 hectares. There is, however, no evidence at all that

---

[1] A draft of this section was circulated as Centre for Russian and East European Studies Discussion Paper, Series RC/D, no. 1, and I am most grateful for the comments received. In particular I would like to thank Professor A. L. Shapiro and his Leningrad colleagues for their help.

[2] *AFZ*, I, no. 201, 180, translated in Smith, *Enserfment*, 40.

such an area was a 'normal' holding, though the amount approximates to the theoretical normal size of holding in many other societies.

By the sixteenth century the *vyt'* (lit. 'a share') was a common measure of peasant holdings for taxation purposes; it varied in size from 12 to 16 chets (6½ to 8½ ha) in each of the three conventional fields according to the quality of the land. It thus amounted to 18 to 24 desyatinas (19½ to 25½ ha) in all.[1] According to Klyuchevskii, on very poor land the vyt might be as much as 20 chets in each field (30 desyatinas altogether: 33 ha).[2] Fletcher's statement that 'the wite conteyneth sixtie chetfird. Every chetfird is three bushelles English, or little less' seems to describe the 20-chet vyt mentioned by Klyucheviskii.[3] The real peasant holding was usually smaller than this taxation unit and was sometimes taken to be half a vyt, though there seem to have been many holdings smaller than this.[4] This unit of 9–15 desyatinas (10–16½ ha) in all three fields is taken as the basis of the following calculation. Incidentally, it may be noted that the basic peasant unit in the Novgorod lands, the *obzha,* amounted to 10 Moscow chets in each field or 15 desyatinas in all, i.e. the same as the maximum assumed for the half-vyt unit in the more southerly Moscow area.[5] It has been pointed out that the Novgorod *obzha* had 2–7 desyatinas (2–7½ ha) sown area.[6]

There seems to have been no fixed relationship between hayland and arable within the holding; estimates are therefore given assuming that two-thirds, one-third, one fifth and, lastly, none of the holding was hayfield. Certainly, it will usually have been possible to supplement any hay from within the holding by some obtained from clearings and patches in the forest. Yields, in both cases, however, are likely to have been low. The rates here assumed are taken from the average hay production for the non-Black Earth areas of European Russia for dry-valley and water-meadow lands in the forest area.[7] Dry-valley meadows and pastures on podzols often have a yield of less than a metric ton per hectare; this is roughly comparable with the 750 kg per ha implied by the frequently used fifteenth–sixteenth-century rate of 10 ricks (*kopny*) a desyatina, since, as Abramovich has shown, these ricks were of about 5 puds in weight.[8] This incidentally illustrates how technical coefficients may be used with reasonable confidence unless there is some reason to believe they have been substantially modified over time.

---

[1] Sreznevskii, *Materialy*, I.455–6; Klyuchevskii, *Sochineniya*, 6.203, 508; Shapiro, *EzhAI* (1960), 208–9; Veselovskii, *Soshnoe pis'mo*, II.441 f.

[2] Klyuchevskii, *Sochineniya*, 6.203. The range 12–16 chets is accepted in *PRP*, IV.95 and V.85. See also Kamentseva & Ustyugov, *Russkaya metrologiya*, 84. Rozhkov, *Sel'skoe khozyaistvo*, 72, took the vyt as 5 desyatinas of good land in each of three fields and this has been accepted by Abramovich, *PI*, XI.369. A peasant vyt of 15 desyatinas (16½ ha) is mentioned in *AFZ*, I, no. 166, 152.   [3] *Of the Russe Commonwealth*, 49.

[4] *PRP*, IV.140.   [5] Abramovich, *PI*, XI.380.   [6] Milyukov, *Spornye voprosy*, 36.

[7] Efron-Brokgauz, *Entsiklopedicheskii slovar'*, XXXII.357; *Sel'skokhozyaistvennaya entsiklopediya*, IV.455–6; *Spravochnik predsedatelya kolkhoza*, II.546.   [8] *PI*, XI.368.

The sown area is taken to be two-thirds of the area of the holding less the hayfield, the remaining third being fallow. This should not, of course, be taken to mean that a regular three-course system existed in a developed form. In many cases as we have seen, the peasant holding consisted of individual fields or closes (*nivy*), but not all of these would be worked every year; it is assumed that such failure to cultivate, to put it in a neutral way, may have affected a third of the arable.

The winter field was sown to rye. The harvest, calculated in terms of threshed grain to seed, varied on old-worked, possibly exhausted soils in central areas at the end of the sixteenth century around 2–3-fold.[1] A village on the more fertile soils of Opol'e, Oboburovo, gave 3.2-fold in 1592. The Patriarch's village of Podberez'e, Vladimir uezd, had a yield of 3.6-fold in 1599.[2] Exceptional yields of 8.5- and 10-fold were recorded on land in Yaroslavl' uezd in 1542–3 and of 1.7 on poorer soils in Shelon' pyatina in 1595–6.[3] Such figures relate to monastic or court lands and it is improbable that peasant holdings would have averaged more than 3-fold.[4] This is the figure which is assumed for the purpose of this calculation.

The seed rate is mentioned in two documents of the late fifteenth and mid sixteenth centuries.[5] The second of these gives much relevant information and in part reads as follows:

And, in all, the village and the hamlet people have 4,150 ricks of hay and in the village and the hamlets there are 134 vyts, each with 6 desyatinas of peasant arable in a field [i.e. 18 desyatinas in three fields]; and they have to till for the Grand Prince in the village at Buigorod 134 desyatinas, a desyatina a vyt, and a desyatina is 80 sazhens long and 30 sazhens wide [170 by 65 metres], and they have to sow 2 chetverts of rye on a desyatina and twice as much for oats, and they have to cart the dung from their tenements to the Grand Prince's arable at the rate of 30 *kolyshkas* on a desyatina, and the *kolyshka* measures 4 spans in length and width and 2 deep.

The Russian span was measured from index finger to thumb, so the *kolyshka* was roughly 71 cm by 71 cm by 35 cm.[6] Until the early seventeenth century the chetvert or chet contained 4 puds (65.5 kg) of rye; this is the unit assumed in these calculations.[7] The manure would amount to roughly 5 cubic metres a hectare; if we assume that 1 cu m of manure weighed 900 kg if well rotted, 300 kg if fresh,[8] this would give something like 4,500 to 1,500 kg per ha.

The spring field was usually sown to oats. The seed rate was twice that for

---

[1] Gorskaya, 'Zernovoe zemledelie', 214. On yield ratios in general, including some Russian material, see Slicher van Bath, *AAG Bijdragen*, no. 10.

[2] Rozhkov, *Sel'skoe khozyaistvo*, 55.

[3] *Ibid.* 55–7. Early- and mid-eighteenth-century yields appear to have been less than 3-fold; see Indova, *Dvortsovoe khozyaistvo v Rossii*, 156–62.

[4] Cp. Mordvinkina, *MISKh*, IV.324.   [5] *AFZ*, I, no. 166, 152; II, no. 178, 176.

[6] Kochin, *Sel'skoe khozyaistvo*, 154, says it was cone-shaped 'about a metre in diameter and half a metre high'.   [7] Kamentseva & Ustyugov, *Russkaya metrologiya*, 88–113.

[8] *Spravochnik predsedatelya kolkhoza*, II, 58.

rye, i.e. 4 chetverts a desyatina. The yield of oats varied from 5.2- to 4.8-fold, exceptionally, and more usually 3-fold or less.[1] It seems to have been about 2.7-fold in central areas in the early seventeenth century.[2] In 1657–61 on an estate of the boyar B.I.Morozov near Nizhnii Novgorod the oats yield averaged 3-fold.[3] Yields therefore seem to be comparable with those of rye, though the exceptional harvests were not so high; a return of 3-fold is therefore assumed for oats, though this may be somewhat optimistic. It is assumed that per unit of volume oats are $\frac{2}{3}$ the weight of rye.[4]

TABLE 8. *A farm unit: estimates of annual production (4 hypothetical farms)*

	A	B	C	D
Tenement area (des.)	9–15	9–15	9–15	9–15
Hayfield (des.)	6–10	3–5	2–3	0
Arable (des.)	3–5	6–10	7–12	9–15
Sown area (des.)	2–3	4–7	5–8	6–10
Fallow (des.)	1–2	2–3	2–4	3–5
Winter field (des.)	1–2	2–3	2–4	3–5
Rye, total harvest (chets)	6–12	12–18	12–24	18–30
seed (chets)	2–4	4–6	4–8	6–10
net harvest (chets)	4–8	8–12	8–16	12–20
Spring field (des.)	1–2	2–3	2–4	3–5
Oats, total harvest (chets)	12–24	24–36	24–48	36–60
seed (chets)	4–8	8–12	8–16	12–20
net harvest (chets)	8–16	16–24	16–32	24–40
Milk (litre), 1 cow		714–955		
Manure (m. tons), 1 cow		3–4		
1 horse		4–5		
Straw, rye (puds)	48–96	96–144	96–192	144–240
oats (puds)	48–96	96–144	96–192	144–240
Hay (puds)	300–500	150–250	100–150	0

Notes: 1. Assumptions: 1 chet of rye weighed 4 puds (65.5 kgs),
   1 chet of oats weighed $2\frac{2}{3}$ puds (43.7 kgs),
   1 litre of milk weighed 1 kg.
   2. Fractions are not shown, since these estimates are approximations.

Rye and oats seem to have been the main crops in the central area, though barley was also grown, as well as small quantities of spring sown wheat and spring rye. Buckwheat was also grown fairly widely by the sixteenth century, though not on a scale to compare with the two chief grains.[5]

[1] Rozhkov, *Sel'skoe khozyaistvo*, 55–7.   [2] Gorskaya, 'Zernovoe zemledelie', 192.
[3] Strumilin, *VE* (1949), 2.52.   [4] See *The Economist guide to weights and measures*, 50.
[5] Krotov, *MISKh*, IV.422–6. Cp. Gorskii, *Ocherki*, 28. Evidently the old saw 'Cabbage soup and gruel is our food' (*Shchi da kasha pishcha nasha*) is unlikely to date from a time earlier than this.

Pulses do not appear to have been very important in quantity. An analysis of areas sown by eight villages belonging to the St Cyril of Beloozero monastery (1604–5) in Moscow and Dmitrov uezds gives the following percentages; winter rye 44, oats 36, barley 15, wheat 3, spring rye 1; peas were only mentioned in two villages. The Joseph of Volokolamsk monastery presented a similar picture as regards pulses, though this estate was livestock oriented and oats was the main crop.[1]

It is assumed that the rye straw weighed twice as much and the oats straw one and a half times as much as the grain harvested. Probably much of this straw would be grazed, since grain appears to have been cut fairly high to judge from illustrations in manuscripts.

Milk production is based on an assumed output of 6–8 Russian pounds a day for a lactation period of 300 days.[2] This is almost certainly an over-optimistic assumption. Cattle at this period, while varying considerably in size, tended to be quite small and probably averaged 200 kg live weight.[3]

The basic consumption figure of 3 chets of rye per adult a year (in figure 2) is based on the following considerations. A will of 1565 mentioned that a man and wife who were allowed a monthly dole should jointly be given each year 'ten chets of rye and eight of oats; and each six chets of barley and a poltina in money, and if there are children, a chet of rye and a grivna each'.[4] This 5 chets of rye per adult, plus an allowance of oats, Rozhkov regarded as a very high allowance. Other wills mention 3 chets of rye and 3 of oats, per person, but this, again according to Rozhkov, was a starvation allowance. He considered an average peasant family (man, wife and two children) needed not less than 10 chets of rye and 4 of oats.[5] This implies little more than 3 chets of rye per adult. In 1614, when the grain tax for the military was instituted, the basic ration was estimated at 6 chets of rye per man per year.[6] This high amount was supplemented only to a small extent by a chet of oat-meal and a chet of grits per ten men per year. Thus, 3 chets of rye a year per adult (little more than $\frac{1}{2}$ kg per day) is on the low side and would have to be supplemented by other grains, say 2 or 3 chets of oats.[7] This is the assumption made here.

Rye was consumed mainly in the form of bread, but was also used in the preparation of a wide range of dishes made with flour and of a slightly fermented beverage, kvas, and beer. Half a kilo a day, therefore, would be by

1 Gorskaya, 'Zernovoe zemledelie', 82; Shchepetov, IZ, 18.128–33.
2 Mendeleev, Sochineniya, 16.251.   3 Tsalkin, MIA, 51.48–50.
4 Cited in Rozhkov, Sel'skoe khozyaistvo, 260. Cp. RIB, II, no. 32 (1549), 32–3, which specifies an allowance of 1 chet of rye monthly to certain beggars at Beloozero.
5 Rozhkov, Sel'skoe khozyaistvo, 261.   6 Veselovskii, Soshnoe pis'mo, 1.163.
7 The same consumption rate was assumed in the eighteenth century by provincial officials in the north since there were vegetables and berries to supplement the diet, as well as the produce of fishing and hunting; but this rate was regarded as half that for the central areas; Rubinshtein, Sel'skoe khozyaistvo Rossii, 243.

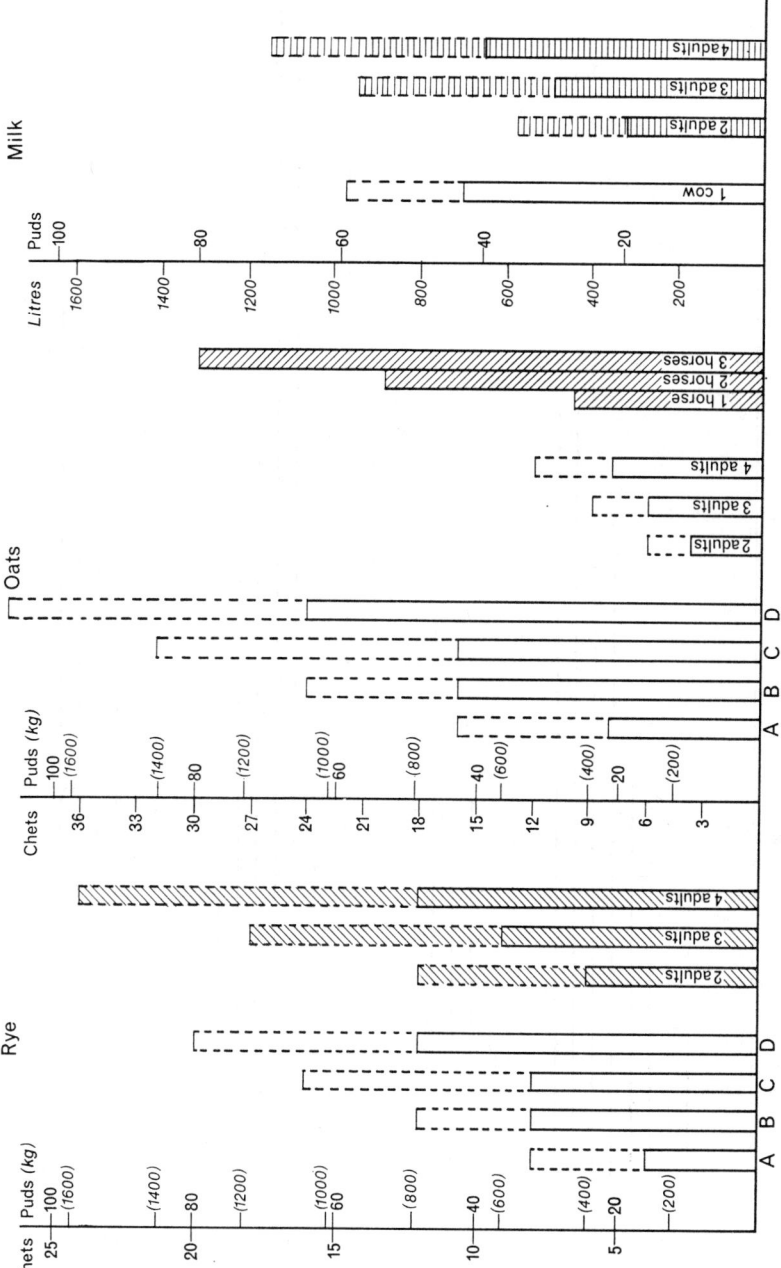

Fig. 2. A farm unit: ranges of net production and consumption by humans.

89

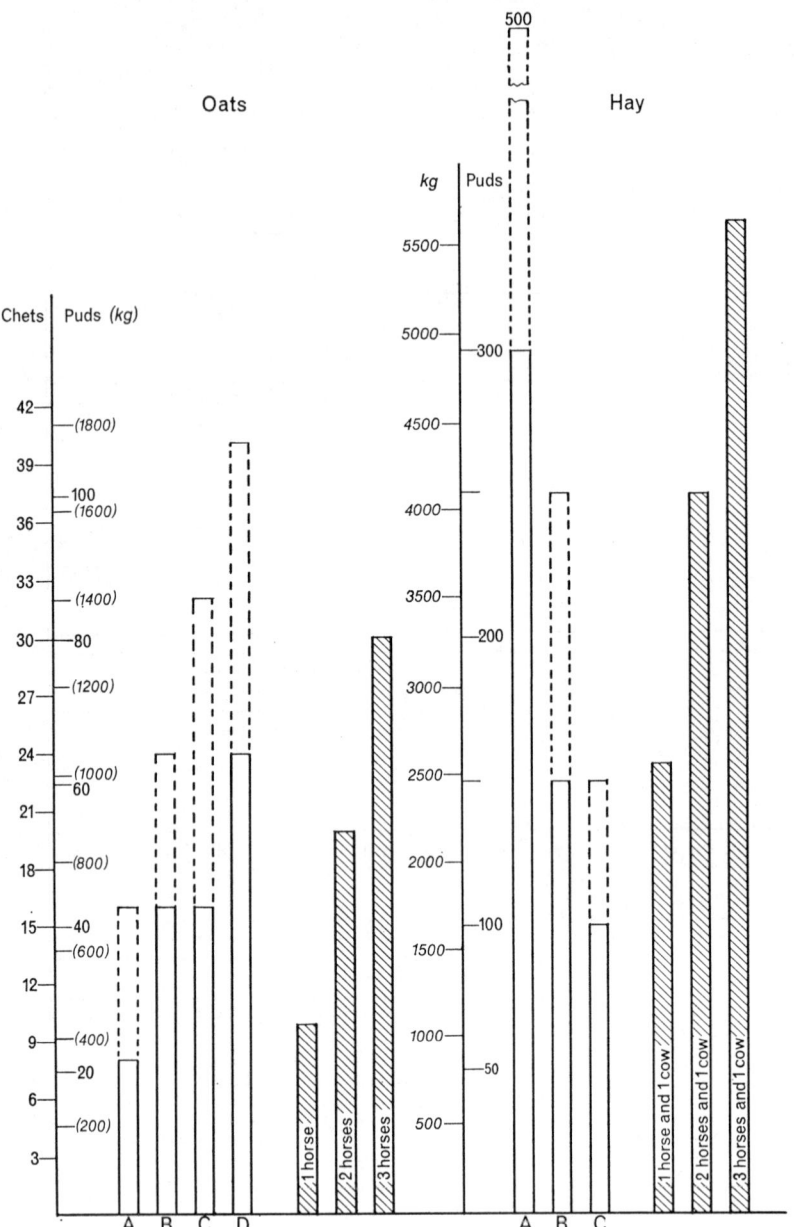

Fig. 3. A farm unit: ranges of net production and consumption by animals.

no means excessive, particularly if losses between field and table are allowed for.

Milk consumption is based on a later nineteenth- or early-twentieth-century figure for Moscow guberniya.[1]

It is assumed that the minimum of livestock were kept, a cow and, for draught, a horse for four desyatinas of sown land, probably the maximum which could be worked with one horse. Table 9 gives estimates of feed requirements for animals from the holdings described in table 8.[2]

In figure 3, finally, an estimate is made of consumption by animals related to output of oats and hay in each of the four hypothetical tenements. No account is taken in this table of other items which would be available, such as chaff and leafy fodder or other supplies which might be won from the forest. Animals were traditionally stalled a total of 204 days. This is within the limits of 200–220 days for the stall period given by the present day handbook for collective farm chairmen.[3] On the St Joseph monastery, Volokolamsk, a horse was allowed 5 ricks (kopny) of hay and 10 chets of oats a year.[4] This rick was evidently the standard measure rick weighing 15 puds or about 245 kg.[5] A monastic horse thus had a ration of hay amounting to about 75 puds or 1230 kg for the year. Hence the complaints, in 1626 at Sol' Vychegda, that 5–6 small ricks of hay (i.e. 25–30 puds) 'will be little feed for one horse for the year'.[6] A cow was allowed one 15-pud rick for 12 weeks, or about 600 kg for the stalled period. Thus, the hay rations for horses in this sixteenth century monastery, and for cows, were half or less of modern amounts, even when these are scaled down to allow for the difference between the weight of ancient and modern animals. The oats ration for the horse was also less than half the modern standard. These figures have been used as the basis of this calculation of actual consumption; it seems improbable that peasant animals would be better fed than monastic ones.

Straw for litter has been put at 2 kg a day for each animal by Mendeleev; a modern text book suggests as an average 2 kg for horses and 3 kg for cows;[7]

---

[1] Thorner, Kerblay & Smith, *A. V. Chayanov on the theory of peasant economy*. 129. Chayanov's figures are based on work by S. A. Klepikov. On p. 160 of the same work Chayanov gives a general estimate for a farm which may be compared with that attempted here; he concluded that an average peasant family of six persons (equivalent to 3.5 adults) required 10 desyatinas of land.

[2] Conversion factors for fodder units are taken from Popov, *Kormovye normy i kormovy tablitsy*, 192, 173, 151; maintenance rations in terms of fodder units are given on pp. 21 and 67 of the same work.

[3] See p. 43.   [4] Shchepetov, *IZ*, 18.112.

[5] Abramovich, *PI*, xi.366. That this identification of the unit is correct can be shown by reference to the proportional relationship between the modern feed rates given in Popov, *Kormovye normy*, 67, 18, 31 and 88; but this check also shows that the unspecified rick used in measuring the ration for sheep was probably the 5-pud small rick (*melkaya kopna*).

[6] *APD*, I, no. 210.   [7] Mendeleev, *Sochineniya*, 16.469; *Osnovy zhivotnovodstva*, 347.

TABLE 9. *Estimated feed and litter requirements*

	300 kg horse			200 kg cow		
	Fodder units	Physical units (kg)		Fodder units	Physical units (kg)	
Oats, stall period						
(200–220 days)	300–330	300–330				
work period						
(145–165 days)	435–495	435–495				
total	765–795	765–795	(440*)			
Hay, stall period						
(200–220 days)	500–550	1,050–1,155		700–770	1,470–1,620	
Other						
(145–165 days)	725–825	1,520–1,730		500–577	1,050–1,210	
total	1,275–1,325	2,675–2,780	(1,230*)	1,270–1,277	2,670–2,680	(500–600*)
Litter, stall period		320–352			480–528	

Notes: 1. Animal weights are based on Tsalkin, *MIA*, 51.48–50, 92–3, 96–7.

2. The rations given in *Spravochnik predsedatelya kolkhoza*, II.754–5, for a 400 kg horse have been scaled down. The totals for oats and hay given may be compared with the feed stocks recommended to collective farms (per horse) (*ibid.* 757–8):

	Fodder units		Kilograms				
	daily	p.a.	straw	hay	concen-trates	succulents	greenstuffs
Centre and N.W.							
RSFSR	8.67	3,164	–	2,900	790	900	4,600
Belorussia	9.06	3,307	400	2,300	1,020	800	4,600

The rations proposed in Popov, *Kormovye normy*, 67, are somewhat less than these.

3. The rate of 3.5 fodder units per day approximates to the rates of 3 units a day for in-milk cows and of 4 units given by Popov (*ibid.* 18, 21). Further information from Popov (*ibid.* 31, 149–53) suggests that from 6 to 9 kg of hay daily, varying according to quality, would be required to maintain this feed rate. See also *Spravochnik predsedatelya kolkhoza*, II.640–1.

* Figure for St Joseph monastery; see text.

I have assumed slightly less than 2 kg for horses and 3 kg for cows. The figure for manure is based on that given for present day animals, but it has been scaled down and checked, as far as is possible, by a calculation based on feed intake and litter.[1]

The estimates summarised here are only estimates. There are numerous sources of possible error. It has, for instance, been necessary to convert measures of capacity into units of weight; but the weight of grain per unit of volume depends on a number of factors which vary and about which we

[1] *Spravochnik predsedatelya kolkhoza*, II.56, 58.

know nothing for the period with which we are here concerned. Such factors are the size and form of the grain, its moisture content, whether the measure used was struck or heaped, and so on. Nevertheless, despite the tentative nature of these estimates, it may be of interest to see how they fit with what is known from this and later periods of Russian history. Therefore, while continuing to stress that the figures are at best tentative approximations, we may make cautious use of them as crude indications of probable sources of strain in the farm economy.

From table 8 and figure 2 it will be seen that tenement A with 3–5 desyatinas (3½–5½ ha) of arable supports two adults, but only in good years and at a low level in rye. In favourable years it might also meet the oats needs of two adults, but with minimum harvests there would be a deficiency. Tenement B with 6–10 desyatinas (6½–11 ha) of arable provides enough rye for two adults, even at a high level of consumption, and enough for three at a low level. The oats surplus would also be enough for three adults if the tenement had only one horse; if there were two horses, however, a maximum oats harvest would only leave enough for the minimum requirements for two adults. Tenement C with 7–12 desyatinas (7½–13 ha) provides enough rye for three adults, or even four in favourable circumstances. If only two horses were kept, there would also be enough oats for three or four adults. Tenement D with 9–15 desyatinas (10–16½ ha) provides enough rye for three or, in most circumstances, four adults, but there would sometimes be shortages of oats if three horses were kept. The family is likely to have included children as well as the two adults which may be regarded as the minimum labour force for a viable unit. It seems likely, therefore, that tenement A would have too little arable land to maintain the family without outside assistance. In fact, evidence to be adduced later suggests that units with 6–12 desyatinas (6½–13 ha) of arable land may have been usual for peasant farms, at least in some areas.[1] Tenement C would probably normally support two or three adults. Tenement D might support up to four adults in normal circumstances. In the 1920s Chayanov took a peasant family of two adults and four children as approximating to 3.5 adults.[2] This suggests that the average peasant farm would perhaps be likely to fall in the range of tenement C. At the yields assumed one cow appears to be able to supply the milk needs of at least three adults, but, as has been pointed out, the assumed milk yield may be much too high. The human diet would, of course, normally be supplemented by supplies of berries, nuts, fruit, fungi, fish, game and such like from the surrounding forest. Unfortunately, it seems impossible to put forward any estimates of the relative weight in the diet of such items resulting from gathering; they are likely to have fluctuated greatly in amount.

If we turn to the consumption estimates for animals (figure 3), we see that in each case there is no shortfall in the supply of oats for the number of horses

[1] See pp. 162–5.   [2] See p. 83, n. 2, above.

probably required by the sown area. Tenement B has enough hay for two horses and a cow in favourable circumstances; but this allows no hay for any other livestock. Tenement C would be unable to provide enough hay for two horses and a cow. Moreover, the larger arable area might sometimes mean that three horses were kept; and, in any case, the hay consumption figures are low by modern standards. The livestock sector posed a major problem in balancing conflicting trends. Within the farm area assumed, as the area of hayfield declined, the arable area increased and would reach a point where an additional draught animal was needed. Thus, the number of animals to be kept increased as the feed base for them within the tenement decreased.

Of course, any supplements of hay and leafy fodder gathered from the forest would help to meet deficiencies. Additional grazing would be available, including that from standing straw left after harvesting, and cut straw and chaff would be included in the animals' diet. Nevertheless, it seems that the livestock sector was inadequately supplied from within the holdings described and would usually have to find additional sources of fodder from the forest in order to survive. The amounts required, and consequently the area of forest necessary to maintain the livestock, would vary; the yields of hay from the holding and from the surrounding forest are not likely to have differed much, if at all.

Overall, then, these tentative figures suggest that the peasant farm unit could provide enough grain for the humans, especially in those periods of the family's life when the burden of children relative to working adults was not too great, but that the livestock sector was likely to be in part, sometimes in large part, dependent on supplies from the forest.

This picture of four hypothetical farms has not been put forward as representing typical farms, but simply in order to try to see as clearly as possible what the basic unit of Russian peasant life was like as regards getting a livelihood and what the rough technical limits were within which it could operate. Obviously size of holding would vary considerably, even within one area at any one period; the proportion between arable and hayfield varied; family size, and numbers and type of livestock varied. Above all, the peasant farm was not completely isolated.

There were social relationships, both horizontally between the farm members and other villagers or peasants in any commune which existed (though, as we shall see, small settlement size probably limited such co-operation), and also vertically with lords, lay or clerical. Apart from this, the farms here outlined had to make use of forest for pasture or hay, of someone's bull for stud purposes; salt had to be got, and the diet of rye bread, cabbage and milk products needed supplementing by fishing, hunting and gathering the abundant variety of food then available in the forest.

The forest, in fact, was more than a setting for the Russian medieval farm.

It was an important supplement to the farm for the diet of both men and especially animals; it provided material for buildings, tools and implements. In addition, where population density was not too great, it provided opportunities for bringing new land into cultivation. This was, of course, a slow process as is shown by the lengthy periods, as much as 20 years in some cases, for which exactions were relaxed from new settlers who undertook to clear the land.[1] Nevertheless, as the family grew and children married, there were normally opportunities for the young people to hive off from the original family and either establish a new holding within an existing community or on an existing estate, or even to start to bring new land into cultivation away from established areas. This phenomenon surely contributed to that lengthy, creeping colonisation by the Russian peasantry of the vast areas of the east European plain and the immense space beyond the Urals, a phenomenon which has sometimes been misinterpreted as evidence for the Russian peasants being nomadic. The alternative growth pattern, that of increasing size of holding with the growth of adult family numbers, seems to occur comparitively rarely. When it does it is often associated with an 'improvement' in the economic well-being of the peasant family which involves their departing from the usual pattern of a nearly exclusive concern with food production. We shall see that the larger peasant families were often concerned with crafts and trades as well as with farming in the narrower sense. For the mass of basic peasant units, however, the family itself, the commune or group of local communes or settlements, would usually provide some possibility of adjusting fluctuations between size of family and size of holding. The land-labour ratio could be modified by adopting strangers into the family, or, in theory, by reallocating holdings (in practice it was the burden of impositions which was reallocated). We know little of either, but more of the first than of the second method. Terms for such adoptees are found in the registers of inquisition. In addition, of course, particular incidents of history, not self-generated by the peasant farm, but impinging on it from outside, contributed greatly to movement and colonisation. Wars, internal and external, plague, famine, and excessive exactions and impositions would all from time to time act as strong and effective incentives to physical and social mobility; but consideration of such relationships between the peasant farms and the larger society must be deferred to a later chapter. Here I have tried to focus attention on some aspects of the basic unit of that society and its internal mechanism.

[1] E.g. *ASEI*, II, no. 92, 55; no. 164, 99. Sometimes, however, the relaxation was as short as three years.

PART II

REGIONS

6

Muscovy

We have so far considered the peasant household unit in static isolation. Now we must consider it in relation to the world around it. This is done here by looking at peasant farming in three regions which extend across central Muscovy.

'Muscovy' is an imprecise term, and some word of explanation may be needed for its use in my title. First, its acceptability to the English reader has been well established since the sixteenth century. Some Russian historians object that it is not the term used by contemporary Russians themselves and partly that it is a disdainful western term. Here it is used to indicate the Russian late-fifteenth–seventeenth-century state centred on Moscow, sometimes called the Moscow State, or the Russian tsardom or state. Second, the term is imprecise since the area under Moscow's rule varied over time, and also that rule itself varied in nature. In the fourteenth and fifteenth centuries Moscow's authority increased and evolved as other principalities were incorporated.

The three areas to be examined here are found on a line about the 56th parallel across what is now European Russia. They are the Moscow area itself, Toropets and Kazan'. Toropets was incorporated as a result of military action against Lithuania at the very beginning of the sixteenth century; Kazan' was incorporated by similar means half a century later. As will be seen, these two areas differ considerably in physical and cultural environment.

The nature of Moscow's rule evolved partly in response to the incorporation of other areas and to the different traditions found there. In essence it was the emergence of a tsardom, of a state, whether we regard it as centralised or not, headed by a monarch at least claiming absolute, autocratic power. This affected both social relations and also administrative and economic arrangements; the latter were not sharply differentiated. Administration overlapped considerably with appropriation, the extraction of surplus in various forms from the peasants.

In the early fifteenth century the core area of what we know as European Russia comprised a number of principalities. These principalities, some of which were apanages (*udely*) of members of the family of the Grand Prince of Moscow, were treated as the hereditary estate of each prince, but the land was classified in different categories and organised in different ways within each principality.

(1) Court lands, held and worked by peasants in the main, but also in part by slaves, mainly on demesne, formed the prince's estate in the narrow sense. This estate, located in different places throughout the principality, provided goods, mainly agricultural produce, and services to maintain the prince and his court officials. The administration of these lands, and the appropriation of the produce obtained from the dependents on them, was organised in court sections (*puti*) headed by great nobles (sometimes *boyare*), with subordinate, often slave (*kholop*) officials; these dealt with separate functions, such as hunting, provision of horses, victualling and so on.[1]

The peasants on courts lands were liable to taxation, mainly tribute, and various other obligations, in kind, in money or in labour. Some held land of, and provided certain services for, the prince – fishing, hunting etc.; this form of tenure has been linked with the later holding of land by service, mainly military.[2] The local units of court lands were responsible as a whole, not as individuals, for meeting tax demands and for the fulfilment of any other obligations. Unfortunately, we know virtually nothing of the peasant organisation for the distribution of such burdens, but it was important both for the peasants and for the court, though in different ways. While the peasant accepted his liability to obligations because he had land, the administrators' view was that the peasant was entitled to land because he was tax-liable (*tyagly*).

(2) 'Black' lands were occupied, originally over extensive areas and thus in more compact blocks, by peasants directly subject to the prince in his capacity as ruler of the state, not as an individual lord. They were administered by the prince's high-ranking officials; these were the prince's officer (*namestnik*) and the volost-head (*volostel'*).[3] These officials were maintained by a 'living' (*korm*) in the form of produce extracted directly from the local population; they were therefore referred to as those maintained by livings (*kormlenshchiki*). Their subordinate officers, some gentlemen, some slaves, responsible for various aspects of taxation and justice (*tiuny, dovodchiki; nedel'shchiki, pravetchiki, pristava*) were also maintained in this way or by a share in the fines and payments exacted from the administration of justice. This system continued into the sixteenth century.

[1] A good description of the Moscow Prince's estate is given in Bakhrushin, *Nauchnye trudy*, II.13–45.
[2] Rozhdestvenskii, *Sluzhiloe zemlevladenie*, 36–8.
[3] In this context the volost was an administrative area of peasant land. Essentially *volost'* meant an area under a single authority. Thus, it could also mean a lord's estate, a manor. Sometimes it referred to a total area, sometimes only to peasant land within that area (see Chicherin, *Oblastnye uchrezhdeniya v Rossii*, 62–3). Other local peasant administrative areas were *stan*, sometimes the equivalent, sometimes a subdivision, of a volost, but always implying central control; *perevara* and *desyatok*, used in Toropets; *pogost* in Novgorod and some other areas; and *guba* in Pskov, for example. In 1966 I was told in Moscow that no work was then being done on the nature or even the terminology of such administrative areas.

Black peasants had their own elected officials, or at least were supposed to have them. They were the hundred-men (*sotskie*) and reeves or elders (*starosty*). At first these seem to have been peasants concerned with self-administration. The actual concerns presumably varied; they included allocation of land or of land rights to newcomers, but there is no evidence for any land redistributions. Increasingly, however, these men tended to be treated by the central administration as its lowest rung. Like the court peasants, black peasants were liable to tax, tribute and justice, burdens for which their officials were made personally liable as central power increased, but which were, of course, distributed internally by means we do not precisely know.

(3) Boyar estates of heritable land (*votchiny*) were scattered over the mass of court and black lands. Each boyar's estate would not necessarily be located in one place, but might consist of manors in various parts. Rozhdestvenskii stressed how difficult it was 'to bind a clutch of villages and hamlets scattered over great expanses, separated from one another by great distances, in to a single economic system'.[1] He went on to point out that these separate groups of villages and hamlets were the prepared economic units which formed the big heritable estates. This seems to me inadequate. The estate or manor was a unit appropriating part of the surplus; it was sometimes a processing and trading, rarely a production unit. When it was a production unit it appropriated labour instead of produce. The basic production unit on these, as on all other lands, was the peasant tenement, especially before the late sixteenth century.

These heritable lands were virtually closed areas from the point of view of the prince's administration. The prince's officials had, in general, no right to enter them for any reason, except for certain very serious crimes, which were specified in the grants of immunity, or if there was a land dispute. Such immunities were only gradually eroded by the prince's administration; they meant that an important sector (for such estates might well be better organised and economically stronger than most) was outside the prince's control.[2]

Within such estates the organisation varied according to size and wealth. The largest might approximate to that on court land with stewards and other officials responsible for supplies to maintain the boyar, his servitors and household. They would also be responsible for the collection of taxation and, sometimes, for the exaction of justice when it was not in the hands of the prince's justices. As on courts lands, again, there would be some peasant organisation at the lowest level. From the mid sixteenth century heritable land was liable to provide military service.

(4) Land was also held by service; gentlemen (*dvoryane*) and junior boyars

[1] *Sluzhiloe zemlevladenie*, 16. He discusses a similar situation in the mid-seventeenth century, *ibid*. 227–8.
[2] On the zig-zag oscillations in policy as regards immunities in the 1490s to 1540s, see Kashtanov, *Sotsial'no-politicheskaya istoriya Rossii*.

(*deti boyarskie*), a lower grade of servitor, might hold lands of a prince or a boyar, or of high ecclesiastics such as the Patriarch or metropolitans, in return for service. Such lands were in theory, and often in practice, not heritable; they were located as far as possible in a single holding, but often allocations were in more than one location; the prime consideration was state requirements for political and military purposes. Not land-holding as such, but a career of service to the crown was the main thing.[1]

The internal organisation and administration of the estate would probably not be as complex as on boyar estates; but, especially when the requirements of military service demanded the absence of the lord for long periods, a steward was likely to be in charge, if the size of the unit enabled this to be done. Any local peasant organisation (of elected reeves, sworn-men (*tseloval'niki*) and representatives of peasant heads of households) was likely to be inconvenient to the lords, since there would be many conflicts of interest, especially where demesne arable land was important and labour rent was exacted. Thus, in the course of the sixteenth century as court lands (including newly captured territories, such as Kazan') and black lands (where peasant organisation was particularly found) were allocated to tenants holding by service, the functions of such peasant bodies were increasingly taken over by estate officials, especially on larger units.[2] The volost elders survived perhaps ten or twelve years after the transfer of land into tenures held by service. The process is reflected in a changed formula of address in documents of the second half of the seventeenth century. The earlier phrase 'reeves and peasants of various volosts and estates held by service' was replaced by 'gentlemen, junior boyars and their stewards and reeves, sworn-men and peasants'.[3]

(5) There were ecclesiastical lands of various types. The estates of the Patriarch, metropolitans and monasteries were comparable with the boyar estates of heritable land; these dignitaries and institutions also granted land in return for service to their own servitors. Parish churches also held land. These ecclesiastical lands were like other heritable lands, largely closed to the prince's administration. Their immunity was reinforced by recognition of the church's right to administer justice in terms of the ecclesiastical law not only to church officers, but to all those on its estates. There were, however, many variations in the applications of this principle. The church, as an institution, was also free of taxation, though on occasion this principle, too, was overridden. Monastic landholding developed particularly in the sixteenth century and monastic estates were often well organised in many respects and economically advanced.

---

[1] Rozhdestvenskii, *Sluzhiloe zemlevladenie*, 225.

[2] Rozhdestvenskii (*Sluzhiloe zemlevladenie*, 267–9) took the view that black lands were somewhat sparingly distributed but he did not specify the proportions of the different sources of land to be held.

[3] Veselovskii, *Soshnoe pis'mo*, I.334.

Overall, then, the land within a principality was not all organised in compact blocks, nor was it all administered in the same way. Different types of land in terms of tenure displayed different arrangements, both internally with those working the land and externally with superior lords. Relations between princes, boyars, ecclesiastical dignitaries, gentlemen and their dependents, slaves, serfs and peasants, were complex and varied. Much depended on custom, local agreement and personal arrangements.

In the course of the fifteenth century, Moscow achieved a dominance which, despite conservative tendencies inherent in the system, resulted in far-reaching changes. First Moscow acquired an enormous territory. At the accession of Ivan III in 1462 the area of the Moscow state amounted to about 0.4 million square kilometres, by his death in 1505 it reached 1.7 million. The total area of the Russian state was about 2.5 million square kilometres by 1533 and reached about 4.3. million by 1598.[1] The tsar had vast quantities of land at his disposal, though since little is known about population growth, we cannot be certain how far, or even in which direction, changes in land available per head went in key areas.[2] The incorporation of the huge, but lightly populated areas of Novgorod's far north, did not mean a commensurate increase in the strength of the state measured in terms of population. The growth in the size of groups of related nobles serving the tsar and other late-fifteenth-century evidence may possibly hint that there may have been a notable increase in population at this time, but land was in general still abundant even at the contemporary level of technology. There is no evidence of land shortage though in many central areas estate boundaries were frequently contiguous.

Second, the increase in the size of the area under Moscow's rule created administrative problems. The acquisition of another principality by Moscow did not at first involve much change in the internal system of the newly incorporated area, but from the standpoint both of the tsar's administration and of the new area, comparable treatment would be attractive.

From the moment of his subordination to Moscow the former apanage prince became accustomed to recognise himself, not as the independent holder of a certain part of the Russian land, which he had now ceased in fact to be, but as part of a numerous class which, led by the Moscow sovereign, ruled the whole Russian land subject to him.[3]

The acquisition of new territory meant information was demanded by the

---

[1] Cp. Kopanev, *IZ*, 64.235, 246. The figures given by Ikonnikov, *Opyt*, II, kn. 2, 1083, appear to be underestimates by a factor of ten or so. The figures given in the text above have been calculated from modern Russian historical atlases.

[2] Kopanev, *IZ*, 64.235, 242, 254, suggests population rose from about 6–7 millions in 1500 to 9–10 millions in the middle and 11–12 millions at the end of the sixteenth century.

[3] Klyuchevskii, *Boyarskaya Duma*, 238.

Moscow departments for the purpose of state administration. Thus, registers of inquisition were evidently compiled for many areas from the 1480s to the early years of the sixteenth century. Registers made after the capture of Novgorod in 1478, of Toropets in 1503 and of Kazan' in 1552 have survived. Such registers are important sources of information; unfortunately, there are none for the core areas of the Moscow state for the mid sixteenth century; for the Moscow area we have registers only from the last quarter of the century.[1] There would be a tendency towards more uniformity since this would simplify administration for Moscow and the locality would wish to avoid burdens not borne elsewhere.[2] The former fragmentation and largely personal *ad hoc* arrangements were often superseded by direct subordination to the centre, but this was done using existing local arrangements and personnel as far as possible.

A balance had to be struck between, on the one hand, retaining the former system, though now subordinated to the tsar's officers, and, on the other, ensuring that there was no resurgence of local centres to challenge Moscow's power. In order to maintain and expand that power, resources had to be gathered from the new acquisitions, but this had to be done as far as possible without arousing resentment, often expressed by peasant flight, by excessive contributions judged in relation either to the past of the particular territory or, increasingly often, to the present of other areas of Muscovy. Peasant complaints in the fifteenth–sixteenth centuries had often been in terms of the situation in the past. In the seventeenth century local complaints of excessive exactions were always in terms of comparison with other areas.[3]

The growth of Muscovy also involved changes in the administration at local level. Even at the end of the fourteenth century local charters were issued in some areas in attempts to control the prince's agents maintained by 'livings'; this was the beginning of a lengthy process of dealing with the alleged excesses of those maintained by livings (*kormlenshchiki*) which only ended with the elimination of this form of maintenance in the sixteenth century. The process was long, complex and pursued a zig-zag course which need not be dealt with here, except to point out that, as the competence of these officers was reduced, the involvement of representatives of the local population, or certain categories of it, was increased. Hundred-men and 'good people' (*dobrye lyudi*), later sworn-men (*tseloval'niki*), had to participate in court cases; and in the fifteenth century the latter became permanent, rather than *ad hoc,* officials. This participation, based on custom, was concerned with aspects of justice and administration not dealt with by the prince's agents and included the collection of imposts. As the state came increasingly to rely

---

[1] Zimin, *Reformy Ivana Groznogo*, 53.

[2] The issuing of a code of laws, the Sudebnik of 1497, is part of this process. See Dewey, *ASEER*, 15.3. 325–38.

[3] Veselovskii, *Soshnoe pis'mo*, II.495; also 307, 311, 323, 326.

on these local officials they were made personally responsible for the sums due and office therefore became an onerous burden. This shift away from reliance on the tsar's centrally appointed agents led to the emergence of a system of police functions based on guba reeves (*gubnye starosty*) and sworn-men in certain uezds in the 1530s, largely to deal with the problem of 'brigandage'. This was probably often popular disturbances, heralding the considerable internal disturbances around 1600. By mid-century the change had become general, and in 1555 'livings' were eliminated. In the 1530s the unit area was usually a whole uezd and all the population was, in theory, involved in electing the guba reeves who, however, had to be of junior boyar rank and were usually those not performing other service. Thus, middle-ranking servitors came to have considerable authority over the local population even though, at this stage, that population was still, in law, free. In the 1550s, the areas were smaller, often a volost, and a new category of reeves, reeves of the land (*zemskie starosty*), were to be elected by peasants only. They were responsible for all local matters (unlike the guba system, which coexisted with the state administration) and were under considerable central control.[1]

These shifting arrangements for local administration reflect the struggle to control the countryside by an emergent tsardom with an insufficiently developed administrative structure, either of central government officers or of gentlemen tenants holding by service. Thus, in the first part of the sixteenth century land grants were made mainly 'to bind and restrain the volost from being taken by a rival prince'.[2] Local participation, in fact, was often only an aspect of the power struggle at princely level. In mid-century, however, the closer control of many localities became possible. It was no longer simply a question of keeping the volost out of the hands of a rival. The tsar's authority had not in any real sense been further strengthened fundamentally, but the fragmentation of the administrative areas resulted in a preference for some centrally controlled officials in order to achieve uniform treatment, so long as the cost was not excessive. The development of grants to be held by service into a fully fledged system resulted in a body of career servitors owing allegiance not to a locality, but to the crown which thus had less need to extend its authority downwards still further.

Evidence for new developments in economic control even at estate level is found at about the same period, The growth of commodity and money-based relations, market links and the general development of monastic estates at the end of the fifteenth century and in the early sixteenth century is the setting for the appearance of monastic account books and a number of similar records.[3]

The increase in land held by service and the emergence of a developed system of such tenures seems crucial for the history of peasant farming. At the

---

[1] The basic charters relating to the guba and other local reforms are to be found in Yakovlev, *Namestnich'i, gubnye i zemskie ustavnye gramoty Moskovskago gosudarstva*.
[2] See Kashtanov, 'Iz istorii poslednikh udelov', 285.   [3] See p. 230 below.

end of the fifteenth century many nobles became impoverished as the custom of partible inheritance, the right of women to inherit land and commemorative donations to monasteries, frequent during periods of calamity, all contributed to the fragmentation of their estates.[1] On the Moscow territory in the fourteenth–sixteenth centuries, for instance, up to 150 remote monasteries and more than 100 monasteries in or near towns arose.[2] Some acquired very extensive estates in the same period. The impoverished nobles found careers serving the Grand Prince, a situation which would be reinforced by any increased rate of growth in the number of servitors there may have been at the time. As Moscow's acquisitions grew, service with its ruler became increasingly attractive. The tsar began to offer comparable conditions of service to such men, and they provided manpower for the army and came to play a greater part than hitherto both on the land and in the state.

The sixteenth century saw a further great increase in state demands, largely associated with the military campaigns being undertaken. The campaigns which finally resulted in the capture of Kazan' in 1552 were a minor effort compared with the quarter century of campaigns against Livonia (1558–83). These drained the country, put a severe burden on the military servitors, and caused much discontent. At the same time some of the factors which impoverished the nobles took land out of the category liable to provide military service. This led to attempts to limit and control ecclesiastical land holding and also, in 1556, to a decree regarding military service from both heritable estates and those held by service.[3] This laid down that there was to be provided

a man on horseback and in full armour, and with two horses for a distant campaign, from a hundred chetverts of good quality, useful [ugozhie] land; and whoever holds land by service and the sovereign gives them a grant, a living, he [i.e. the tsar] also makes a money payment for the people to be provided; and whoever holds land, but does not pay service from it, those men are to pay money for the people [they should provide]; and whoever provides, for service, people surplus to the land, more than the people he should provide, such men are to have a large grant from the sovereign, they are to be given two-and-a-half [times] in money for the surplus people.[4]

From the time of Ivan IV to the end of the sixteenth century, Muscovy's total military force rose from 70,000 to 109,000 and its gentlemen and junior boyar element from 17,500 to 25,000.[5] During this period, the differences between heritable tenure and tenure by service narrowed, the latter becoming in fact normally heritable.[6] Grants were made in the mid sixteenth century to those willing to hold by service at rates which took account, in varying degrees, of the availability of estates, the quality of service, and the office and

[1] Rozhdestvenskii, *Sluzhiloe zemlevladenie*, 62–9, 145.
[2] Ikonnikov, *Opyt*, II, kn. 2, 1103; Zverinskii, *Material*.
[3] *PSRL*, XIII.1, 267–9. The decree itself is not extant.
[4] Another English version is in Vernadsky, *Source book*, 1.141–2.
[5] Hellie, *Enserfment*, 267.   [6] Got'e, *Zamoskovnyi krai*, 390–2.

rank of the tenant.[1] Probably land quality, evaluated by the officers of inquisition on the basis of local information, had been taken account of earlier; the innovation was that standard rates were now laid down for it.[2]

The amount of land was estimated in notional chets of good quality land considered 'live' (*zhivushchaya*), that is with a population of dependent peasants on and working it, thus making it fully viable and hence tax-liable. By the late sixteenth century, owing to internal disturbances and much increased state demands, there was a lack of such land, and by the end of the century in some areas at least, a shortage of any land for distribution to tenants willing to hold by service. This shortage existed alongside the universal and constant existence of much 'empty' (*pustaya*) land. This land, as we will show, was not necessarily literally empty, devoid of population and unworked, but was exempt from taxation because of its condition (which would sometimes include total abandonment, but would often be some stage between that and viability).[3] There was often not so much a real lack of land as inconvenient location and poor distribution of the resources required.[4]

As will be seen, in the Moscow registers of inquisition, there are scarcely any traces of peasant black lands which, together with court lands, were a major source from which allocations were made to servitors. Tenants in this area sometimes received only two or three chets of 'live' land per hundred chets of grant.[5] 'Empty' land, was thus included in grants, but at modified rates. In the grants (*dachi*) made in Kazan' in 1565–8, for instance, each chet of long fallow counted as two-thirds of a chet of good quality arable, and overgrown and tillable oakwoods as one half.[6] Despite the vast expanse of land acquired by the capture of Kazan', it was rare for a tenant to receive half his notional allocation (*oklad*).

Thus, especially in the centre and in other areas where there came to be few black peasant lands, the development of a regular and uniform system of allocations had made great progress. These allocations were always noted in terms of three fields. It seems that it was usual to expect one field, the fallow field, to be measured and the other fields to be estimated on the basis of any earlier documentary evidence available in local reports, and checked by visual estimates by the officers carrying out the survey.[7] This notional amount was then equalised between the three fields, one-third of the total arable area being recorded with the phrase added 'and in two at the same rate'. The inquisitions which note these allocations, however, were cadastral surveys carried out primarily for fiscal purposes; they aimed to disclose the resources

---

[1] Rozhdestvenskii, *Sluzhiloe zemlevladenie*, 240.

[2] Veselovskii, *Soshnoe pis'mo*, II.349.

[3] See especially the evidence from Toropets: Smith, *Forschungen zur osteuropäischen Geschichte*, 18.125–37.

[4] Rozhdestvenskii, *Sluzhiloe zemlevladenie*, 271.   [5] Veselovskii, *Soshnoe pis'mo*, II.400.

[6] TsGADA fond 1209, no. 152, passim.   [7] Sedashev, *Ocherki i materialy*, 27–8.

available to the state. These resources were summarily estimated in terms of the large Moscow sokha, a unit of 800 (notional) chets of good quality arable on lands held by service, but 600 such chets on monastic lands. These units came into use in the central areas of Muscovy just after the mid sixteenth century, and applied to court lands as well. As court and peasant black lands were distributed to tenants to be held by service, these units were used to estimate their wealth and consequent tax liability. From shortly after 1550, therefore, there was an increased uniformity especially in recording service land. But it was a uniformity not based on precise measurement of arable land. 'Every normally running economic unit consisted of a combination of various appurtenances; the composite parts could not but influence the whole set-up of the unit', wrote Veselovskii; and he rightly stressed that arable was an important, but not the only, element.[1] Sometimes the non-arable elements were estimated in terms of chets of arable land, as in Moscow (see below). The unit of measurement was arable land, but, apart from the sokha, scythe and axe continued to play their parts, and measurements of chets of arable land must not be assumed to represent the physical tilled area.

Military and associated demands from the enlarged state called not only for men, but also for many other resources, mainly provisions, horses and, increasingly and especially important, money. This contributed to, and was affected by, price rises for farm produce. It has been calculated that, in general grain prices rose more than four times during the sixteenth century, meat and livestock prices roughly doubled.[2] The late sixteenth century sees a shift both to money rents and to the growth of demesne production by means of labour services (barshchina).

Two great crises contributed to the changes then taking place. First, the Oprichnina (1565–72), a privy organisation of Ivan IV in his internal struggles with factions among the great nobles, resulted in extensive devastation in many central and northern areas of the state, greatly contributing to a decline in the amount of taxable land and to the rise in prices of agricultural produce. The major Tatar raid of Devlet Girei in 1571 also caused considerable destruction in and around Moscow. Second, although from around 1580 there had been an economic upswing, the turn of the century saw a series of disasters collectively known as the Time of Troubles. The Rurik dynasty came to an end in mysterious circumstances. There were three successive years of famine (1601–3), peasant disturbance, a major revolt led by Bolotnikov, invasions by Poles and Swedes resulting in an enormous territory being occupied, much destruction, and great confusion in questions of land holding. These were the circumstances in which the enserfment of the Russian peasant was taking place. By the end of the sixteenth century, the tax-liable peasant was virtually

---

[1] *Soshnoe pis'mo*, I.394; cp. I.22, 35, 361, etc.

[2] Man'kov, *Tseny*, 40–1, 53. Cp., however, p. 143 and table 12 below on price changes in the Moscow area.

bound; 'to till the land and, in general, to run his farm had become for him not a right, but an obligation'.[1] This process was finally enshrined in the law code of 1649.

The unit of measurement for allocations of land to be held by service was arable land, but despite attempts to introduce identical units and greater uniformity over an increasing area of the state, the estimate of the officer of inquisition remained important. In essence, the exactions of state taxation continued to be 'according to their strength' (*po sile*) measured in notional units of arable land.[2] Moreover, attempts to achieve regularity met with differing success by area and type of tenure. In general, the central areas around Moscow were most readily controlled, remoter areas might preserve their particularities longer. The tsar's officials directly controlled the court lands which were uniformly administered. The black lands continued to use their own internal arrangements to distribute the tax burden according to strength. The peasants on both these categories of lands thus came into contact with the tsar's officers, and especially where the black lands still formed extensive blocks, as in Toropets, they retained many local features and were able to continue to bargain to some extent about their assessments. On lands held by service, however, this was not so. The tsar's officers left the internal control of estates held by service to the estate-holders; their registers of such estates were not checked by reference to the bound peasants working such land. This will have been the case especially in a captured, non-Slav area such as Kazan', where there was no tradition of former peasant black lands. In all these categories, therefore, the measurement of land was a notional measurement of wealth and thus, to some extent, imprecise, though for differing reasons; it must not automatically be taken as an accurate indication of real areas of arable land.

Nevertheless, the material provided by the sixteenth-century registers of inquisition and similar documents should not be dismissed out of hand. It is, of course, the only such material we have; but, more important, we should consider some of the evidence of changes affecting farming at that period to see whether this material may not be used to make approximate calculations of certain kinds. There is evidence in some areas, Moscow, for example, of an increase in village size; in the central area villages seem in general to be much larger than those in Toropets. Sometimes we find the phrase 'let into the arable land' (*pripushcheno v pashnyu*) associated with such larger settlements with perhaps ten or more tenements; this extension of arable land is probably linked with a regularisation of field layout. We have already seen how, mainly in the central area and on wealthier and better-regulated estates, such as those of monasteries and princes, a three-field layout spread during the late fifteenth

[1] Veselovskii, *Soshnoe pis'mo*, II.354.
[2] Veselovskii laid much stress on the subjective nature of these estimates, but this view was strongly and, I believe, rightly challenged by D'yakonov, *RIZh* (1917), 1–2.54.

century, and may even have been beginning to extend to larger and richer peasant farms. The use of a three-field notation by the sixteenth-century officers was, therefore, probably soundly based in most cases. It should be stressed, however, that this layout did not imply any periodic redistribution of strips in each of the three fields. Sixteenth-century references to redistribution involved taxation, not land.[1] In peripheral areas, however, where population densities were particularly low and there was non-nucleated settlement and abundant forest land available to black peasants, the individual close, rather than the manor's field, is likely to have continued to play a basic part. This had been so in Toropets in the first half of the sixteenth century. Yet even here it is unwise to dismiss the notational system as virtually meaningless. If bargaining and agreement between black peasant and tsar's officer, inherent in estimation 'according to their strength', took place, it meant there was some sort of peasant check on the estimation; and although power was on the side of the officer, the peasant was not totally powerless. The solidarity of peasant communities is well known; and the officers had few ways or resources to check what they were told. Although the scattered nature of settlement was not conducive to joint activity and resistance, the customary reliance of the officers on local information, as well as their instructions to decide according to their consciences and not to impose excessive burdens, put a certain curb on their demands.[2] Even in areas such as Toropets, therefore, some reliance may be placed on the figures as approximate indications of wealth.

The sources which have been used for the three case studies we are about to examine are not of the same type, nor are they always contemporaneous with one another. For Moscow we have collections of deeds and charters which vary widely in date, as well as some incomplete and poorly published late sixteenth century registers of inquisition. For Toropets we have a single register of inquisition nominally of 1540; the Kazan' area is described on the basis of a survey carried out in 1565–8 after the Russian conquest. There was a choice between attempting a fairly generalised survey of peasant conditions in the Moscow state, or accepting the limitations of available sources and trying to find out what they can tell us about peasant conditions in certain areas. The latter course may help to make clearer the nature of the sources and the variations in the history of different parts of European Russia.

---

[1] *APD*, I, no. 1 (1583), 4. The fiscal distribution may have been the model for that of the land, according to Veselovskii, *Soshnoe pis'mo*, II.458.
[2] Veselovskii, *Soshnoe pis'mo*, II.85.

7

Moscow uezd

The Moscow uezd extended on either side of the river Moskva roughly from the point where the Istra joins the Moskva in the west to the confluence of the Otra in the east. Thus, the area did not include the whole length of the Moskva; the upper reaches of the river were blocked by the Zvenigorod lands, the lower by Kolomna. The uezd, however, included on the south a major tributary, the Pakhra, together with its tributaries, the Desna and the Mocha; this basin was the early southern limit. The northern part included the Istra, Vskhodnya and Yauza, and extended over the upper reaches of the Klyaz'ma and its tributary, the Vorya.

The soils of the region are mainly turfy podzols. The eastern part of the uezd has sandy loams, while in the south-east to the right of the Moskva there are clays; left of the Moskva are sands. The region has few lakes. Marshes occur mainly along the Yauza and the Pekhorka.[1] It is likely that before much forest had been cleared the water-table was higher and the lakes and marshes were, therefore, somewhat more extensive than at present. From the area of Ryazan', to the south-east, a tongue of more fertile grey forest soils stretches out towards Moscow and crosses the Oka; these soils are of reasonable quality for cultivation. This is a region of mixed woods, broad-leaf varieties and spruce and, where clearances have been made, with secondary woods, mainly birch or mixed trees. To the north of Moscow the coniferous forest zone starts, at first interspersed with deciduous woods, such as oak. The river valleys have thin soils of sand or sandy clays where pines or pine-spruce woods grow. Even today 'the areas of tilled land are often of limited size due to the hillocky relief, the considerable density of the river network and, in the valleys, considerable marsh and forest'.[2] As in former times, the low-lying areas, including the marshes and flood meadows along the rivers, are important for hay and pasture. The climate is classed as moderately continental with minimum average monthly temperatures of $-8°C$ to $-14°C$ and maximum of $17.5°C$ to $18.5°C$; 120 to 150 days in the year are frost-free and spring and autumn are quite long for Russia, which means there is less seasonal pressure of work on the land than in some other parts.

Solov'ev stressed the central position of Moscow between the rivers Volga,

[1] Semenov, *Geografichesko-statisticheskii slovar'*, III.333.
[2] *Pochvenno-geograficheskoe raionirovanie SSSR*, 83.

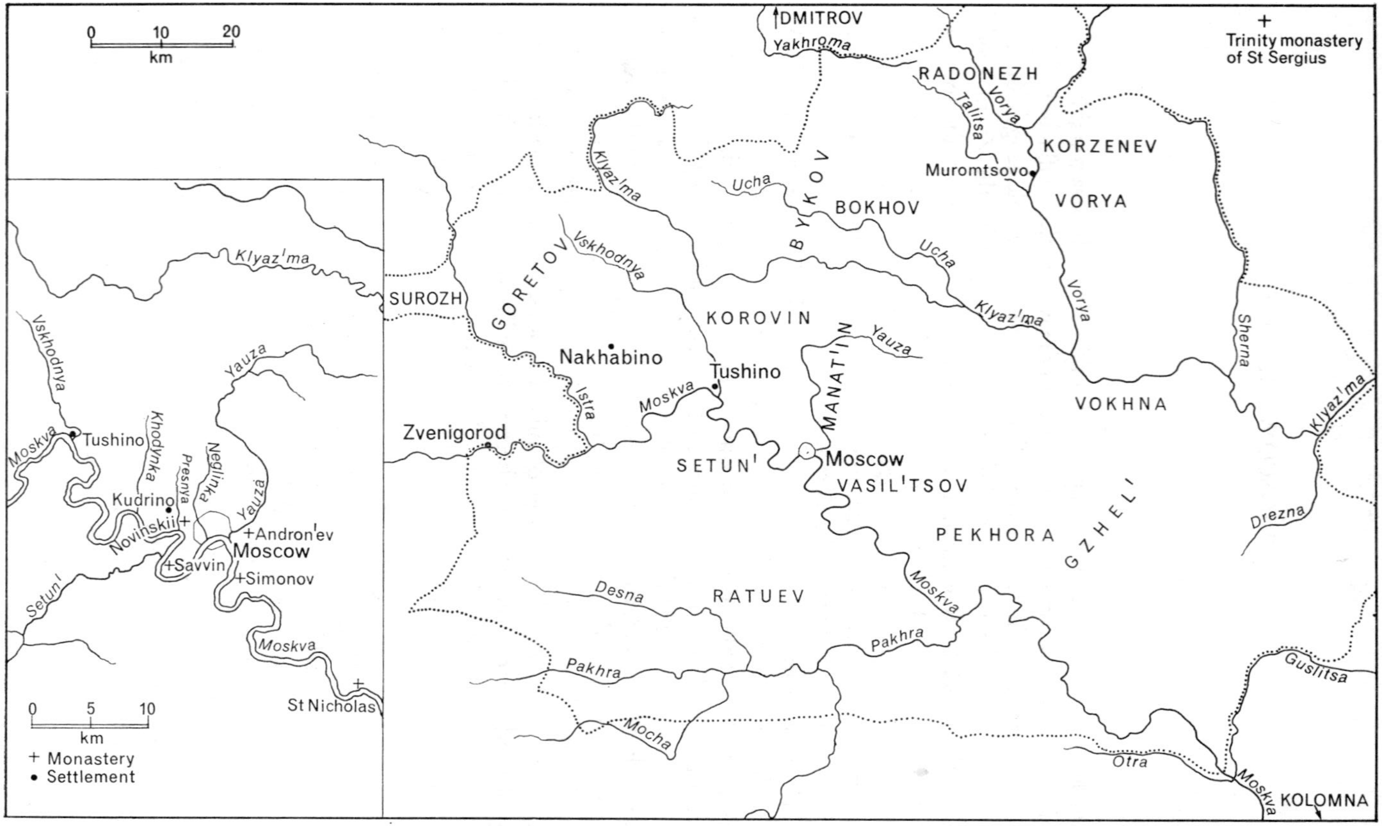

DMITROV
Yakhroma
RADONEZH
Trinity monastery of St Sergius
Talitsa
Vorya
KORZENEV
Muromtsovo
VORYA
Ucha
BYKOV
BOKHOV
Klyaz'ma
Vskhodnya
GORETOV
SUROZH
Ucha
Klyaz'ma
Vorya
Sherna
KOROVIN
MANAT'IN
Yauza
Nakhabino
Tushino
Istra
Moskva
VOKHNA
Klyaz'ma
Zvenigorod
SETUN'
Moscow
VASIL'TSOV
GZHELI
Drezna
PEKHORA
Moskva
Desna
RATUEV
Pakhra
Pakhra
Mocha
Guslitsa
Otra
Moskva
KOLOMNA
0 10 20
km
Klyaz'ma
Vskhodnya
Yauza
Tushino
Moskva
Khodynka
Presnya
Neglinka
Yauza
Kudrino
Novinskii
+Andron'ev
Moscow
+Savvin
+Simonov
Setun'
Moskva
St Nicholas
0 5 10
km
+ Monastery
• Settlement

Map 2. Moscow uezd.

Oka and upper Dnepr, between north and south, between Slav and Finnish peoples.[1] Moscow is also at an interface of forest zones and at a linguistic interface. Indeed, such factors seem to have received more attention from many Russian authors than the obscure problem of how the region was settled by the Slavs.

It seems probable that the general region of the Oka and upper Volga was being colonised in the late eleventh to mid twelfth centuries; as a consequence of continual Tatar raids, the eastern part of this general area may have gradually declined somewhat in importance and the more westerly parts came to be somewhat more important. Because of dense woods towards the Klyaz'ma–Drezna junction, Moscow itself was rarely subject to Tatar raids, at least from the east.[2] By the fourteenth and fifteenth centuries, colonisation of the Moskva basin was fully under way.[3] The nature of this colonisation, however, is obscure. We know too little to say how far there was a natural creep of peasant farmers hacking themselves out clearances in the forest and then being taken over by princes and monasteries asserting overlordship, and how far there was something like a policy from above of deliberate encouragement of settlement. Both processes seem to have been involved. To the south and south-west of Moscow, a densely wooded area even at the end of the fifteenth century, there were a number of free settlements, that is settlements under the prince's protection and granted privileges, such as exemptions from specified obligations, usually for a term of years.[4] From the 1320s the Moscow prince had become responsible for the payments made to the khan, and this involved various agents (*chislyaki, ordyntsy, delyui*) whose lands were also in this area (*chislyatskie zemli*). So there was both princely encouragement for colonisation and close princely control here of any pre-existing population in the fourteenth century. At the same period the prince had a number of villages near Moscow: Kolomenskoe, Nogatino, Bityagovo, Yasenevo, Ostrov, Konstantinovskoe, Orininskoe, Novoe on the Khupavna, Koponya, Naprudskoe, Deguninskoe, Semchinskoe on the Zhabna, Krasnoe, etc. The numerous extensive holdings of heritable lands of the great men of the area seem mainly to have been found to the north and west of Moscow, probably the turfy podzol area of earliest settlement.[5] Perhaps, therefore, the early settlement of the area was largely due to the natural creep of peasant colonists, later made dependent by lords, or themselves becoming lords. The more exposed internal frontier to the south and south-east, where there was a challenge from the princes of Ryazan'

---

1 *Istoriya Rossii*, 1.75–6. On the importance of the rivers for trade, see *Istoriya Moskvy*, 1.32f.

2 Tikhomirov, *Rossiya v XVI veke*, 106–7.

3 Lyubavskii, *Istoricheskaya geografiya Rossii*, 38–9, 152–5; Cherepnin, *Obrazovanie*, 166; Veselovskii, *Podmoskov'e*, 17. Cp. Got'e, *Zamoskovnyi krai*, 192.

4 Tikhomirov, *Rossiya v XVI veke*, 109.

5 Veselovskii, *Podmoskov'e*, 14. Howes, *The testaments*, 365f., cites the settlements bequeathed by the Grand Princes, but, unfortunately, does not plot them on a map.

and Kolomna, tended to be colonised as a result of princely policy. The southern border, moreover, continued to be exposed to Tatar raiders fording the Oka even in the sixteenth century.

Again, we do not know how far the Moscow uezd was wooded in the fourteenth and fifteenth centuries. By examining the figures given in sixteenth-century registers of inquisition Rozhkov came to the conclusion that the uezd then had little forest, 10–15 per cent and sometimes as little as 2 or 3 per cent of total land area in some parts.[1] This seems to be nonsense; the total area accounted for is far from the whole (probably only allotted or claimed forest was noted) and it is probable that most of the difference would be wild forest. However, it is certainly true that the rise of Moscow to the headship of the Russian lands in the fifteenth century is accompanied by more frequent contiguity of estate borders; this, in itself, however, tells us little of population density and the consequent likely extent of forest clearance. All that can be said at present is that cultivated area even then was small relative to total area; forest, both virgin and of secondary type, was overwhelmingly dominant.

There is some early placename evidence for forest (*bor*) by the walls of the Moscow Kremlin. In the early fourteenth century east of the town, from the mouth of the Yauza towards the south and south-east, the forest was bespecked with clearings of the small settlements of the prince's bee-men and fowlers; elsewhere 'there was forest everywhere except where marshes prevented it from growing'.[2] In the late fourteenth century along the Setun' and its tributary the Ramenka the forest was impenetrable, untouched. A century later, near Moscow, between the Presnya and the Khodynka peasants of Kudrino village were felling forest and making clearances which gave rise to disputes over boundaries. There may also have been some natural clearances; the name *Khodynskoe pole* (Khodyn prairie), for instance, suggests this.[3]

In the last quarter of the fifteenth century some grants of privilege included provisions aimed at forest protection; this evidently reflects the rapid and extensive internal colonisation then taking place.[4] In the course of the centuries of continuing colonisation human interference, especially the clearance of forest for tillage and the pasturing of cattle, changed the nature of the forest. Stands of maple, ash and several varieties of elm are mentioned in early documents; these species are now found only in parks in the Moscow area.[5] Conifers were displaced by the more easily started deciduous species, especially birch and aspen.

Dense woods evidently lasted into the seventeenth century; to the north of

---

[1] *Sel'skoe khozyaistvo* 7–9, 478–9. Such conclusions have been rightly criticised by Got'e, *Zamoskovnyi krai*, 165, 179; Kochin, *PI*, II.145–86.

[2] Veselovskii, *Podmoskov'e*, 20.

[3] Semenov-Tyan-Shanskii, *Rossiya*, I.46; Semenov, *Geografichesko-statisticheskii slovar'*, V.511.

[4] E.g. charters to the Trinity monastery of St Sergius in 1479, 1485 and 1490.

[5] Veselovskii, *Podmoskov'e*, 20.

Moscow, there was then still a relative abundance of forest animals, such as elk, for the tsar's hunt which was further organised and became especially developed towards the middle of the century.[1]

The process of internal colonisation coincides in the late fifteenth century with the evidence for the spread of three-field layout on larger units such as the monastic estates.[2] These permanent arable fields involved a complete elimination of the tree cover for ox ploughing. By the early sixteenth century pine and spruce of a size suitable for building was rare in the area. Later in the century there were complaints of shortage of timber for fuel and small jobs. As the nature of the forest changed, as a result of human activity, so did its fauna. For instance, beavers had been widespread in the Moscow area in the fourteenth century, but an early fifteenth-century grant of fisheries and beaver runs on the Vorya appears to be the last mention of them there; farther east they survived somewhat longer.

In the south-east, marshes, as well as the political situation, protected the forest from being exploited as intensively as in other parts. Along the Guslitsa, and especially its tributary the Shuva, as well as along the Vokhnya, Rogozhnya and others, the woods continued to survive. They provided an abundance of high-grade squirrel pelts in the fourteenth and fifteenth centuries.[3] It was here, too, that peasant farming without immediate lords survived.

The main documentary sources for the history of farming in the Moscow uezd are of two types; both, unfortunately, give relatively little information on peasant farming. First, there are collections of charters; virtually all of those before 1505 which are extant have now been published in a series of volumes during the post-war period.[4] These collections represent princely and monastic archives; there are very few private documents. The Moscow area was overwhelmingly a region of noble and ecclesiastical land-holding in the period for which we have documents; the question of monastic land-holding around Moscow, however, seems to have been inadequately studied.[5] Secondly, there are some registers, or rather sections of registers, of inquisition.[6] Unfortunately, some of those which might have been of the greatest use for our limited purpose have not survived. For instance, there were registers compiled for crown estates in 1628–9 which are not extant; these

---

[1] Tsvetkov, *Izmenenie lesistosti*, 8, 12–13; Kutepov, *Tsarskaya okhota na Rusi*, passim.

[2] See p. 243. [3] Veselovskii, *IGAIMK vyp. 139*, 96; *Podmoskov'e*, 24.

[4] *DDG*; *AFZ*, I–III, *ASEI*, I–III. Later material is only partially published.

[5] Ivina in *AE za 1966*, 14.

[6] Got'e, *Zamoskovnyi krai*, 7–36; gives a useful survey of these. Those relating to the sixteenth century were published by Kalachov, *PKMG*, I. 1; the index volume with a preface by Chechulin is also of great use. Unfortunately, the standard of this publication is so bad that it is impossible to compile a meaningful summary table as for Toropets and Kazan'. A useful list of registers of the 1580s–1590s for all areas is given in Koretskii, *Zakreposhchenie*, 304–19.

might well have contained more information on peasant farming than registers describing heritable estates and those held by service; in the latter the main concern was often the allocation of land to servitors, not its internal organisation and working.[1] Moreover, the Moscow registers are dull documents; they lack many of the interesting features of some other registers, such as those for Tver', let alone the one for Toropets. Even so, it seems possible to draw some conclusions from the charters and registers about peasant farming in the area.

Only occasionally is it possible on the basis of documents of later date to say anything specific about the active colonisation of the Moscow area in the fourteenth century. A court case of the late 1460s, for instance, quotes documents probably going back to the 1380s.[2] At that time the Grand Prince Dmitrii Donskoi exchanged certain lands with a monk; he took a village 'with its hamlets and clearances, with the forest and with everything that was administratively subordinate to the village, wherever the plough went and the sokha, the axe and the scythe'. In exchange he gave a small monastic retreat (*monastyrek-pustynka*) 'which the abbot Afonasei established on my land on the shore by Medvezh'e [Bear] lake with lakes Verkhnii and Nizhnii and the bee-tree hamlets' one of which also had fowling runs.[3] 'And whatever people Sava has to live on those lands those people do not have to pay any of my tribute' nor a series of other obligations, including labour duties, 'nor to be liable with the stan [an administrative district] for any customary dues'.[4] This area, about 18 km to the north-north-east of Moscow, on the road to Shchelkovo, in the general direction of the Vorya–Klyaz'ma junction, seems in the late fourteenth century to have been thinly settled.[5] Here a monk could establish a retreat; the forests and marshes gave game and wild bees. For prince or monk to exploit nature's abundance, however, hands were needed; but we are here given no details of the people the monk Sava 'had to live on those lands' or of how they were recruited. A document of 1453 records the purchase by the abbot and an icon painter of the St Simon monastery of five wastes: 'Odintsovo, Kharino and Stupino, Malakhovo and Koltovo'.[6] In the late 1460s a court case refers to peasant officials, a hundred-man and two tenmen, one in charge of bee-trees, and to 'the whole Pekhora volost of peasants on this land'.[7] By the late fifteenth century, then, there were peasants organised in independent administrative units on land of the Grand Prince. Moreover, one of the named peasant representatives, Isachko (or Isak) Bashlov, a black-

---

[1] Got'e, *Zamoskovnyi krai*, 26–7.   [2] *ASEI*, II, nos. 381, 340.

[3] Such retreats were clearances made by individual monks from a mother-monastery, in this case the St Simon monastery, founded in 1370. See Antonova, *TrODRL*, XXII.190, 192

[4] *ASEI*, II, no. 340. Cp. *AFZ*, I, no. 71 (1465), which, however, does not include tribute in the immunity.

[5] Antonova, *TrODRL*, XXII.194.

[6] Antonova, *TrODRL*, XXII.193; *ASEI*, II, no. 352, and see also no. 381.

[7] *ASEI*, II, no. 381, p. 380.

smith, seems to have set up a clearance about the middle of the century. It is, perhaps, also noteworthy that the court case was brought by the peasants against the monastery holding these lands, but there is no explicit mention of arable land. Here we have a glimpse of slow, continuing colonisation; conflict over land, the growth of administrative control and the monastery succeeding in establishing its claim, according to these late-fifteenth-century documents, to black peasant lands: the peasants themselves were still able to take suit.

A charter of the Grand Prince of 28 March 1447 established a rent of half a ruble, in place of tribute and certain other obligations including the liability, together with 'the black people', to commune payments and labour rent, from 'whatever people they have to live' on the monastic lands of St Simon in Moscow uezd.[1]

In 1489–90 a grant was made on behalf of the Grand Prince to Stepanko Doroga who petitioned that he wished 'to settle and live on that waste [Kozlovskoe]; and that waste, they say, is to be rented, it has lain waste about fifty years and is overgrown with big trees, nor is there any tenement on it, neither stake nor arable'.[2] Stepanko (the diminutive form of his name may indicate that he was a peasant) was granted the right to settle and live on the waste which had evidently been devastated during the internecine strife and epidemic disease of the second quarter of the fifteenth century. He was freed from tribute and other customary dues for six years. 'And when he has sat out his term of six years he is to give the Grand Prince from that waste a rent of a quarter of a ruble year by year at Christmas, at the Palace.' Apparently, however, this grant should not have been made; within a few years a court case established that this land had already been given to a monastery.[3] Again, there is reference to a peasant hundred-man, as well as to 'all the peasants of Korzenev volost' and the knowledgeable men (*znakhori*) and old-established peasants (*starozhil'tsy*) who gave testimony and indicated the bounds – one of them, incidentally, claiming to remember eighty years back.

It was also at about this time that we see competition for work hands reflected in a charter of immunity issued by the Grand Prince to a monastery in 1481.[4] This freed from tribute and a range of other obligations 'whatever people they have to live in that village and its hamlets, or whatever people they summon to themselves to live in that village and its hamlets, from other principalities, but not from my, the Grand Prince's, heritable estate, old-established peasants and newcomers'. Nevertheless, the document also suggests the area, although on the Yauza, was still quite wild. The dependants are freed, among other things, from the need to provide guides or guards for nobles, officials and other travellers and from being called on for service in hunting bear and elk, both animals only found where woods are relatively

---

[1] *ASEI*, II, no. 349.   [2] *ASEI*, I, no. 541. The renting of land will be discussed below.
[3] *ASEI*, I, no. 595 (1495–9).   [4] *ASEI*, I, no. 492.

undisturbed by human occupation. Here internal colonisation quite close to Moscow was evidently still in progress in the late fifteenth century.

Evidence in a court case in the late 1490s illustrates some of the general problems of finding and keeping labour to work the land.[1] A land-owner had died in 'the great plague' leaving an infant son; as a result his land became waste. 'But, lord, when Fedor Neplyui had become a man he petitioned the Grand Prince, lord, and received a charter from the Grand Prince Vasilii to his heritable estate, the waste Korobovskaya land. And Fedor Neplyui, lord, gave the Grand Prince's grant to my father, Gavril S'yanov, for him to summon people to that Korobovskaya land. But, lord, there was brigandage and much theft on the road and people did not settle on that land.' The reference here is probably to the last great plague which came to Moscow in 1425 and continued till 1427.[2] Under Vasilii II (1425–62) the fighting between the princes resulted in considerable internal disturbances which doubtless facilitated the spread of disease and criminal activity, particularly along the roads.

In 1494 a local prince granted a monastery 'a mill on the Klyaz'ma for rent and Chern'tsovskaya and Podkinskaya wastes which the millers tilled; and whatever millers they have to live at that mill, and peasants at those wastes, they have no need to pay...tribute...' and certain other obligations.[3] The monastery was to pay yearly 'for the mill and the wastes a rent in place of tribute and all customary dues of four rubles on St Peter's day'.

On 10 November 1491 a peasant who wanted to work a waste, formerly a hamlet, on the banks of the river Moskva on land of the metropolitan's Novinskii monastery was given a grant by a steward.[4] The peasant wanted to 'establish himself a tenement on that waste, to fell and colonise [rozselivati] the forest'. He was freed for four years from joining the monastery's peasants in paying tribute, the post-horse obligation, from work for the monastery, communal obligations and commune payments. The cellarer of the Trinity monastery of St Sergius made a similar grant to a peasant on 2 June 1499 in the Vereya area.[5] The peasant and four named sons were settled 'at a site (stanishcho) in the forest at the junction of the Vorya and Talitsa. They are to fell the forest, establish tenements, fence the gardens and clear the meadows'. To this end they were freed for six years from any obligation to work for the monastery; 'but when their six exempt years have passed they are to be liable with their fellow peasants, and like the other peasants, are to work for our monastery'. This peasant family was evidently fairly substantial, probably because five males gave it an advantage over others. Another charter of the same period notes that a boundary is now delimited 'by the new patch of cleared forest'.[6]

---

[1] ASEI, II, no. 410; cp. no. 411.
[2] PSRL, xxv.246–7; Tatishchev, Istoriya Rossiiskaya, v.233–4.   [3] ASEI, I, no. 575.
[4] AFZ, I, no. 34; cp. Veselovskii, IGAIMK, 139.70. See also Appendix I.
[5] ASEI, I, no. 623.   [6] ASEI, II, no. 411 (1494–99), p. 438.

A document of 3 June 1496 is an agreement between an officer of the stables and a monastery; the officer took over 'the metropolitan's manorial settlement Turab'evo on the Klyaz'ma in Bokhov stan, Moscow uezd, 'with all that has been subject to that manorial settlement from of old, forests, meadows and hayfields, for my lifetime; and I have taken that manorial settlement empty, without people, and I, Fedor, am to control and till that manorial settlement for myself for my lifetime, to fell and colonise the forest and to summon people to that manorial settlement'.[1] After his death the settlement, with any money, grain and livestock in it, was to revert to the monastery. Another case of a monastery tenant, Boris son of Nikifor Pavlov, records that, about 1477–84, he 'tilled those lands, felled the forest and set up buildings on them; and we, lord, have now taken those lands of ours from Boris to our monastery lands because they have come to the monastery lands, with all that has been subject to them from of old...'[2] Evidently in this case the initial forest clearance led to the expansion of the arable area which finally came in contact with the monastery lands and was then incorporated in them. The tenant was granted a life use in return for expanding the monastery's arable area.

Thus, about peasants, we have a few glimpses of their colonising activities in these documents, and also clear references to the existence of peasant administrative bodies on the black lands of the Grand Prince. Due to the early establishment of land claims by prince and monastery, however, in this area peasants often sought permission to develop a farm, either because of the years of remission from obligations granted in such cases, or, because it was easier to re-establish a work unit on an abandoned waste, rather than attempt to establish one from the start in wild forest. The two factors were sometimes combined. For their part, of course, the lords were anxious to encourage colonisation as a means of increasing the number of dependants. Even the documents we have then, despite their being almost without exception from princely or monastic archives, show something of peasants felling and colonising the forest. Clearly, too, there were more extensive areas of peasant black lands, mainly east of Moscow (Pekhora, Vokhna, Korzenev, etc.), where dense forest long continued to exist, though explicit information on these is extremely scarce in these documents.

In the main, however, the sources make us much more aware of lords, lay or clerical, attempting to develop the resources of the areas within their claimed jurisdiction and, at least from the late fifteenth century, increasingly competing for labour resources in order to do so. Dependants may only be recruited from other principalities. After the peasant has come 'to live' on an estate, i.e. to make a livelihood, to win a living there, and his period of remission from obligation has passed, he is liable to pay his dues in money,

---

[1] *AFZ*, I, no. 40.   [2] *ASEI*, III, no. 44.

produce and labour as, evidently, was the normal lot of all peasants.[1] Tenancies are also sometimes granted for a life.

It is interesting to note that in these fifteenth-century charters the liability to obligations is defined as 'according to their ability' (*po sile*),[2] and not, as in the later registers of inquisition, at certain set rates. Presumably, 'ability' was assessed largely in terms of local custom; but it may also be taken to imply that within any one area there was some generally acceptable norm, though with a customary range. The amount of arable land, taking quality (probably in terms of yield) into account, was the most likely standard measure; not an accurately measured area, of course, but an estimate, often in terms of seed used. This seems to be implied, for example, by the statement that 'if any hamlet has more arable land and appurtenances they (the officers of inquisition) impose more produce-rent (*korm*) and exactions on that hamlet'.[3] The appurtenances were usually valued in terms of chets of arable land.[4]

Unfortunately we do not know what lay behind attempts to settle these forest lands. The chief impulse at peasant level seems likely to have been population growth, however seriously modified by other factors. The nature of the Russian family and the farm household unit, as well as small settlement size, were also crucial to the mechanism put in motion by that impulse.

The early-fifteenth-century growth of population may have been adversely affected by plague, though such disasters are likely to have struck towns and settlements on and in contact with roads and trade-routes more severely than the remoter small settlements scattered over the countryside. There had been other disasters as well. Edigei's attack of 1408 was particularly severe and caused considerable losses in the Moscow area.[5] Then, from 1418 there were a series of plague years, culminating in the Great Plague of 1425-7. About the same time a number of early winters and floods caused harvest failures and famine.[6]

The desertion was so great that in many places no-one remained to say who had formerly held settlements. Land-holders who survived abandoned their lands for want of hands to work them, sold them cheap, or gave them to monasteries. It was from this time, i.e. the second quarter of the fifteenth century, that the rapid growth of monastic land-holding began.[7]

Even if depopulation due to plague and emigration due to social disturbances were the proximate causes of some late-fifteenth-century social changes in-

---

1 For a discussion of the term 'to live', see Smith, *Enserfment*, 59, 65; Cherepnin, *Obrazovanie*, 212–16 has a more extended analysis; cp. his article in Novosel'tsev *i dr, Puti razvitiya feodalizma*, 219.

2 *ASEI*, I, nos. 291 (1461), 305 (1462), 401 (1471).   3 *AAE*, I, no. 150, 121.

4 See Veselovskii, *Soshnoe pis'mo*, I.13, 22; see also pp. 107 and 124–5 above.

5 *PSRL*, xxv.238–9.

6 Kahan, *Jahrbücher fur Geschichte Osteuropas*, 16.353–77, surveys such natural calamities.

7 Veselovskii, *Podmoskov'e*, 17.

volving increased concern by lords for land settlement, and, sometimes, for land reorganisation into larger work units, the more remote causes still remain hidden.[1]

Given farming on closes by small families in a relatively abundant forested area, partible inheritance at peasant level could contribute to colonisation and the expansion of the total arable area to keep pace with the population growth; here the family and marriage pattern contributed to extensive economic growth. It was a different situation at higher levels of society. Here, quite apart from such accidents as plague and famine, the custom of partible inheritance practised by the boyars and other notables led to the fragmentation of the large estates of the early fifteenth century which sometimes became absorbed in the growing monastic lands. In the late fifteenth century the system of distribution of lands to be held by service spread, though it only reached its full development in the mid sixteenth century. These new gentlemen holding by service (*pomeshchiki*), the survivors of the old holders of heritable lands, and especially the monasteries, consolidated their holdings and came to claim greater control over the peasant population of their estates. This is the background for the spread of a three-field layout in some cases, the amalgamation of arable land, the increase in the amount of demesne arable land, and also the enserfment of the peasantry. These may be seen as attempts in very unfavourable circumstances at a somewhat more intensive economic growth; but in this case partible inheritance and the insecurity of careers in the service of the Moscow ruler acted as a considerable brake on growth.

The general picture we have of fifteenth-century colonisation continued on occasion even in the late sixteenth century, but it was gradually dying out. In 1576-7 a junior clerk of the Great Treasury took, from the Trinity monastery,

the manorial settlement Davydovskoe and its hamlets, to hold that manorial estate for my lifetime, and to till the arable land; and in that live manorial settlement to establish a monastic tenement in my lifetime, to open up the tilled arable land [*pashnya rospakhati*] and summon peasants in the hamlet, and to establish live hamlets in my lifetime, and to hold in Davydovskoe manorial settlement and in the hamlets every sort of appurtenance, hayfields and meadows and forests.[2]

On the tenant's death, the settlement and hamlets with all their appurtenances, 'with the grain, standing and in the ground, and all their property' were to revert to the monastery; the widow, children and other relatives were to have no claim. Such monastic conditional tenures, especially those for a life, declined in the seventeenth century.[3] Colonisation continued, of course, but mainly in other areas, such as the steppe borderlands; and this involved an outflow of peasant population from the central area around Moscow.

[1] See Alef, in *Forschungen zur Osteuropäischen Geschichte*, 15.37-40.
[2] Sadikov, *Ocherki*, no. 22, 483.   [3] Got'e, *Zamoskovnyi krai*, 426.

Basically, even into the sixteenth century, then, peasants and some lords were clearing and cultivating the forest in small units, hamlets often of very small size. These patches of forest clearance were not worked by means of a developed three-field system, even though there were usually three courses in rotation: winter-sown rye, spring-sown oats and fallow. The arable land was often a close (*niva*), not a field (*pole*); the difference in terminology is significant. This system, based on family labour, and with an extensive and intermittent use of forest, though at first giving high yields often came into conflict with the more closely regulated system of princely and especially monastic estates on which villages and larger hamlets, together with larger fields and a more controlled farming, sometimes developed.

Before the major changes of the sixteenth century the greater part of the peasant population lived 'in small hamlets scattered on glades cleared of forest. We should consider a hamlet of 1–3 tenements typical for that time. Hamlets of 4 and more tenements were exceptions.'[1] But hamlets were administratively subordinate to villages, and in the Moscow area, where black peasant lands were early incorporated into estates, especially ecclesiastical ones, village size was of great importance. The contrast in size in the late fifteenth to early sixteenth centuries was considerable in certain cases. Partial surveys of monastic land and of the metropolitan's land, a total of 21 villages, gives an average of 27 tenements (and 27 males) a village, but only 3.3 tenements (and 3.4 males) per hamlet at the end of the fifteenth century.[2] The average number of tenements in settlements of all types of the Trinity monastery was 5.6 in the 1590s, fell to 5.2 in 1616, but in the period of recovery from the Time of Troubles, around 1625, reached 9.9.[3] Thus, on this wealthy monastic estate even average settlement size was high and increased rapidly by 1625. The sporadic information we have suggests that hamlets rarely had more than 4 tenements; villages, on the other hand, mostly had between 15 and 30 tenements in Moscow uezd.

Villages and manorial settlements had dependent settlements and lands, usually hamlets and wastes, which were often cultivated or at least used for pasture or gathering. Even an isolated church on the Pruzhenka, with a holding of 6 chets per field and 20 ricks of hay, had, 'on the other side of the Pruzhenka, Popova waste, and Bobriki waste with 12 chets of tillable arable land in a field and in two at the same rate, 17 desyatinas of tillable forest, 3 desyatinas of marshy forest and 21 ricks of hay on the Pruzhenka'.[4] A deed of 1526 records the purchase of the hamlets and clearance and notes that it is 'with all that has been subject to that hamlet from of old, arable land, hayfield and forest and all items of income, in all three fields, wherever the sokha, the

---

[1] Veselovskii, *Podmoskov'e*, 39; Cp. a similar statement by him in *IGAIMK*, 139.78.
[2] Based on data in *AFZ*, I, no. 39, 54–5; *ASEI*, I, no. 649, 565–71, Rozhkov, *Sel'skoe khozyaistvo*, 348; Got'e, *Zamoskovnyi krai*, 214–15. See also Romanov, *TrLOII*, 2.469.
[3] Based on data by Got'e, *Zamoskovnyi krai*, 214–15.   [4] *ASEI*, III, no. 52 (1498).

axe and scythe have gone'.[1] Here, land designated in terms of the three-fold formula is recorded in a three-field notation. Both are conventions; the former indicated the functional aspects of a viable farm unit, the latter was an attempt at a common measure of wealth for purposes of tax assessment.

The problem of what a field was, and the complex problems of what 'three-field' was, raise questions both of layout and of the nature of work, as distinct from tenurial, units. We have seen that there is evidence in the late fifteenth century for changes taking place in the nature of layout at least on some monastic and princely estates.[2] In certain cases these changes apparently involved the creation or development of demesne arable. This was probably related to the relative size of estates or, rather, of the manorial units comprising estates and to the general economic, including the labour, situation.

Estate size of itself, of course, tells us nothing about organisation and layout, about work units rather than management units. That the officers of inquisition and other writers of deeds regularly used a three-field notation by the sixteenth century by no means proves the widespread existence of a three-field layout even in the relatively well settled Moscow area. Even when a regular three-field layout existed in a village the supplementary tilled land of the forest, the wastes and other clearances, would remain important; the hayfields and forest were crucial to maintain the livestock on which arable tillage relied. This symbiotic situation is depicted, in fact, in the earliest Russian estate maps which have survived; these are from the seventeenth century.[3]

Concern for the organisation of arable land was demonstrated in 1474 when the Grand Prince exchanged some lands with the Trinity monastery.[4] The Ploshchevo arable was now either contiguous with the lands of the Trinity village or at least located where it had been made administratively subordinate to it. It may, thus, refer either to the physical organisation of the monastic three-field layout of the demesne arable or to the monastic administration exacting obligations from this cultivated land. No doubt both processes increasingly often coincided. Glinkovo village, towards the same northern borders of Moscow uezd, early in the sixteenth century had two fields in Kinela stan, Pereyaslavl' uezd, the third was in Radonezh; this third field part of the village had 29 tenements listed and the village had 4 or 5 hamlets subordinate to it.[5] Here the populous and well-regulated estate of the Trinity monastery shows one situation. The lands of a settlement belonging to the Grand Prince, probably an outlying hamlet or waste, might in practice become part of a monastery's three-field organisation; on such estates a 'field' was clearly not always necessarily just a single block of arable land,

---

[1] *AFZ*, I, no. 61; cp. nos. 64, 66.   [2] See p. 109 above and appendix 1.

[3] Goldenberg, in *PI*, VII, 296–347; see also plates 10 and 11.

[4] *ASEI*, I, no. 424; also note on page 622. See appendix 1, no. 2.

[5] *ASEI*, I, no. 649 (1503/4–1540s), 567. See also the seventeenth-century estate maps mentioned above.

though in the late fifteenth century, such larger and better organised estate units seem to have been concerned with improved organisation of layouts. A somewhat similar situation, which may help to show how such changes might come about, was recorded in a court case in the 1490s.[1] A very large exchange of lands took place in 1520 between the Trinity monastery and the Grand Prince, for similar reasons. The monastery exchanged 2 villages and 67 hamlets for a *pogost* and 71 hamlets.[2]

During the period of forest colonisation and given that cultivation was at least at times somewhat irregular, for example when slash and burn was practised, or when there was a form of infield–outfield, there must have been many occasions when the right to use, and even the nature of particular land areas, were in dispute. Moreover, such patches of forest clearance would sometimes find themselves, given further clearance from the manorial centre, no longer surrounded by forest, but in contact with the demesne arable lands as these were expanded in the late fifteenth century.

Only occasionally do our sources suggest regular three-field layout, however.[3] There is, for example, one reference to a 'middle field'.[4] The meaning of 'three-field' thus seems unclear; but the size of some monastic and princely villages, though still associated with relatively numerous small hamlets with their wastes and other forest appurtenances, probably often involved an increase in the amount of demesne arable and a more regular three-field layout.

A charter of 1498–9 illustrates some of the complexities of tenure and layout. It does not enable us to give a clear picture of the nature of these fields or of their size, but it shows that they were named after particular settlements and included hayfield.[5] Other examples of similar date also show 'fields' as particular locations and as not necessarily being entirely arable land.[6] That fields may have been, even in the late sixteenth century, at least in certain cases, taken as indications of location, not of an open field, seems to be shown in the case of one hamlet where 'the churchyard of Paraskeva called Pyatnitsa has been newly set up on a field; the church is of wood, with cells; it stands without services; a poor monk [*starchik*] lives there and there are two cells and a priest's tenement empty; Ivanko Oksenov, a peasant of Klimova hamlet, built that churchyard'.[7]

For the late sixteenth century the general picture is somewhat more complicated. Then, too, the three-field notation continued in use; but now the

---

[1] *ASEI*, II, no. 404; see appendix I, no. 5. See also *ASEI*, III, no. 44 (1477–84).

[2] Veselovskii, *IGAIMK*, 139.98.

[3] See, for example, appendix I, no. 10.   [4] *ASEI*, I, no. 630 (about 1490s–1510s).

[5] *AFZ*, I, nos. 51, 52; see appendix I, nos. 11, 12.

[6] *AFZ*, I, nos. 46 (1506–7), 49 (1495–1511).

[7] *PKMG*, I.1.86. St Paraskeva or Pyatnitsa (Friday) (there are several identifications of this saint) was the object of a cult popular among peasants in the sixteenth century but originating much earlier; see Dyuvernua, *Materialy*, 173; Chicherov, *Zimnii period*, 56–62.

notation was not only universal in the registers of inquisition, but also modified by the reforms of the 1550s. The extension of the use of the Moscow large sokha-unit as the measure for tax assessment, the regularisation of allowances for quality of land, uniform demands for military service by size of estate measured in notional chets of arable land – these attempts at standardisation throughout Muscovy were based on notional chets, that is estimation of wealth in terms of arable land conventionally measured in terms of a three-field layout. But by this time a three-field arrangement of some sort was doubtless quite widespread in reality, at least at manorial centres on estates and even, to some extent, at peasant level, where forest had been extensively cleared. Therefore, although the figures reported do not indicate the precise amount of actual arable land, they attempt to measure wealth in terms of arable land (for purposes of taxation) in a situation where that aspect of the estate was becoming more regulated and used as an indicatior for the other sectors. In this sense, these conventions in part reflect reality.

In Gor'etov stan', for instance, Osan Zavesin had a tenement 'on the third field' of a manorial settlement beyond the Klyaz'ma which he rented.[1] In Surozh stan', Dmitrokova hamlet on the Molodilna stream had '26 chets of tilled arable land and 10 chets of fallow a field; the third field is in Ruza uezd; hay 15 ricks, firewood 3 desyatinas'.[2] Novoe waste, formerly a hamlet, had 'two fields in Zvenigorod uezd and the third field in Moscow uezd'.[3] These three examples were all heritable lands; the fields may have been still administrative units rather than the parts of a three-course rotation, but they seem to indicate real locations and the irregular amounts look genuine.

There is evidence for the amalgamation or consolidation of arable land, moreover, a development linked with the organisation of three-field layout. Evidence for this is found in the charters of the 1560s.[4] An estate, evidently fairly compact, of the tsar's treasurer, N. A. Funikov-Kurtsev illustrates the process: Salarevo village had eight dependent hamlets (as well as three wastes and a clearance) and three of the hamlets were 'let into the arable land' of the village.[5] In Manat'in, Bykov and Korovin stan two wastes were brought together as regards the arable land which amounted to forty-five chets of medium quality.[6] The phrase 'let into the arable land' (*pripushcheno v pashnyu*) occurs fairly frequently.[7] The land involved was usually a waste, or more rarely a site being incorporated into the arable land of a manorial settlement or other estate unit.[8] In the eighteen instances of incorporation found in the Moscow registers, sixteen wastes and three sites were 'let into the arable' of eight heritable, five monastic or other ecclesiastical estates, and three estates

---

[1] *PKMG*, I.I.142. [2] Ibid. 100. [3] *Ibid.* III.
[4] Sadikov, *Ist arkhiv.* (1940), III, document no. 14 (1567), 205; no. 29 (1568), 227.
[5] Kobrin, *Ist arkhiv.* (1958), 3, document no. 4 (1569), 158–9. [6] *PKMG*, I.I.204.
[7] *Ibid.* I, 63, 129, 130, 131, 138, 175, 180, 183, 189, 203, 257, 263, 264, 277.
[8] *Ibid.* 43, 53.

held by service. It is not clear precisely what the phrase indicates. While it might conceivably mean that certain wastes, sites, etc. were now to count as part of a 'field' for purposes of exactions or estate administration, it seems much more likely that two contiguous units of arable land were now to be worked as a single block (this seems to be the case on p. 119 above).[1] It suggests an extension of demesne arable land around the central settlement and the replacement of the former system (infield–outfield, or close farming) by a three-field organisation with closer manorial control.

The amalgamations indicated by the phrase 'let into the arable land' are not the realisation of a deliberate policy by landowners of enlarging settlements from the late fifteenth century and throughout the sixteenth century (Veselovskii's view). They are the enlargement of fields (at least as administrative units, but usually in a physical sense) within a manor or estate, usually heritable or ecclesiastical ones, in the process of adjusting the man–land ratio in specific circumstances.[2] These circumstances in the second half of the sixteenth century in Moscow uezd included social, political and economic factors. The extensive growth of land-holding by service, the growth of serfdom and villeinage, the great increase in state demands, some growth of the market and of the use of money, the consequential disturbance of existing arrangements, much physical destruction and abandonment of land due to the Oprichnina, peasant ruin and flight, all played a part.

The dominance of the natural economy was being eroded, a fact reflected by both the great increase in state taxation and the considerable rise in prices for agricultural and other produce in the course of the sixteenth century.[3] The autarky of the household unit was weakened; this had considerable effect on the peasant's social standing and his farm economy; this will be discussed below.

The question of the desertion of land deserves a word. In Moscow uezd in the second half of the sixteenth century much land had been abandoned and was overgrown. Frequent entries are found similar to that against a total for Bokhov stan: '...and 4 other wastes, but there was no one to ask the names of the wastes'.[4] In Surozh stan there were 'empty lands which were estates held by service by boyars and gentlemen and junior boyars, but now lie waste and no one holds them'.[5]

There were several elements involved in this desertion and the documentary evidence alone is an inadequate basis for final conclusions. Till at

[1] Cp. the evidence for a similar process, though with a different terminology, in the late fifteenth century, p. 123 above.

[2] Veselovskii, *IGAIMK*, 139.131, and chapter V in general; Romanov, *TrLOII*, vyp. 2, esp. 455, 461.

[3] See Veselovskii, *Soshnoe pis'mo* and Man'kov, *Tseny*.

[4] *PKMG*, I.1.27; cp. 9, 31, 32, 33, 37, 38, 39, 48, 53, 59, 78, 120, 201, 202, 224, 230, 235, 236, 244, 272, 274.

[5] *Ibid.* 116.

least the mid sixteenth century the general pattern of settlement, even in the central area around Moscow, was one of small, non-nucleated hamlets. The two, three or four tenements of such hamlets undertook forest clearance; cultivation by such units would not always involve a regular rotational sequence on constant fields or a three-field layout. Even a great monastic estate on the outskirts of Moscow at the end of the sixteenth century combined regular three-field by the main monastery buildings with cultivation on peasant holdings at some distance.[1] These small peasant units probably tended, in the main, to cultivate by extensive, rather than intensive, means. There is no hint of any open field system of redistributed strips. Nevertheless, the forest clearance and degradation involved was enough in this zone of superfluous moisture and poor soils to result in leaching. This would itself tend to contribute to the retention of forms of extensive cultivation. A shift to more intensive forms, such as a three-field arrangement, would only aggravate the problem of falling fertility and, at times, lead to some desertions, as a result of soil exhaustion.[2] Thus, apart from the social, political and economic factors mentioned above, there were inherent aspects of the system of peasant farming which contributed towards an extensive system of cultivation. In some cases this might involve desertion.

The extent of land abandonment is difficult to measure precisely.[3] Table 10 summarises the available evidence for Moscow uezd; it shows some of the difficulties in interpretation. The proportion of settlements classified as 'empty' is many times higher than the proportion of tenements which were 'empty'. In part, no doubt, this is accounted for by Goehrke's procedure of counting all wastes and sites (selishcha) mentioned as desertions. He did this as an approximation believing that 'most wastes and sites were, in fact, true land abandonments. Thus, in each case the percentages of abandoned lands are fixed as a maximum which, as a rule, is in need of a corrective down grading.'[4]

The very much smaller proportion of empty tenements, however, perhaps indicates how great a corrective might be required in some cases. The data is evidently far from complete; but it is probably noteworthy that the 1586 material (line 4), the only reasonably complete item with an average of about 4 tenements per settlement (a convincing figure for this area), has the lowest percentage of desertions in terms both of settlements and of tenements. The 1592-3 data (line 6) in part confirms this picture, but both items relate to ecclesiastical estates which may have been better able to resist at least some adverse circumstances. The proportion of desertions of settlements after the Oprichnina, therefore, was perhaps as high as 50 per cent, but certainly much less than the 90 per cent or more indicated by lines 1 and 2. Line 8 perhaps

---

[1] See below and appendix 2.   [2] Goehrke, Die Wüstungen, 220–1.

[3] Rozhkov, Sel'skoe khozyaistvo, 305–6, adopts the unsatisfactory procedure of relating the number of wastes to the number of clearances referred to in deeds.

[4] Goehrke, Die Wüstungen, 97.

indicates the maximum peak of desertions after the Time of Troubles.

In fact, of course, the nature of the registers from which this material is derived doubtless distorts the picture. 'Empty' in most cases was used to indicate that, for some reason, the item was not fully viable and was, therefore, to be treated as not liable to taxation. In many instances this was no doubt due to the abandonment of the tenement. Settlements, of course, were less liable to move immediately from a fully viable (and hence tax-liable) situation to one of complete abandonment; for them there were many steps between these two situations, but most of these might cause them to be classified as 'empty'. It is probably in terms of some such explanation that one should seek for the downgrading required for the inflated figures of desertions; but this, even if correct, does not enable us to put forward any corrective at all precisely. Thus, we need not accept that 90 per cent or more of Moscow uezd settlements were abandoned in the late sixteenth century. It is clear that many, perhaps half or more of all settlements suffered sufficient economic setbacks for the officers of inquisition to classify them as tax-exempt, but we do not know how many were abandoned and deserted in the full sense.[1]

TABLE 10. *Desertions in Moscow uezd*

	Date	Areas	Type of estate	Settlements		Tenements		Arable
				total no.	empty no. %	total no.	empty no. %	land (chets)
1	1574	6 stans	service, heritable, ecclesiastical	541	494 91.3	—	— —	1,256
2	1576–8	3 stans	service	199	192 96.5	>27	>2 7.4	>90
3	1585	—	all	2,697	2,012 74.6	2,999	151 5.0	15,918
4	1586	—	ecclesiastical	473	225 47.6	1,967	44 2.2	7,552
5	1594	3 stans	ecclesiastical	18	12 66.7	—	— —	310
6	1592–3	—	ecclesiastical (Trinity monastery)	533	253 47.4	1,585	— —	8,997
7	1616	—	ecclesiastical (Trinity monastery)	369	278 75.3	477	— —	—
8	1624–6	—	ecclesiastical (Trinity monastery)	567	412 72.6	1,528	— —	—

Source: Goehrke, *Die Wüstungen*, 98, 104, 124.

The political and social events of the late sixteenth and seventeenth centuries, the main immediate causes of desertions, took place when internal clearance

[1] See also the section below on Toropets, for which we have more detailed evidence on desertion than for Moscow.

and colonisation had come to an end. The technical possibility of continuing extensive cultivation was now restricted by the limits imposed by tenurial rights and, in some cases, by the extension of demesne arable land; at the same time, however, new territories in the Volga area and the south-east steppe frontier became available for colonisation.

It should be remembered that throughout the sixteenth century, even on the largest and best organised estates of wealthy monasteries, the work units of land were a combination of demesne arable land at the manorial centre with a wide variety of small scattered arable units in peasant hands in the surrounding forest. The symbiosis of close and field continued; but the key question of the balance between the lord's central demesne arable land and the arable land of the peasants, remains unanswered. The disasters of the late sixteenth century, the destruction and desertions, are likely to have increased the amount of land (but not cultivated land) per man and this shortage encouraged transfer to a somewhat more intensive three-field arrangement. At the same time, the disasters put many difficulties in the way of such a transfer. The patches and clearances in the forest continued in the majority of cases to play a crucial part, as sources both of grain and of peasant work hands. The tension, at the extremes, between close farming and demesne arable farming was increased.

The information on settlement size reinforces this general point. While in Moscow uezd village size was often quite large by Russian standards, hamlet size did not differ much from that of other regions, as we shall see. The small, non-nucleated settlement, usually of no more than four tenements, was the type in which the majority of the population in the uezd lived; it was probably here that most land was tilled and most grain was produced. But all this is largely surmise; we have little hard evidence.

Koretskii has, indeed, pointed out that 'the lord's arable land was a normal phenomenon in the first half of the sixteenth century on both estates held by service and heritable ones in the central uezds of the Russian state'.[1] One might query how 'normal' it was, but such arable certainly existed. He adduces evidence to show that lord's arable land, sometimes from the late fifteenth century, was also found on estates of the metropolitan and monasteries, and on court lands (one example, not in Moscow uezd), on all lands other than those of black peasants. Instructions to peasants in the 1550s in the central uezds all included references both to payments in kind and also to tillage duties.[2] This arable land was worked not only by villein labour, servile dependants of the lord, but particularly by peasants, even in the first half of the sixteenth century; Koretskii also points out that the 'boyar's labour rent' is mentioned in the 1550 Law Code.[3] This is no doubt basically

---

[1] *Zakreposhchenie krest'yan i klassovaya bor'ba v Rossii*, 18.   [2] *Ibid.* 19–23.

[3] Rozhkov, *Sel'skoe khozyaistvo*, 133, 138–43; *Sudebniki XV–XVI vekov*, 173; translations in Dewey, *Muscovite judicial texts*, 71; Smith, *Enserfment*, 93.

true.[1] Koretskii himself, however, added a word of caution.

We spoke of the lord's arable land and labour rent existing for the first half of the sixteenth century in all types of tenure, and noted their tendency to develop; but in the second half, and especially at the end, of the sixteenth century, widespread lord's arable land and labour rent is characteristic; this begins to occupy a predominant place in the system of peasant obligations.[2] . . . In the difficult years of the Livonian war a mass transfer of peasants in court villages to tilling a set number of desyatinas [*desyatinnaya pashnya*] takes place to ensure grain for the forces.[3]

In fact, the increase in labour rents, despite some earlier evidence, occurred mainly from the 1570s and 1580s when they may have become more important than other peasant obligations in Moscow uezd.[4] This process was encouraged, if not caused, by rising prices for agricultural produce (see table 12, p.143) and the need for closer control of labour; the burdens of war and the devastation wrought by the Oprichnina stimulated peasants to flee to monasteries, other powerful lords, or even to the newly acquired lands along the Volga and elsewhere.[5]

There is a further problem. This is the question of the location of the lord's arable land. The obligation to till a set number of desyatinas, for instance, might, in theory at least, be fulfilled by work either on the peasant holding or on the lord's demesne arable land located at his manorial settlement or village. There certainly were cases when tillage by desyatinas meant tillage on the central manorial demesne.[6] Nevertheless, the view that 'everywhere, whether peasants tilled arable lands for the monastery by vyts (the so-called tillage by desyatinas) or compulsorily [*vzgonom*], i.e. collectively, monastic tillage was always concentrated on the village fields' is not as clearly indicated by the Moscow registers as Veselovskii suggests.[7] He stated that 'the majority of peasants, living in hamlets, fulfil this obligation "from a distance" [*naezdom*], i.e. riding to work from their hamlets'.[8] If this were so, it is difficult to see what distinguished the two forms. Was tillage by desyatinas performed by a single peasant, while compulsory tillage was performed by a group?

Fedorovich held that 'usually the peasants of the village or hamlet worked it [the monastic arable land], but sometimes peasants were assembled "compulsorily" from all the estate settlements to work some large portion concentrated at one of the monastery's settlements'.[9] The 1590 charter of the

---

1 But see below for seventeenth-century evidence of a continuation of the use of villein labour.
2 Koretskii, *Zakreposhchenie*, 25.   3 *Ibid.* 31.   4 Stashevskii, *Opyty*, 43.
5 Makovskii, *Razvitie*, 472. On the colonisation of the Volga area see Peretyatkovich, *Povolzh'e v XV i XVI vv.*, 273f. Colonisation of the Perm' and Kazan' areas seems to have been virtually simultaneous; Lyubavskii, *Istoricheskaya geografiya*, 207.
6 An example, in Rostov uezd, is given by Got'e, *Zamoskovnyi krai*, 487–8.
7 *IGAIMK*, 139.141; cp. a similar view in Got'e, *Zamoskovnyi krai*, 500.
8 Veselovskii, *IGAIMK*, 139.141.
9 Fedorovich, *Uch. zap. Kazan. GU*, t. 90, kn. 6, 1110–1; cp. Got'e, *Zamoskovnyi krai*, 500.

Novinskii monastery (appendix 2), which details peasant obligations, like other such documents, in fact distinguishes between tillage (§§ 3, 24) and compulsory tillage (§§ 21, 23). There are regularities in the formulae used even though the meanings are not explicitly given. Tillage, not otherwise qualified, is not found together with payment of any produce rent in the form of grain; yet cartage of grain to the monastery is included in the obligations; this suggests the grain was produced on the peasant farms. Tilling compulsorily, however, was accompanied by paying produce rents in grain; in one case a chet of rye and two of oats (i.e. a *yuft'*); in the other, 4 chets of rye per vyt. Moreover, in the last instance, it was stated that there were 'four desyatinas of monastery arable land which the peasants mow and till compulsorily'; this is the only explicit reference to monastic arable land. There is no mention of cartage of grain, apart from that paid as produce rent. Thus, compulsory tillage here took place on monastic arable evidently located near the manorial centre; this would, of course, be organised by the steward probably on set dates and so, as Veselovskii wrote, be collective. Reference to tillage alone, however, indicates cultivation on the peasant's own farm; the amount specified, one-and-a-half desyatinas per vyt, indicates one chet per course per vyt. Since the peasant unit was normally some fraction of a vyt, this obligation does not appear to be very onerous.

On larger or better organised monastic estates grain, then, was grown on demesne, but at least some was grown on arable units outside the main block of arable land, however organised. Manorial settlements sometimes had fully dependent, servile villein labour; this was supplemented by dependent peasant, i.e. serf, labour in the main working on its own holdings around the central settlement. At Nikonovo the monastery 'children' (the monastic equivalent of lay villeins) tilled six chets of good and seven of medium arable; there were also four chets of good quality peasant arable land, but 'the monastery takes no income from the peasants because they have been newly established'.[1] This was the reason, presumably, for the relatively small amount of peasant arable land, their tenements were not yet fully viable.

There were, thus, various methods of working monastic arable land.

First, it was an obligation of monastery peasants who should till, sow and gather a specified number of desyatinas according to the size of their portions. In the seventeenth century this obligation fluctuated in the range of two to four desyatinas per vyt of peasant land. Second, monastic arable land was maintained compulsorily [*vzgonom*] by means of peasant and labourer hands, i.e. by common efforts, obligatory help. Third and last, the children [*detenyshi*] tilled arable land for the monastery.[2]

It seems only necessary to add that the first method was evidently employed on peasants' own holdings and that peasants might have to fulfil more than one type of labour obligation.

---

[1] *PKMG*, I.I.60–I.   [2] D'yakonov, *Ocherki iz istorii sel'skago naseleniya*, 307–8.

After the 1584 decision rescinding the grants freeing monasteries from the main tax payments (*tarkhany*), there was an incentive for monasteries to expand the demesne arable area; while peasant land was tax-liable, the more powerful monasteries were able to obtain exemptions from the decision for their demesne arable land, at least by the end of the sixteenth century. For instance, in 1594 Trinity lands in Goretov stan had 'to pay the Sovereign's taxes from eleven live chets, apart from that the children till'.[1] Such circumstances modified the proportion of demesne arable to the arable land in peasant hands, at least on monastic estates. Unfortunately, we do not have enough evidence to give a precise picture of the changing proportions of demesne arable land to land held by peasants in Moscow uezd. On monastic land it looks as if the concessions gained at the end of the sixteenth century and in the immediate aftermath of the Time of Troubles encouraged a higher proportion of demesne. On the largest monastic estates, such as that of the Trinity monastery of St Sergius, the old combination of servile and serf cultivation, both on monastic demesne and on peasant farms continued. In 1617, for instance,

the Sovereign ordered the live monastic arable land, together with that which the children [*detenyshi*] till for the monastery, in Moscow uezd and elsewhere to be free of taxation...As for the peasant and servant arable land, with all that arable which the peasants and servants till for themselves, the Sovereign ordered them to have the sokha-unit people, the money income and every sort of petty communal obligation, as before. And the clerks...sent a pay-register [*platezhnitsa*] [derived] from the various registers of inquisition, [listing] in those monastic lands the arable land, what the children till for the monastery, and whatever arable land the peasants till for the monastery by themselves [*na sebya*]...[2]

Soon after this, however, the introduction of a new tax basis eliminated this advantage. This new unit was the 'live chet' (*zhivushchaya chet'*), based on the number of live, i.e. viable and hence tax-liable, tenements; the area of arable land was no longer considered. The new unit took account in a rough and ready way, and with many variants in practice, of both peasant and labourer tenements in certain ratios (see table 11).[3] Monastery children (*detenyshi*) were probably excluded.[4]

Special circumstances might ease demands. 'The peasants till for the monastery, mow and cut firewood; but the peasants give no money rents, grain or any other stocks because the village is on the high road and provision of horses for the Sovereign's riders and for the monastery's affairs is had from them.'[5]

---

1 *PKMG*, I.I.280.
2 Sadikov, *Ocherki*, no. 54, p. 529; cp. D'yakonov, *Ocherki iz istorii sel'skago naseleniya*, 311–12.
3 Veselovskii, *Soshnoe pis'mo*, II.486–502. See also Blum, *Lord and peasant*, 233.
4 D'yakonov, *Ocherki iz istorii sel'skago naseleniya*, 304.
5 *PKMG*, I.I.69; cp. 60–1, quoted above.

In other instances, too, peasants gave no money or grain income to the holders of heritable estates, but sometimes tilled arable land for them, mowed hay and cut firewood.[1] Such labour obligations seem to have been commoner than produce rents in grain on the late sixteenth century lay estates, but they may have been performed on peasants' own holdings.[2] Moreover, the question of relative labour inputs remains totally unresolved. We know nothing of the amount of time spent on work on the demesne land, nor of when labour had to be supplied; and this last was an important consideration when the work-period was severely restricted by climate.

On lay estates too, then, the lord and his 'people' (i.e. villeins) are found in the central settlement, the peasants (i.e. serfs) are in the surrounding hamlets and wastes.[3] The lord's village and the peasant's hamlet were the typical Moscow settlements of the fourteenth–sixteenth centuries.[4] In at least one instance there was a distinction between out-servants who worked the fields and in-servants who worked in the house; a settlement had one household (*chelyadinnoi*) tenement and four people's, i.e. villein, tenements.[5]

TABLE 11. *District categories, tenements per live chet*
(*decrees of 1620–8*)

Category	Estates			
	Service and heritable		Ecclesiastical	
	Peasant	Labourer	Peasant	Labourer
I	2	2	2	2
II	2	3	2	2
III	3	2	2	2
IV	4	2	3	2
V	4	3	3	3
VI	5	3	4	2
VII	8	4	6	3
VIII	12	8	9	6

Source: Veselovskii, *Soshnoe pis'mo*, 11.489.

The lands of settlements were divided, among heirs, for example. This suggests a comparatively stable, rather than a nomadic, population. One manorial settlement, Stepanovskoe, was divided into halves, one of which was subdivided into three lots.[6] A third and a sixth of this settlement are also mentioned but the precise relationships of the family members are obscure.[7] A purchase was made from four relatives, probably cousins, in the 1520s of three strips of the land.[8] These were evidently located in a single block, again

---

[1] *Ibid.* 77. [2] *Ibid.* 59, 62, 64, 77. [3] *Ibid.* 173. [4] Veselovskii, *Podmoskov'e*, 39.
[5] *PKMG*, 1.1.186. [6] *AFZ*, 1, no. 65 (1525–6). [7] *AFZ*, 1. no. 66. [8] *Ibid.* no. 69.

suggesting that here at least there was no developed three-field layout with strips allotted in each of the subdivisions of the fields. This process of division sometimes resulted in estates held by service being so fragmented in the seventeenth century that some tenants were entirely without peasants.[1] Small pieces of land such as these are likely to have been not only tenurial but also in effect work units.

Small compact blocks are likely to have been particularly characteristic of peasant holdings rather than the larger, and sometimes more organised, estates of military servitors, of princes, and especially of monasteries. Such peasant holdings, of course, were not always the result of fragmentation resulting from partible inheritance, but were especially characteristic of the initial phases of forest colonisation and of the farming of closes. For instance, at Men'shee Red'kino a peasant of the St Simon monastery had sown one-and-a-half chetverts of rye.[2] This was by Medvezh'e (Bear) lake in Pekhora volost where peasant proprietors, such as Isak Bashlov, a blacksmith, had in the middle of the fifteenth century been setting up clearances and working some single tenement hamlets in the dense forest with its bee-trees, marshes and fowling runs.[3] Perhaps a little later, but before 1475, a monk who had a number of dealings in rye left the Trinity monastery 'four closes [nivy] of rye on the Dubnichnoe field'.[4] This once again shows that the term 'field' must be looked at carefully.

It has already been mentioned that land was worked 'from a distance' during the continuing internal colonisation of the fifteenth century.[5] Wastes and sites were also worked.[6] References to turnip patches (repishcha) again suggest something like an enclosed work unit rather than a common field; moreover, they may possibly hint at the use of fire as a means of tillage.[7] Other such hints are the use of the term for cut forest (secha) and a record of a court case in which it was alleged that the defendant 'cut the forest this spring over the old boundary, the monastery's land, and sowed it'.[8] It is just possible that something other than slash and burn, cultivation is referred to here, but it seems unlikely; fire would be the quickest means to clear trees and sow in one season. Hayfields were, apparently, occasionally tilled.[9]

Various cultivated or cultivable lands beyond the central fields are often referred to in the late sixteenth-century registers. Units of tilled forest, most frequently recorded in the range 2–4 desyatinas, sometimes as small as $1\frac{1}{2}$ or as large as 7 desyatinas, were found at wastes 'beyond the fields' (po zapolyu).[10] A huntsman of Prince Andrei Telyatevskii lived in a tenement 'along the end

---

[1] Got'e, Zamoskovnyi krai, 400.

[2] ASEI, II, no. 371 (1462–85).   [3] ASEI, II, no. 347; no. 371 and note on p. 561.

[4] ASEI, I, no. 450 (1475). The precise location has not been identified.

[5] ASEI, II, no. 404 (1490–8); also no. 385 (c.1460s).

[6] ASEI, I, nos. 232 (1440s–50s), 142 (c.1430s).

[7] ASEI, II, no. 425 (1501–2); AFZ, I, no. 47 (1496).   [8] ASEI, I, no. 595 (1495–9).

[9] ASEI, II, no. 411 (1494–9).   [10] PKMG, I.1.20, 24, 27, 28, 29, 39, 40, 41, 42, 48.

of the field'.[1] There were 'remote arable lands' and arable land tilled 'from a distance'.[2]

At the same time peasants, among others, were renting similar lands, that is paying quit-rent (*obrok*) for them. This payment was less than the amount due in taxation from 'live' land. Such rent-liable lands, therefore, though granted without term (unlike hired land (*naem*) which was for a limited term) were in an intermediate situation between viable, 'live', fully tax-liable land and 'empty', tax-exempt land. They were consequently at risk, since anyone willing to pay the full amount of tax might take them over.[3]

Peasant land-renting was quite considerable in some areas, especially renting by the peasants of certain lords. The peasants listed as renting these empty lands, however, have to be supplemented by others who are not always mentioned in the registers. For example, Gavrilko, a peasant of Bogdan Belskii, rented 7 wastes in Bokhov stan.[4] The total given, however, states that Gavrilko rented these lands 'with his fellows'.[5] In a court village of Ievle volost Vlas Ivanov 'and his fellows' rented land.[6] In other instances, although only one man is named, the text refers to peasants in the plural.[7] There are examples of the same peasants being mentioned in different lists.[8] Moreover, the forms of some names are not those usual for peasants, although the individuals are specified as such; there is Patra, son of Volk, for example, while two 'people' (i.e. villeins) of Shemet Ivanov, a clerk of the stables, are given as Ivashko, a typical peasant name form, and 'Zhuk, son of Danil Poznyakov'.[9]

All this suggests that this renting of land may have been undertaken, not so much by individual rank and file peasants, but rather by peasant officials acting on behalf of a group, possibly a tax commune or partnership of peasants of a particular estate. Such groups are implied by the phrase 'with his fellows'. Among these men there may have been some peasants of higher status known by patronymic forms; they would be likely to be found among peasant officials, but it is also possible that sometimes they were acting on their own behalf. The view that this land renting is evidence of rich, entrepreneurial peasants seems overstated.[10] Such land renting may be an indication either of wealthier peasants being able and anxious to cultivate additional land, or of pressures obliging groups of peasants on certain estates to seek additional lands to cultivate. In the former case, the process may be seen as analogous to the amalgamation of arable land we have already seen on certain estates.

---

[1] *Ibid.* 59. [2] *Ibid.* 13, 56, 57, 58, 90, 91, 94. [3] Veselovskii *Soshnoe pis'mo*, II.465–6.
[4] *PKMG*, I.I.224. [5] *Ibid.* 225. [6] *Ibid.* 261; cp. 271. [7] *Ibid.* 269, 270–1.
[8] *Ibid.* 259 and 268, two entries on 269, two on 271.
[9] *Ibid.* 269, 262. Much weight should not be given to the use of patronymics, however; there was no consistency in their usage. See Tupikov, *Slovar'*, 24.
[10] Makovskii, *Pervaya krest'yanskaya voina*, 153–4.

There is less evidence about livestock husbandry than about arable farming.[1] The main draught animal was the horse, but the documents do not mention it. It was probably the only draught animal on peasant farms, but on monastic estates oxen were also used.[2] Horses were also used for riding, and provision of horses was a common obligation imposed on peasants.[3] Probably on peasant holdings the same animals were used for both purposes. Though no doubt peasants often went on foot, there is evidence that they rode.[4] There is also evidence for trading in horses with Nagai merchants, perhaps from the late fifteenth century, certainly in the sixteenth century, but nothing is known of the numbers or extent of this trade.[5] When horses were sold, a branding tax was exacted and monasteries were sometimes granted the right to exact this themselves.[6]

The cow was almost as essential as the horse, yet, again, there seems to be no direct evidence about peasant cattle. Monastic obligations included the delivery of sour cream,[7] butter and sour cheese, though the last item may have come from sheep or goat milk.[8] The metropolitan and monasteries had specialist cowhands.[9] Sheep, and probably goats, were almost certainly found on most peasant farms, but the documents give no details except that sheep-skins, sometimes tanned, were among peasant obligations and might be taken in lieu of flax.[10]

Poultry, too, are only implied by eggs listed among the items of produce peasants had to deliver.[11]

Numerous references to hayfields and meadows in the charters and registers show the importance of hay as feed. Virtually every record of a holding mentions hay or hayfields. The water-meadows of the river valleys seem to have been of particular importance.[12] Such meadows, as well as sites, wastes and clearances in the forest, might be worked from a distance – a fact which suggests that, at least for some of the larger holdings, there was pressure on feed supplies, if not an actual shortage. Concern was shown to protect meadows and pasture lands from trampling.[13] Fenced droveways were some-times made by peasants to enable their livestock to reach water without

---

1 See Rozhkov, *Sel'skoe khozyaistvo*, 72.
2 *ASEI*, I, nos. 307 (1462–6), 492 (1481), 575 (1494); *PKMG*, 1.1.82, 247, refer to ox-men (*voloviki*).
3 See e.g. *AFZ*, I, no. 26 (1526); *ASEI*, I, no. 307.
4 *ASEI*, I, no. 414 mentions that peasants rode to church; the term *naezd* indicates land worked from a distance involving riding.
5 *ASEI*, I, no. 1 supplement on p. 26; Kashtanov, *Ocherki*, 437–8.
6 *ASEI*, I, nos. 309 (1462–6), 416 (1473).   7 *PKMG*, 1.1.54, 58, 62, 63, 64, etc.
8 *PKMG*, 1.1.63, 64.   9 *AFZ*, I, no. 2;4 *PKMG*, 1.1.53, 61.
10 *AFZ*, III.1, no. 23; *PKMG*, 1.1.63, 76.
11 *AFZ*, III.1, no. 23; *PKMG*, 1.1.54, 58, 62, 63, 64.
12 E.g. *ASEI*, I, nos. 554, 555 (1490s).
13 *ASEI*, I, no. 493 (1481). *Sudebniki XV–XVI vekov*, 27 (§61 of 1497 law book).

damaging crops.[1] Thus, even though the charters and registers do not include much information about peasant livestock, we seem justified in assuming that most tenements had some livestock; at least a horse, a cow and, probably, a few sheep and poultry.

Evidence on gathering is a little more frequently found, but is still inadequate for anything like a full picture to emerge. Even in the early sixteenth century, that is after the colonisation process had fully developed and estates had claimed much of the land, but before the disasters of the second half of the century led to any population stagnation or decline, there seems to have been abundant forest in most parts of the uezd. Only at some large settlements is there likely to have been any shortage of forest. The registers of the late sixteenth century note carefully not merely the amount of forest of various types allocated to servitors, but also the extent and nature of forest which had grown on deserted lands lost to cultivation. Entries such as 'coppice forest suitable for stakes, poles and beams' described the overgrown land.[2] Moreover, throughout the registers, whenever obligations from peasant tenements are specified, tillage, hay and firewood seem to be regarded as fundamental.[3] In such cases the three-fold formula of earlier times reappears, as it were transformed from the plane of the primary producer to that of the lordly consumer.

Our sources tell us nothing of many aspects of the forest as a source of food at peasant level; nothing on berries or fungi in the Moscow uezd has been found, for example. Apple trees, presumably wild, are mentioned mostly in describing the bounds of estates in the fifteenth to early sixteenth centuries.[4] In the late sixteenth century, however, in Manat'in stan there were apple orchards shared by tenants holding by service.[5] In Kinela stan in the seventeenth century, a clerk holding of the Trinity monastery had an orchard with 30 apple trees.[6] There is also one mention of a nut tree, evidently wild, near a hamlet's fence.[7]

There is more material on fish, and some on game and wild bees, though most of it relates to the holdings of lords rather than peasants. Fish was, of course of great importance to monasteries since it could be eaten on Christian fast days. Monasteries sought and were granted exclusive rights to fish in rivers.[8] Lakes are also mentioned as supplying fish.[9] These included pike, bream, carp, perch, roach and others it has not been possible to identify.[10] Nets,

---

[1] *ASEI*, I, no. 431 (1474–5).   [2] *PKMG*, I.I.20 and many other entries.
[3] *Ibid.* 59, 62, 68, 70, 77, 85, etc., cp. *AFZ*, III.I, no. 23.
[4] *AFZ*, I, nos. 44a (1486), 46b (1497); *ASEI*, II, nos. 409 (1494), 411 and 412 (both 1494–99), 429 (1504); *AFZ*, I, no. 44 (1510–11).   [5] *PKMG*, I.I.35, 39.
[6] Got'e, *Zamoskovnyi Krai*, 459. In nearby Pereyaslavl' uezd in 1620 the tsar had two orchards with 240 apple and 8 pear trees; *ibid.* 458.   [7] *ASEI*, II, no. 412, (1494–9).
[8] *ASEI*, III, no. 41; I, nos. 40, 376; Kashtanov, *Ocherki*, no. 4, p. 346; no. 26, p. 386; no. 28, p. 390.   [9] *AFZ*, III.I, no. 61 (1620).
[10] *Shchuki golovy, shchuki kolodki i ushnye.* These may, in fact, be pike or parts of pike for particular dishes.

including drag nets (*bredniki, sezhi*) were used as well as traps (*kuritsy*) which were probably used for taking fish under the ice.[1] Fishing expeditions were regularly sent out by some monasteries in the fifteenth century.[2] By the late sixteenth century, however, there are numerous references to ponds in some settlements, not only monastic ones, and it is made quite clear that by this time fishing in the Moscow uezd had, on the more advanced estates, changed from gathering activity to deliberate cultivation.[3] The pike and perch of lake Isakovo in Vokhna volost were fished by peasants 'for themselves'.[4] At other small settlements there was probably a continuation of such ancient activities; for instance, at the two carp ponds of Bulatnikovo waste or hamlet.[5] At the court village of Cherkizovo, however, pond-men (*prudchikovy*) are noted, as well as a carp pond 'and two more beyond the wood'.[6] At a monastic settlement there were three carp ponds and '3 fish nurseries [*sad*] and in them is every sort of noble fish'.[7] At a monastery's manorial settlement on the Klyaz'ma was another fish nursery; 'they put every sort of fish in it, they fish in the Klyaz'ma and carry them off to the monastery'.[8] This development of fish cultivation seems to be analogous to changes in the sixteenth century in Poland and Czechoslovakia.[9] By the seventeenth century peasant fishing rights were considerably restricted by the growth of estates.[10]

The taking of game is hardly mentioned in the sources. An immunity of 1481 freed monastic dependants in a village near Moscow from the obligation to turn out for bear and elk hunts.[11] The metropolitan was granted a similar immunity which distinguished the battue and the chase; his peasants were not liable to 'go to the palisades for elk, for bear or for wolves, nor to the field for elk. And our huntsmen and kennelmen are not to ride in to them nor to station themselves on them, nor are they to have to feed my dogs...'.[12] The royal hunt in fact seems to have been quite well developed in the late sixteenth century. The kennels were evidently located mainly in Bokhov stan, north of the Klyaz'ma, judging from the several references to holdings of kennel-men in the area.[13]

Peasants, however, are likely to have been more concerned with the taking of smaller game than bear or elk, but there seem to be no references to hare

---

[1] *ASEI*, I, no. 376; II, no. 384; III, no. 41; III, no. 38 mentions fishers under the ice on the Sherna. [2] *ASEI*, I, no. 41; also Kashtanov, *Ocherki*, no. 47, p. 421.

[3] For references to ponds see *PKMG*, 1.1.55, 61, 79, 80, 82, 129, 167, 180, 189, 277–8 (two ponds, 'but the water has now flowed out').

[4] *Ibid*. 93.   [5] *Ibid*. 189.   [6] *Ibid*. 167. Sadovniki are also noted (see below).

[7] *Ibid*. 82. Sadovnik usually means gardener, but in the above context it probably refers to fish breeders.   [8] *Ibid*. 67; 219.

[9] See B. K. Roberts' review of works by W. Szczygielski in *AHR*, XVI.II.161–3.

[10] Got'e, *Zamoskovnyi krai*, 532.   [11] *ASEI*, I, no. 492.

[12] *AFZ*, III.I, no. 11 (1564 with confirmations in 1569, 1574, 1584, 1587). See also nos. 37 (1599) and 48 (1605).

[13] *PKMG*, 1.1.206, 207, 210, 211, 217. A few others were found in Pekhora stan and one in Sherna; *ibid*. 282–3, 256.

or other small game. Hunting rights had probably long been restricted by this time. Beavers, for instance, are only mentioned in connection with grants to monasteries.[1] In the east of Moscow uezd, Vokhna volost had been an area of the prince's beaver men.[2] In the seventeenth century even the sale of wild hawks had been forbidden; a peasant arrested in 1629 for selling a hawk he had caught three days before pleaded ignorance of 'the sovereign's prohibition'.[3]

In the south-east, along the Guslitsa and some tributaries, the taking of squirrel for pelts remained important in the fourteenth and fifteenth centuries. Fowling nets (*perevesi*) were used in the fourteenth and fifteenth centuries in clearances and by marshes; they were favoured for taking duck and may possibly sometimes have been used on peasant holdings.[4] They do not seem to occur in the sixteenth century registers, though there is a reference to net-men (*tenetchiki*) (probably men who kept the nets used in battues) on an estate held by service and formerly held by kennelmen.[5] This tends to confirm the growth of lordly hunting at the expense of that at lower social levels; it may also in part reflect the changing nature of forest resources owing to human activity.

There are few references to wild bees in the Moscow uezd, again possibly due to human disturbance of the forest. Man and his forest clearance was a greater danger to the bees than even the honey-loving bear. In the fourteenth century, along the Medvenka and Vyazemla, minor tributaries of the Moskva, was an area of bee forest; bees were found also in Radonezh, Korzenev and Vorya stans. On the Pakhra, around Dobryatino village, was a huge bee-forest of the prince. In the first quarter of the fourteenth century there was bee-forest just to the east of Moscow itself; under Ivan Kalita it was in the charge of a certain Vasil'tso from whom the later Vasil'tsevo sto, the department responsible for the sovereign's honey and mead supplies, derived its name. In the early fifteenth century the prince still had his bee-men and grouse-men here, but in the sixteenth century there is no mention of them; only place-names such as Bortnoe (*bort* meant a bee-tree) and Teterevniki (grouse-men) remained.[6] In Bokhov stan towards the middle of the fifteenth century a landowner made a gift to a monastery which included forest and wild bees.[7] The only other references found all relate to that wild area about Medvezh'e (Bear) lake already mentioned; there were evidently a fair number of bee-trees in the area since there was a ten-man responsible for bees; this suggests an area of relatively undisturbed forest.[8] There is no mention of bees in the

[1] *ASEI*, I, no. 40 (1423); Kashtanov, *Ocherki*, no. 3 (1432–45), 345.
[2] Veselovskii, *Podmoskov'e*, 39.   [3] Kutepov, *Tsarskaya okhota na Rusi*, 238.
[4] *ASEI*, II, nos. 340 (c. 1380), 347 (1445–53), 355 (1453–62), 371 (1462–85), 381 (c. 1465–9), 418 (1498–9).
[5] *PKMG*, I.1.265.   [6] Veselovskii, *IGAIMK*, 139.95–6; *Podmoskov'e*, 24–5.
[7] *ASEI*, II, no. 342 (c. 1440–44).
[8] *ASEI*, II, no. 381 (1465–69); cp. nos. 347, 355 (both mid 15th century).

sixteenth century registers which have survived for the Moscow uezd, nor in our seventeenth-century sources.[1]

Few crafts and trades are mentioned in the sources. Among the witnesses giving evidence in a court case at the end of the fifteenth century were the good men 'Prokosh the fisherman, Oleshko the carpenter'.[2] A blacksmith was rare and perhaps likely to be a substantial man; one in the late fourteenth century was engaged in opening up new land in the forest.[3] An early sixteenth century list of certain monastic lands specifies only one blacksmith (but two jesters!)[4] For Radonezh stan there is an immunity to the Trinity monastery granting new peasant tenants remission, among other items, of the obligation to make saddles (*sedelnoe delo*); possibly there may have been a local craft in Kiyasovo (or Kesova).[5] This village was on the road to Moscow.[6] A monastery peasant living in a village was described as a shoemaker, but this is likely to have been a rare trade in the villages, though prominent in Moscow itself.[7] One curious aspect of trades linked with horses is that a series of documents, all deeds of full bondage, were written by post-horse clerks.[8] This seems to have been the result of an innovation in procedure in Moscow apparently about 1480 by Ivan III.[9] The post-horse clerks were probably used for this purpose as they were responsible for collecting the customs duty on transported goods (*tamga*) and the duty paid on such documents was treated as an item under this head.

The fifteenth- and early-sixteenth-century charters give us no evidence on textile production, though obviously domestic production of linen and woollen objects was common. The later registers include textiles among items of rents in kind; two sorts of linen cloth are mentioned and some interesting equivalents noted.[10] '...if in any year canvas is not taken, 3 altyns a canvas are taken; but if in any year canvas and money in lieu are not collected, a tanned sheepskin each is taken from the peasants instead of canvas.'[11]

No detailed description of a peasant unit has been found, but there is one for a post-horse station established on the Dmitrov high road by the Ucha in 1510–11.[12]

The land for the tenement is 30 by 20 sazhens [64 m by 42 m]. On the tenement is a building of two living huts [*izby*], one of 3 sazhens [6½ m], the other of 2½ [5½ m], and a room under a hayloft. The hut with the hayloft over is 2½ sazhens and the other hayloft with the room below is 2½ sazhens, and between the haylofts is a stable and a

---

1 Chechulin, *PKMG*, I, ukazatel', xxiv; Got'e, *Zamoskovnyi krai*, 532.
2 *ASEI*, II, no. 411 (1494–99).   3 See above, p. 116; *ASEI*, II, no. 381.
4 *ASEI*, I, no. 649, 565   5 *ASEI*, I, no. 175 (1444–45).
6 *ASEI*, I, p. 604. See also no. 494 but no. 649 gives no hint of such a trade.
7 *ASEI*, II, no. 404, 415. Rabinovich, *O drevnei Moskve*, 286–8.
8 *ASEI*, III, nos. 397; 402, 403 and 417 by the same man; 418; 421, 422, 423, 425, 427, 430 and probably 434 and 438 all by the same man; 446, 450, 451.
9 Gurlyand, *Yamskaya gon'ba*, 45–7; Kolycheva, *AE za 1961*, 63–4.
10 *PKMG*, I.1.63.   11 *Ibid.* 76.   12 *AFZ*, I, no. 44.

hay jetty of 4 sazhens [8½ m]. The two huts, a hayloft and the room under it are new, but the hayloft, stable and hay jetty are noticeably repaired from the old tenement. The buildings are covered with new bast and shingles. The yard has a wattle fence with gates of squared timber with bolts on both sides and covered with boards. Sixty chets of land have been allocated and 60 ricks of hayfield... The station is 30 verstas [32 km] from Moscow and the same from Dmitrov.

This holding was, of course, more substantial than, but doubtless similar to, the usual peasant farm. The provision of hay along the river was here of especial importance because of the nature of the holding; usually it would have been the arable land which was of greatest importance.

Peasant farms similar to this but smaller, usually with only one adult married couple and their children, were the basis of all estates in Moscow uezd, as far as can be seen. Whether rents were exacted in the form of produce or in labour, it was mainly peasant serfs, not villeins or hired labourers who were the producers. Moreover, as we have seen, there appears to be a possible exaggeration of the importance of demesne arable land at manorial centres. We lack the data which might have allowed us to sketch the proportions between demesne at manorial centres and the arable land in peasant hands, and to describe adequately the labour force, the labour inputs and the work units of land. Nevertheless, it seems that we have enough evidence for Moscow uezd to make some allowances for the distortions introduced by the nature of our sources. We can see that even on the wealthiest and most organised estates of the great monasteries, and given particularly favourable conditions, relatively large central fields organised in a three-field rotation were combined with peasant holdings, probably of a wide variety of size and, presumably, of organisation. On smaller or less well-regulated estates the probability of the peasant units being important, perhaps even dominant, is much more likely. Within any estate, therefore, for a full picture we need more information about the peasant lands, unfortunately often neglected and overlooked by our sources.

A similar point also applies when we consider the burden of obligations on the peasant. The late-sixteenth-century registers of inquisition give little information on the produce and money rents paid by peasants.[1] Moreover, the information given on obligations is not consistent with that given elsewhere. Obligations on the Trinity monastery estate do not specify grain rents. The information, moreover, is inconsistent; for example, the numbers of tenements do not always agree with those given earlier in the register. The section of the register detailing obligations states there are 34 peasant tenements at Nakhabino; the earlier account of Nakhabino, however, lists 60 peasant tenements. Moreover, the number of peasant tenements per vyt, the tax unit, varied from 0.2 at Muromtsovo to 7 at Tushino; the inclusion of

[1] *PKMG*, I.I.54, 58, 61–4, 68, 76–7, 83, 86–95, 280–1, 283–4, 286.

labourer tenements in the calculation makes no significant difference. Amounts of produce exacted per vyt also vary greatly. Nine dengas per vyt appears to be the standard rate paid to the steward, but, in addition, some settlements pay quite considerable money rentals (*obrochnye dengi*). Quit-rent (*obrok*) here appears to be a commutation for labour obligations at the rate of 5 altyns a vyt, more than three times the rate for payments to the steward. On this great monastic estate labour rent was evidently a highly valued item. Overall, however no clear pattern emerges either of the proportions between the various forms of obligations or of the amounts per tenement of various items. Some of the latter appear unlikely; 13,000 eggs from 34 peasant and 10 labourer tenements at Nakhabino, for example.

The entry for Vokhna volost, if related to the number of tenements in the volost, indicates very light burdens of produce rents (20 yufts of grain for five or six hundred tenements), but money payments (90r.) amounting to 5 or 6 altyns a tenement, more than payments to the steward on the Trinity settlements, but much less than the money rentals. In 1574 Vokhna was not recorded in vyts, presumably because the state had no need of these units used by the peasant community (*mir*).[1]

The information available is extremely limited, but suggests that obligations usually were not very heavy, though the amounts and proportions varied between the different types. We have too little material to generalise: one point, however, may be added about labour rents. We have seen that they were highly valued on the Trinity estate and that we do not know their proportion relative to produce and money rents. In terms of peasant labour, however, labour rents would be particularly onerous since the days demanded were likely to fall at peak work periods. This would be detrimental to the peasant farm's own calendar of work, and the period involved would be extended by travel time to and from the centrally located demesne arable land. Perhaps this helps to account for the relatively high commutation payment for labour rents noted on the Trinity estate.

In the late sixteenth to early seventeenth centuries there were great upheavals and considerable changes in many aspects of Russian economy and society. Farming was not exempt from these changes and the Moscow area was particularly involved in them. We have already seen some changes in both cultivation and gathering. Cultivation of the basic grain crops at some manors came to take place on consolidated or enlarged fields incorporating former closes, outfields and wastes. On many such estates labour rents grew, overshadowing money rents of the period before the Oprichnina. At peasant level three-field organisation spread, displacing the former farming of closes. There were limited attempts to improve livestock, virtually restricted to the level of the tsar's court.[2] Gathering, too, underwent some change: we have seen evidence of fish breeding and the development of orchards, but these

[1] Veselovskii, *Soshnoe pis'mo*, II.458.   [2] Merder, *Istoricheskii ocherk*, 5–7.

changes from gathering to cultivation were evidently limited to the lordly levels of society.

Prices of agricultural produce fluctuated considerably in the sixteenth century with major peaks in the late 1540s and early 1570s. The view that grain prices rose consistently 4–4$\frac{1}{2}$ times in the sixteenth century, does not appear to be demonstrated for the Moscow area; certainly there was no substantial upward trend in the second half of the century.[1] The objection that the monastic account books used by Man'kov gave prices distorted by

TABLE 12. *Moscow grain prices (dengas per chet) and average index for Russia (1598–1600 = 100)*

Date	Rye Price	Rye Index	Oats Price	Oats Index	Wheat Price	Wheat Index	Barley Price	Barley Index	Buckwheat Price	Buckwheat Index	Peas Price	Peas Index
1550	48	139*										
1556	22	126		160								
1557	40	106*										
1564		60	12†	60								
1569	23	76	30	125								
1570	200	519	100	350								
1575		77		110				42	30	42		
1576		55		80		50		40				
1579		77		54								148
1581		87		100	53	78				100	36	
1583	40	121	20	112		117		176				216
1585		129		163	92	148				117		213
1586	45	118	28	173	80	118	36	88	30			211
1587	50	158		142		211		156				288
1588	50	182		164	78	143		193		167		300
1589	90	171		105		136		138		98	84	264
1592		117	27	126	68	100		112		101		
1593	30	102	21	105				69		56		156
1594	18	62		87						68		126
1599		92	21	95		100		100				
1600	40	100	19	100		100		100				
End XVI					20							

Source: Man'kov, *Tseny*, 104–11, 118–21.
* Volokolamsk and Moscow towns.   † Moscow area.

local market factors is doubtless valid, but there is not enough available material to propound an alternative generalised series.[2] General evidence suggests a decline in marketed grain during the period of the Oprichnina, but that the grain market recovered and developed at least in Moscow both in the 1590s and after the end of the Time of Troubles. In the 1550s Chancellor wrote that

[1] See table 12. Man'kov, *Tseny*, 40. Prices are comparable since the coinage was stable from 1535 to 1606, and Man'kov adjusted his series prior to 1535.

[2] Abramovich, *Ist. SSSR* (1968), 2.117. Hellie, *Enserfment*, 98, wrote of 'extraordinary high level of prices in 1589, 1590, and especially 1591', apparently solely on the basis of Abramovich's prices from a monastic estate in Vologda.

the ground is well stored with Corne, which they carrie to the Citie of Mosco in such abundance that it is wonder to see it. You shall meet in the morning seven or eight hundred Sleds comming or going thither, that carrie Corne, and some carrie fish. You shall have some that carrie Corne to the Mosco, and some that fetch Corne from thence, that at the least and well a thousand miles off...[1]

Availability of local grain supplies and a grain trade will have cushioned the impact of adverse factors. 'The development of the grain market extensively and in depth proceeded more intensively in the central agricultural zone where', as Man'kov stressed, 'the central figure in the economy was the cultivator.'[2]

More striking, of course, were the social changes culminating in the series of disasters in the early years of the seventeenth century known loosely as the Time of Troubles. The violence and destruction in the Moscow area had struck both at the great and the 'poor peasant and country folk'.[3] Land went out of cultivation. The consequent fall in quantities of available and especially of marketed produce contributed to the sharp increase in prices. Many peasants fled. Taxes in money rose from 2 dengas or so per chet in the 1540s to about 15 dengas at the time of the Oprichnina and at least 27 dengas in the 1580s.[4] This great increase in state exactions was much more of a burden than the obligations owed to lords, especially since the former were at least claimed in the form of money. This drove peasants into increased contact with the market and was no doubt a contributing factor to the great increase in villeinage at this period. The relative shortage of hands, together with the much increased state tax demands (and requirements in men and materials), contributed to a complex of readjustments: the further development of enlarged demesne arable land; increased control of labour, at least at manorial centres; the growth of labour rents and of villeinage (kholopstvo); and, finally, pressure on the government to restrict peasant movement.

In the 1580s 'forbidden years' were declared in which the peasant right to move was temporarily rescinded.[5] At this period there was probably a peak in the number of peasants becoming villeins, a procedure permitted without payment by the peasant under the 1550 Law Code.[6] This provided some source of labour for those lords with greater resources; but it imposed strain on smaller lords, who found military service increasingly burdensome, and on peasant communities (where these still existed) since they were faced with rising taxation to be met by fewer tax-payers.

The 1590s saw some recovery from the slump of the previous two decades,[7]

---

[1] Hakluyt, *Principal Navigations*, 2.225.   [2] Man'kov, *Tseny*, 30. See also p. 146 below.

[3] Johann Taube, Elert Kruse. Cited in Adelung, *Kritisch-literärische Übersicht*, 269.

[4] Based on Rozhkov, *Sel'skoe khozyaistvo*, 222–33; Blum, *Lord and peasant*, 228–9; see also table 25, p. 233 below.   [5] Koretskii, *AE za 1966*, 306–30; *Zakreposhchenie*, 108.

[6] Paneyakh, *Kabal'noe kholopstvo*, 34f.

[7] Got'e, *Zamoskovnyi krai*, 234–40. A severe reduction of cultivated land per peasant household (Hellie, *Enserfment*, 97) is not clearly demonstrated.

but it has been suggested that the fall in prices indicating this recovery was due not to the increase in tilled area, but to a fall in demand for agricultural produce. In the 1590s we see a decline of population in the artisan quarters of towns, the market rows and trading settlements, i.e. a flight from heavy state taxation by those in towns liable to tax.[1] In 1592 further restrictions were imposed on peasant movement.[2] This presumably was intended to redress the imbalance as regards availability of labour between the greater and the lesser lords, as well as to gain support among the lesser servitors for Boris Godunov's political ambitions. A decree of similar nature was issued in 1597 but extended the right of search for five years to peasants who had run away.[3]

In 1601 there was much rain, one source alleges 70 days, and the grain was 'green as grass' when, on 15 August, there followed a great frost which destroyed all the crops including both winter and spring sown grains.[4] Famine rapidly set in; it was vividly described by a foreign eye-witness.

At that time, by God's will, there was a dearth and famine throughout the land of Moscow so great that no historian has had to describe its like before. The famine time described by Albert, Abbot of Staden and many others, cannot be compared with this disastrous time of famine. It was so terrible and so great was the disaster over the whole country. Mothers ate their children. All the peasants and the rural population in general, those who had cows, sheep, horses and chicken ate them paying no attention to the fast; (the peasants) gathered various plants, such as fungi [campernoellie, duvels-broot] and many others and ate them with tremendous greed; they also ate chaff, cats and dogs. People making use of such food had stomachs which became as fat as cows. Such people suffered a piteous death. In the winter they suffered from giddiness and fell unconscious on the ground. On all the roads people lay, dying from hunger and their corpses were gobbled up by wolves, foxes and other animals.

It was no better in Moscow itself. Bread had to be brought secretly to the market, otherwise it would have been seized. People were appointed supplied with sledges and carts (to carry out the corpses). Daily they carried out a multitude of corpses to the pits dug in the open land beyond the town.[5]

Price data reflect the fact that those not growing their own grain and, of course, those with fewer resources, were immediately endangered. In the central area of Russia, however, the impact of the dearth and famine was partly cushioned by local sources of supply, probably greater stocks and a more developed market.[6] Prices rose as much as 15 times in two years; that

[1] Makovskii, *Razvitie*, 226.   [2] Hellie, *Enserfment*, 97–8, 321–2.
[3] Hellie, *Enserfment*, 105. Translation in Smith, *Enserfment*, 99–101.
[4] *Skazanie Avraamiya Palitsyna*, 105; PSRL, XIV.1.55. For a survey of evidence for other areas see Koretskii in *VIRA*, 7.219f.
[5] Massa, *Skazaniya*, 77. For some Russian accounts see Popov, *Izbornik*, 190, 219, 414.
[6] See Boris Godunov's charter on ending grain speculation, *LZAK*, IX, otd. III; reprinted in *Vosstanie I. Bolotnikova*, 68–73.

they rose 80–120 times, as stated by Klyuchevskii and Koretskii, appears very doubtful (table 13).[1] In areas of dearth it was the townsfolk, especially the poorer ones, and 'needy' peasants who were most exposed since they depended on the market.

TABLE 13. *Famine grain prices (Moscow dengas per chet)*

Date	Location	Rye	Oats	Barley	Wheat
c. 1600	Moscow	40	19		20
Spring 1601	Centre	30–32	20	24	
Autumn 1601	Sol' Vychegda	200	120	160	
Nov. 1601	Sol' Vychegda, proposed controlled price	100	50	80	
1601/2	Vaga★	320	200	288	300
Spring 1602	Novgorod	360	150		
1602	Pskov	400	260	400	
1602/3	Pskov	600	260	500	
1602	Pomor'e	600			
1602	Kazan'	300–360			
1602	Moscow	600			

Sources: Man'kov, *Tseny*, 111;
Koretskii, *VIRA*, 7.222, 226;
Vosstanie I. Bolotnikova, 68, 71;
RIB, XXII.393;
PL, 2.265;
Tikhomirov, *IZ*, 10.94;
Mat. po ist. SSSR, II.605.
★ Of other foods, turnips were 48 dengas a chet and cod 50 dengas a pud.

In 1602 the crops again failed as a result of late frosts.[2] There were said to be 50,000 dead of famine in Moscow in seven months. During two years and four months of famine the dead buried in three great pits at Moscow were

[1] Klyuchevskii, *Sochineniya*, 7.187, 472 (on p. 216 he refers to a rise of 360 times); Koretskii, *VIRA*, 7.227. This calculation is founded on an alleged normal base price of 4½–7½ dengas a chet given in a seventeenth century chronograph; Popov, *Izbornik*, 190, 219. Man'kov's price for rye at Moscow *c.* 1600 was 40 dengas a chet. Klyuchevskii himself, however, pointed out the difficulties in determining 'normal' prices and the obscurities surrounding these chronographs (*Sochineniya*, 7.187, 452–3) and in his article on the ruble in the sixteenth–eighteenth centuries he accepts 22 dengas as a quite modest average price for rye in Moscow in the second half of the sixteenth century (*Ibid.* 212) and specifically mentions a rise from 20 dengas to 3 rubles from 1601 (*Ibid.* 216, 471–2). Clearly, there were great fluctuations in price, but the chronograph figure seems quite exceptional.
[2] Tikhomirov, *IZ*, 10. 94.

said to number 127,000.[1] The dearth and famine were accompanied by disease.[2]

These natural disasters were aggravated by social conditions; lords, unable or unwilling to feed their dependants, set their villeins free or drove them out; villeins and peasants, ruined and starving, were forced to flee or to resist.[3] The main stream of runaways flowed out of the Moscow area to the less devastated Seversk and Polish towns along the Desna and the Sein towards the south-west frontier; 20,000 villeins fled here under Boris Godunov.[4] The tsar issued a decree in August 1603 authorising documents to be granted to any villeins dismissed from households.[5] Resistance, however, culminated in 1603 in a mass rising by villeins and peasants in the Moscow area; this was led by Khlopko (a name itself cognate with the term for a villein). In fact, the main forces of this rising were villeins, those driven out of the great households of the capital as a result of the famine.[6] Peasants, servitors and the men of the artisan quarters were evidently not prominently involved. The rising was put down after a major fight in the autumn of 1603.[7]

The following year the first of the major pretenders to the Russian throne entered Russia and drew support from the local population in the rich agricultural area of the Seversk and Polish towns and from those from the central area who had taken refuge there. False Dmitri I entered Moscow in 1605, supported not only by better-off peasants (*muzhiki*) and many of the townsfolk, but also by military servitors, including some high officers who had gone over to him.[8] Within a year Dmitri had been killed by a rising led by the Moscow boyars, though including other groups of the population, and Vasilii Shuiskii had been declared tsar.

In October 1606 a rebel force led by a former villein, Bolotnikov, approached Moscow. This rising had started in the south and was led mainly by dissatisfied petty military servitors. Bolotnikov was able to unite with a second group, led by Pashkov, and to besiege Moscow for two months, his force augmented by peasants from villages near Moscow.[9] Grain prices rose by a factor of three to four during Bolotnikov's siege.[10] This was a much more serious rising than Khlopko's since it extended over a considerable part of

---

[1] *Skazanie Avraamiya Palitsyna*, 106. This is reasonably close to Margeret's 120,000; *Estat de Russie*, 72. Bussov mentions the incredible figure of up to half a million; the fact that this would be greater than Moscow's population would have to be accounted for by an influx of starving people from the surrounding countryside. See also Bakhrushin, *Izv. AN SSSR, seriya ist. i filosof.* (1947), IV, 3, 201–2.

[2] Karamzin, *Istoriya*, X, note 451; XI, note 170; Koretskii, *VIRA*, 7.229–30.

[3] *Skazanie Avraamiya Palitsyna*, 483.   [4] *Ibid.*; Koretskii, *VIRA*, 7.231.

[5] *PRP*, IV.375.

[6] Koretskii, in: *Krest'yanstvo i klassovaya bor'ba*, 210–11; Makovskii, *Pervaya krest'yanskaya voina*, 275.

[7] Kusheva, *IZ*, 15.91.

[8] Belokurov, *ChOIDR* (1907), 2.6. *Vosstanie I. Bolotnikova*, 331–2.

[9] *Istoriya Moskvy*, I.307.   [10] *Ibid.* 308.

Russia; it is often termed a peasant war. It seems doubtful whether this is entirely accurate. There are at least differences of emphasis among Soviet scholars. There is considerable agreement about the importance of the deteriorating real and legal situation of both peasants and villeins as a contributing factor at this time. Smirnov stressed the importance of villeins in the revolt.[1] Makovskii has argued that the social forces involved were, first, the artisan quarter people, then the black peasants, the petty- and middle-rank servitors of the frontier zone in the south, and the enserfed peasants.[2] Apart from the paucity and deficiencies of the sources as regards the social composition of the revolt, there is a possible difficulty over the word 'peasant'. Peasant (*krest'yanin*) had long indicated a dependant, a social and legal category, as well as an occupational category. Thus, the use of this term in the sources did not always indicate a man working the land; it might mean, for example, a dependant working in a craft or trade and sometimes living in town. The 'New Chronicler', for instance, mentions in a section headed 'On the beating and ruin of serving people by their villeins and peasants', that people 'suffered not by the faithless [i.e. non-Russians], but were abused and killed by their own slaves and peasants'.[3] 'In 1606/7 the boyars' people [i.e. villeins] and peasants gathered together, the Ukrainian artisan quarter people, musketeers and cossacks joined them', it goes on. Even though at its height the revolt no doubt involved considerable numbers of peasants on the land, its hard core was composed of the dependants both slave and serf of lords (especially the magnates with large households and larger, more closely controlled estates), townsmen engaged in crafts and trades and numerous petty servitors. This contributed to conflicts of interest and a fissiparous tendency in the movement which resulted in treachery at some crucial stages.

In late November 1606, during fighting on the outskirts of Moscow, for example, Pashkov deserted to the government forces and Bolotnikov had to retreat southwards; in the southern regions resistance long continued.

This breathing space for the capital gave time and the opportunity for the tsar to try to deprive the rebels of villein support and also to strengthen his own forces. The decree of 7 March 1607 forbad 'voluntary villeins' (*dobrovol'nye kholopi*) being issued with deeds of obligation (*kabaly*) against their will and, thus, aimed at winning them away from support for the rebels.[4] Two days later a law was issued which laid down that peasants recorded in the registers of 1592/3 were to remain subject to those who then held them; a right to reclaim runaway peasants and slaves within fifteen years was also granted.[5] This aimed at gaining increased support of servitors against the rebels.

[1] *Vosstanie Bolotnikova 1606–1607 gg.*, 106.   [2] *Pervaya krest'yanskaya voina*, 468–72.
[3] *Vosstanie I. Bolotnikova*, 84, 342.
[4] *PRP*, IV.376–7; *Vosstanie I. Bolotnikova*, 210–11, 398.
[5] *Ist. arkhiv* (1949), IV.72–87; Hellie, *Enserfment*, 108–9. Translation in Smith, *Enserfment*, 103–7.

In the summer of 1607 a new pretender, False Dmitri II, appeared and established himself at Tushino with Polish and Lithuanian support. A confused situation followed, with many changing sides, some on several occasions, between Moscow and Tushino. This lasted until 1610 by which time the Poles had declared war and invaded Russia. In the struggle for Moscow, Shuiski was overthrown by military servitors and the Moscow townsfolk. The boyars proposed acceptance of Wladislaw, son of the Polish king, in order to exclude False Dmitri II who found potential support with the villeins and lower strata of the townsfolk.[1] The Poles entered Moscow. There were soon complaints about their offensive behaviour and that they 'took plenty of all sorts of goods and victuals without paying'.[2] For their part, the Russians pushed up prices as much as tenfold.[3]

Resistance to the Poles, which had become effective in some northern parts of Russia in 1608–9, now developed into a widely supported movement which included gentry, junior boyars, musketeers, cossacks, as well as peasants and men of the artisan quarters.[4] In order to swell the ranks, treatment as free cossacks was offered to 'boyars' people [i.e. villeins], bond and old-established, from the Volga and elsewhere'.[5] This national levy reached Moscow in the spring of 1611. Like the Bolotnikov revolt, this was a heterogeneous movement with a lack of internal cohesion. Some leaders of the resistance devoted part of their energies to seizing goods and estates for themselves and their followers. Finally, a second national levy rallied in Nizhnii Novgorod late in 1611. This had a somewhat lower-level social composition than the first levy and was more successful in uniting resistance to the Poles. The Poles wrote pointedly to Pozharskii, one of the leaders of the levy, 'You had better despatch your people to their sokhas'.[6] This levy freed Moscow late in 1612, but marauding bands continued to operate in some areas. The following year an Assembly of the Land elected Mikhail Romanov the first tsar of the new dynasty. Polish attacks were renewed in 1617 and 1618, but were unsuccessful and an armistice was agreed. The Russians were obliged to cede to the Poles a large slice of territory in the west, including Smolensk.

This series of events resulted in temporary chaos and also put immense strains on a devastated country with its administrative system in disarray. As Veselovskii stressed 'the tense struggle of the state with its internal and external enemies demanded huge resources precisely when the destruction continued and demanded new surveys'.[7]

Estate size in the Moscow area in the seventeenth century after the Time of Troubles is shown by the results of enquiries on landholding by gentlemen holding by service in the 1630s. Military tenants made these reports (*skazki*)

[1] *RIB*, XIII.1187.   [2] *Istoriya Moskvy*, I.329; *RIB*, XIII.211–12.
[3] *Istoriya Moskvy*, I.332.   [4] *Istoriya Moskvy*, I.334, 339.
[5] *SGGD*, ch. 2, no. 251.   [6] *Istoriya Moskvy*, I.350; *RIB*, I.326–9, 337.
[7] *Soshnoe pis'mo*, II.187.

in response to enquiries during the general attempt to sort out the confused land-holding situation in the aftermath of the Polish invasion. In broad terms, they provide a rough check on the materials for the pre-invasion period derived from the registers of inquisition. They are likely to be reasonably reliable; denunciation of false information would result in the estate being forfeit to the informer. Stashevskii estimated that the average gentry estate, often spread over 3–5 uezds in different corners of the state, was 494 chets. 'Such scattered gentry land-holding, first, hindered large-scale agricultural units being formed; second, it prevented the holder from being able to run his economy with equal zeal and success in all his numerous petty estates.'[1] According to Stashevskii 303 land-holders had 134,487 chets of land and 6,540 peasants and labourers (*bobyly*) (not separately designated in the source). This gives an average of 444 chets, somewhat less than the overall average already mentioned, and about 21 peasants and labourers to work the land.[2] This would mean about 20 chets a field, or 30 desyatinas in all, per working man; that is larger than the usual sixteenth century taxation unit, and perhaps at least twice the size of actual peasant units. This may be accounted for by a high proportion of demesne arable land or may reflect a continuing shortage of hands in the Moscow area in the 1630s. In any event, a complex organisation on units which often had only 4–7 work people (21 persons scattered over 3–5 uezds) appears a remote possibility, In fact, the evidence for more complex organisation of demesne continues to come mainly from the larger monastic and heritable estates.

According to Got'e, in the seventeenth century on court lands, there was both growth of cultivation by desyatinas and commutation to money payments. There was tillage by desyatinas near Moscow itself (here evidently cultivation of demesne arable land); produce rents in grain (*posopnyi khleb*) came from distant villages and volosts; more remote areas contributed money payments.[3] Unfortunately, no more detailed information appears to be readily available on Moscow uezd in the first half of the century.

On estates held by service, and on heritable ones, lord's arable land was widespread, but the amounts were so varied that it is difficult to say much; 'on large boyar estates the owner's arable land often measured hundreds of desyatinas; for rank and file tenants holding by service it was frequently no more than a few chets a field'.[4] Got'e wrote that, as regards Moscow uezd, 'throughout the century there was quite extensive lord's arable land dis-

[1] Stashevskii, *Zemlevladenie*, 20–2. Similar points are stressed by Kochin, *Sel'skoe khozyaistvo*, 348, 364–5.

[2] *Zemlevladenie*, 31. Checks show, unfortunately, that Stashevskii's arithmetic was probably in error. The figures he gives for the distribution of numbers account for at least 7,000 workers, but only 293 owners. His stated average number of peasants and labourers is 24. The differences are not so great as to affect my argument.

[3] *Zamoskovnyi krai*, 488–9.

[4] *Ibid*. 493. None of Got'e's data relates to Moscow uezd.

tinguished by extremely varied size for individual [villages?] near Moscow'.[1]
Apparently, many estates in the 1620s had small amounts of arable land (1–10
chets) and were entirely tilled by villein labour.

For monastic lands, again, there seems too little evidence to draw a clear
picture. In Goretov stan, in 1624–5, the Trinity monastery had 50 chets of
monastery arable and 106 chets of peasant arable, with 20 peasant households.
In Radonezh stan in the immediate vicinity of the monastery comparable
figures were 216 and 880 chets.[2]

Got'e has argued that the seventeenth century overall saw an increase in
arable land per tenement, but a decline, except on court lands, per man.[3] The
data he himself gives, however, seems too confused to enable such a clear
picture to be drawn, at any rate for Moscow uezd. The available evidence
shows very small amounts of arable per tenement for the sixteenth century;
the procedure of averaging data from disparate sources for such an extensive
area might in part account for this improbable result. The data for Moscow
uezd taken alone suggest that the average amount of arable land per tenement
was 8 or 9 desyatinas towards the end of the sixteenth century; it declined
dramatically, at least on the Trinity monastery estate and almost certainly also
in the western and northern areas (not in Moscow uezd) which suffered
particularly during the Time of Troubles. By the 1620s, however, the northern
stans had the former amount per tenement. The figures for the 1620s show
considerable variations. In part these may be due to monasteries having a
considerable labour force, much of which was formed of monastic dependants
and servants, not peasants, especially during the Time of Troubles. By the
mid 1620s some lay and ecclesiastical estates had amounts of arable land
comparable with those before the crisis at the start of the seventeenth century.
It thus looks as if something like the old man–land ratio was re-emerging in
Moscow uezd in the 1620s. Unfortunately, the form of the Moscow inquisitions is such that farm size can only be surmised from these averages; we have
no direct and explicit evidence.

Information on mills for grinding the grain produced on the arable lands
is somewhat sporadic. Most of those noted were water-mills and belonged to
monasteries or ecclesiastics. Even in the early nineteenth century, Moscow
guberniya had only 29 windmills, but 652 water-mills.[4] Sometimes the
amounts of quit-rent or income noted is very high – 25 rubles, in one case
even 30 rubles, and it seems likely, therefore, that the officers of inquisition
only noted the larger and more profitable mills; churn-mills were seldom
deemed worthy of note since they were fairly commonplace and no indication
of especial wealth. The only mill specifically noted as being in peasant hands
was associated with a court village and produced flour for the tsar.[5]

The cost of milling a chet of grain in the late sixteenth and early seventeenth

[1] *Ibid.* 496.   [2] *Ibid.* 498, 512.   [3] *Ibid.* 513–14.
[4] Lyubomirov, *Ocherki po istorii russkoi promyshlennosti*, 227.   [5] PKMG, I.I.279.

centuries, was 3 or 4 dengas, so the maximum here may indicate an expected annual output of at least 1500–2000 chets, say 5–7 chets a day, about 10–15 times the output of a simple churn mill.[1]

Flour and other milled products, such as oatmeal, groats (especially buckwheat) and malt, were necessities in every household. Most were no doubt produced within the household. As we have seen, even the great monasteries seem to have milled mainly for consumption, not for the market.[2] Nevertheless, in Moscow itself in the seventeenth century, there was evidently a considerable sale of flour-based products.[3] Evidently for the first time in 1601, a famine year, an edict was issued on the weight and price of loaves.[4] Subsequently in the period after the recapture of Moscow, a number of such edicts was issued; they laid down detailed regulations for the control of the weight and price of various sorts of bread for as many as twenty-six prices of rye flour and thirty of wheat.[5] Among the related materials are petitions from bakers complaining that the rates had been set too low; this suggests that, despite the seemingly unrealistic detail of the schedule, the measure had some effect.[6] The attempts to regulate bread prices and quality, however, are of most importance in themselves. They show an awareness of the need to control the market for an essential food item not merely in the famine conditions of 1601, but also over a period when the region was recovering from the disasters of the Time of Troubles. Moreover, the bakers employed wage labour, hired hands (*naimity*) who 'bake pies and rolls, mix them in the dough troughs, roll out and knead the dough and set them in the oven'.[7] Here we have the development of an urban commodity production in one of the basic food-stuffs shortly before Olearius commented on the numerous bakers, bread and flour stalls in Moscow.[8] There seems to be no such evidence for any other part of Moscow uezd. The town, with its considerable population, perhaps more than 100,000, many probably not permanently resident, provided a sizeable and exceptional market.[9]

In Moscow itself by the late 1630s there were probably a few thousand persons engaged in crafts and trades, many in occupations concerned with agricultural raw materials or objects of gathering. The material derived from

---

[1] Cost are given by Gorskaya in *MISKh*, VI.52. See above p. 210.   [2] P. 39 above.

[3] See Makovskii, *Pervaya krest'yanskaya voina*, 159–60, on peasant traders in grain in Moscow in 1601.

[4] Massa, *Skazaniya*, 81.

[5] See appendix 3 for a translation of one such Assize of Bread. Edicts and related materials for the period 1624–31 were published in *Vrem. MOIDR* (1849), kn. 4, *materialy*, 1–60; there is a reference to an assize in 1619; *ibid.* 30. See also Klyuchevskii, *Sochineniya*, 7.180–3.

[6] *Vrem. MOIDR* (1849), kn. 4, materialy 30, 32, 35 etc.   [7] *Ibid.* 43.

[8] Baron, *Travels of Olearius*, 115.

[9] *Istoriya Moskvy*, 1.446, states that Moscow's population rose to 200,000 people in the seventeenth century, but adds (p. 453) that there were severe fluctuations throughout the century.

a list of 1638 gives some rough indication of the distribution of such occupations (table 14).[1] In many cases, however, the terms used are obscure; in particular, it is often difficult to be sure whether the trade is concerned with processing or selling the particular product; often, no doubt, it may have been concerned with both aspects.

TABLE 14. *Tradesmen in Moscow* (1638)

DEALING WITH CULTIVATED PLANTS

GRAIN AND ITS DERIVATIVES

13 granary men (*zhitnik*)
2 grain dealers
10 millers
1 master miller

food
19 flour men
1 toasted-grits man (*pryazhnik*)
24 pancake men
53 bread men
1 bread woman
76 roll men
2 roll women
6 gingerbread men
1 bun man
35 piemen
3 fool men
1 pudding man (*kuteinik*)
10 food men (*kharchevnik*)

drink
10 brewers
30 maltsters
1 mash man
1 distiller
38 kvass men

OTHER PLANTS

food
19 vegetable gardeners (*ogorodnik*)
2 vegetable garden foremen
3 cabbage men
11 fruit men (*ovoshchnik*)
1 apple man
2 pear men
7 nursery men (*sadovnik*) (may include fishermen)

[1] Belyaev, *TrMORVIO*, I.xviii–xxiii.

1 hemp oil man (*oleinik*)
24 oil men (*maslenik*)
1 hops man

textiles (There were also 122 master tailors and almost 100 others in the clothing trade.)
1 retter
15 canvas men

DEALING WITH GATHERED-PLANT PRODUCTS

food for humans
7 nutmen
3 cranberry men

animal feed
2 hay men

raw materials
1 woodworker (*derevshchik*)
3 woodcutters
124 carpenters (*plotnik*)
1 bast shoe maker
2 splint sellers

DEALING WITH LIVESTOCK

horses
1 horsedealer
175 grooms, horse-harness men etc.
3 shoeing blacksmiths (but 98 blacksmiths)

cattle
5 herdsmen overseers (*prikashchik stadnyi*)
8 cowherds
1 milkman

sheep
1 felt maker
10 drab-cloth men (*sermyazhnik*)
3 wool-stocking men
13 sheepskin men

pigs
1 bristle man

meat
72 butchers

hides
2 fleecers (*strigol'nik*)
1 tanner
3 leather workers
1 hide man (*kozhannik*)

1 morocco-leather man (*irkha*)
125 shoemakers
7 solemakers

DEALING WITH WILD ANIMALS

3 hunters
3 kennelmen

fur
16 beaver men
3 sable men
65 furriers
19 furcoat men (incl. sheepskins)

fish
32 fishmongers
1 fish factor
1 fisherman
9 herring men
8 smelt men

birds
4 gerfalcon men
2 pigeon men
2 fowl sellers? (*kuryatnik*)

bees
9 mead men
35 candle makers

These approximate figures may appear very low; in reality they would doubtless have to be supplemented by many peasant trading trips to Moscow. Nevertheless, they clearly indicate the importance in Moscow of foodstuffs, especially dealings with grain and its derivatives (338 persons). Other cultivated plants were much less important (71 persons), and of these vegetable oils were the largest group (25). Textile raw materials derived from plants were poorly represented (though tailors and other clothing workers, many no doubt using hempen textiles or linen, were a large group).

Gathering plants was evidently of little importance as a source of marketed foodstuffs (10), but was an important source of raw materials for a wide range of objects (131). Horses were much more frequently implied than other livestock; this was probably accounted for by the presence of the court and central administration. The considerable group of leather workers of different types (140) shows that livestock were important. Moreover, the archaeological evidence that the inhabitants of Moscow wore leather footwear, rather than bast shoes, is here confirmed. The number of butchers (72) suggests, however, that cattle and perhaps especially pigs were much more important than the predominance of horses in the figures appears to indicate. Wild animals provided both luxury objects (furs (103)) and an important source of protein in the diet (fish (51)). Bees were also quite important (44).

We have already seen that shortly before this the bakery trade had developed to the point where government intervention had taken place. Much of these supplies will have come from within Moscow uezd, even though some, especially some materials for textiles and clothing, came from outside the uezd, at times from considerable distances.

We may briefly sum up. Peasants colonised the Moscow area originally by a natural creep due to increasing population mediated by what may be called close farming, that is farming largely on natural glades and man-made clearances in the forest supplemented by a variety of gathering activities. Colonisation was encouraged by the custom of partible inheritance. Our sources, however, give a clearer picture of lords, lay and clerical, undertaking and encouraging attempts to attract settlers and establish tenements. This active colonisation of the central area around Moscow, which later became Moscow uezd, was taking place mainly in the fourteenth and fifteenth centuries; at the same time, the estates of the prince, of the great nobles and of the monasteries, and the smaller estates of tenants holding by service, came to dominate the area, even though, in places, much forest remained virtually untouched. Black peasant lands were a major source of such land mobilisation; they were early swallowed up by estates and had virtually entirely disappeared in the uezd before the seventeenth century.

In the late fifteenth and especially in the first half of the sixteenth century the system of distribution of lands held by service grew; the military servitors claimed increasing control over the peasants on their lands. On larger estates, especially princely and ecclesiastical ones, on which larger settlements were found, a developed three-field system emerged in the same period; demesne arable land, tilled by villeins or enserfed peasants, or usually both, became more important. References to three fields, however, do not always imply a three-field system; they may indicate an administrative function of a manorial or estate area, especially before the reforms of the 1550s. By then some form of three-field organisation was common, at least at villages (considerably larger than hamlets in the Moscow uezd) or other manorial centres on estates. In some cases the central arable area was expanded in the second half of the sixteenth century to take in former outlying patches of arable land; moreover, demesne arable land seems to have become more widespread in the same period. This increased concern of lords for agricultural production, when the colonisation had been completed, was encouraged by rising prices. The relative importance of production on demesne arable land compared with that on peasants' own holdings is obscure, as is the relative weight of labour inputs. Both the desertions of the late sixteenth century and also the small size of early-seventeenth-century estates hindered widespread demesne farming; but such factors also contributed to the predominance of great monastic estates, since the latter were often able to continue to gain both land and labour.

Throughout the period there was a changing symbiosis of close and field. Small areas of arable land worked by some estate holders, both the closes of the early colonisation and the later outfields and wastes which surrounded the manor's central fields, were comparable to peasant work units. These remoter patches of various types were, from the point of view of production for direct consumption at least, better suited to survive disasters such as plague and desertions due to state interventions (burdensome taxation, the oppression of the Oprichnina, the disasters of the Time of Troubles) than the better organised, more advanced, but also more exposed, estates. By the late sixteenth century peasant flight had become virtually a transfer of labour between lords, if within the area; but flight to newly conquered areas to the east and south had also become possible. Nevertheless, despite the crises of the late sixteenth to early seventeenth centuries, the former man–land ratio was being successfully re-established by the 1620s. Sufficient labour remained in the area despite the fact that, from the state's point of view as reflected in the registers of inquisition, the overwhelming majority of the arable land had ceased at times and for a variety of reasons to be viable and so tax-liable.

Finally, in the period after the Time of Troubles, Moscow itself, with a large population, including a few thousand engaged in crafts and trades, provided a sizeable market for foodstuffs and other agricultural produce. These supplies came both from the Moscow uezd and also from more distant areas. The local supplies did not all pass through the market; many of the great men, officials and ecclesiastics, were maintained largely by their own estates in the uezd. Such supplies were produced both on demesne arable and on peasant holdings; much, too, continued to be the result of gathering. Marketed produce also came in part from peasant production. Thus, even in Moscow, where there was an exceptionally large trade in agricultural produce by the mid seventeenth century, peasant farming remained important; the legal enserfment of the peasants did not diminish that importance. Peasant farming remained important not merely as a source of labour for work on demesne land, but also in terms of produce rents which included items both grown on peasant household units and gathered from the surrounding forest.

8

Toropets

The town of Toropets is in the area of the old principality of Smolensk, due south of Novgorod, almost midway between Pskov and Smolensk, influenced by both, but somewhat to the east of a straight line joining these two towns. It is near the headwaters of the Western Dvina which flows along the eastern and southern border of the Toropets uezd. A block of wooded hills extending from Kholm form the division between the basins of the Western Dvina and the Lovat' to the west. The region has a moraine landscape with numerous lakes most of them strung out and linked by streams; many settlements were situated by lakes. Of 159 lakes in the uezd, 139 are less than two miles long. There are, however, two large lakes, Zhizhets and Dvin'e, and around these is an extensive marshy area; in summer this dries out to a large extent, though not completely. The Sparrow Hills, about 140 to 200 feet high, run in a south-westerly direction through the uezd, north of the town; they are often steep and clayey. In the north there is little water and only a few small lakes, but in the south, between the upper Kun'ya, a tributary of the Lovat', and the Western Dvina there is a plateau with a number of lakes. To the east, towards the Western Dvina the land descends in a series of steps with many lakes, marshes and wet valleys. There is a noticeable difference in the landscape on either side of the river Toropa (which presumably gave its name to the town and hence to the uezd). On the right bank, to the west of the river, in the area of the Kodosna, Zhizhitsa and the Western Dvina, the land descends in broad steps. The soil is everywhere sandy and covered almost completely with huge pine and spruce forests. On the left, the ground undulates, water tends to run off and the lakes are linked by small streams; here the soil is mainly clay. Even in the late nineteenth century sixty per cent of the uezd was classed as forest; conifers were dominant, especially in the north, but birch, alder and aspen were also present.[1] Up to the sixteenth century the forest cover is likely to have been similar in kind, but much more extensive; this was an area almost entirely covered with virgin forests and swamps.[2] Even at the end of the seventeenth century, in 1686 and in 1692, concessions were being granted to foreigners to extract mast timber.[3] One effect of this dense forest would be that the water-table was somewhat higher than at the present time.

[1] Semenov, *Geografichesko-statisticheskii slovar'*, IV.231; V.196–7.
[2] Poboinin, 'Toropetskaya starina', I, 93, 264.   [3] *Ibid.* 76, 101.

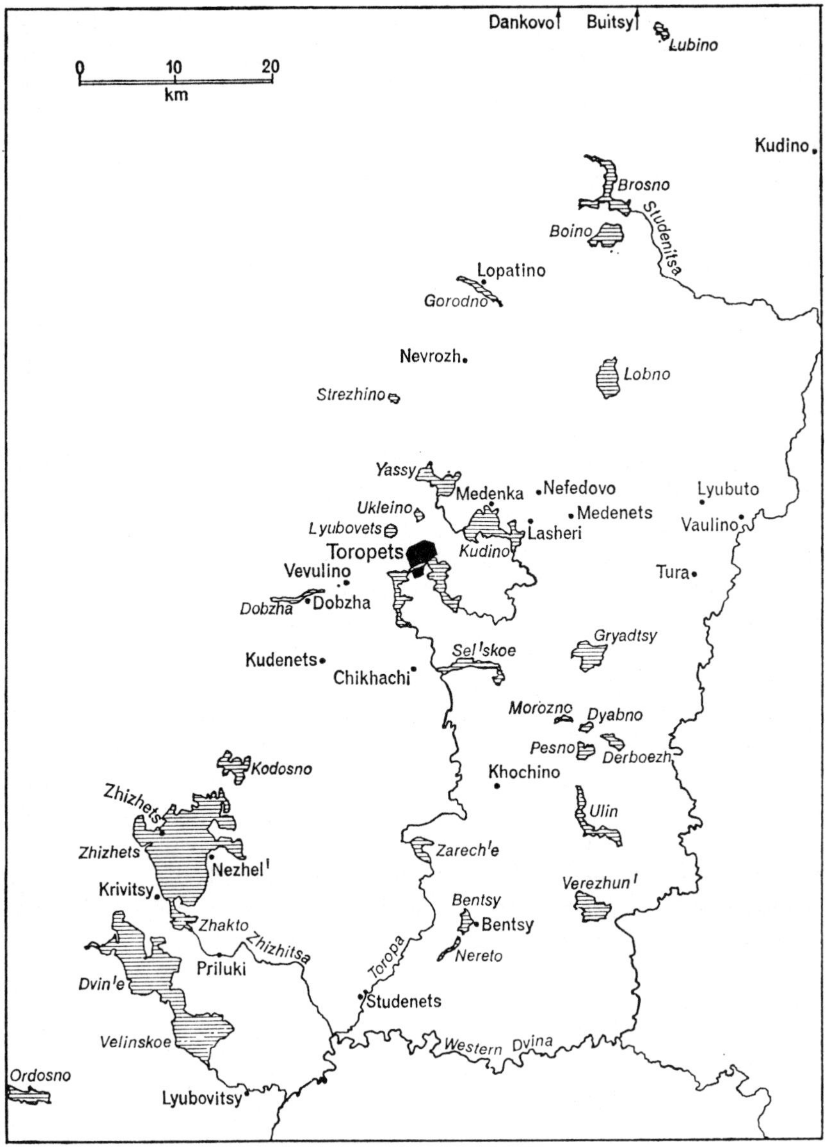
Dankovo
Buitsy
Lubino
0 10 20
km
Kudino
Brosno
Studenitsa
Boino
Lopatino
Gorodno
Nevrozh
Lobno
Strezhino
Yassy
Medenka
Nefedovo
Lyubuto
Ukleino
Medenets
Lyubovets
Lasheri
Vaulino
Toropets
Kudino
Vevulino
Tura
Dobzha
Dobzha
Gryadtsy
Sel'skoe
Kudenets
Chikhachi
Morozno
Dyabno
Pesno
Derboezh
Kodosno
Khochino
Zhizhets
Ulin
Zhizhets
Nezhel'
Zarech'e
Krivitsy
Verezhun'
Bentsy
Bentsy
Zhakto
Zhizhitsa
Nereto
Toropa
Priluki
Dvin'e
Studenets
Velinskoe
Western Dvina
Ordosno
Lyubovitsy

Map 3. Toropets uezd.

It was not only the physical environment of the area which retained old forms. During the time the district was under Lithuanian control the Trans-Dvina area, as it was known, retained old customs, such as the *veche* or town's meeting, and some ancient forms of tenure, partly because it was an easterly territory of Lithuania, remote from the centre, and partly because the Lithuanians tended to avoid innovations if possible.[1] The area was on the frontier between the growing Moscow state and Lithuania. In the second half of the fifteenth century, after Novgorod had been incorporated in the Moscow state, the Moscow government began to exert pressure along the northern border of the Toropets lands. Towards the end of the century, under Ivan III, this pressure increased and was extended so that minor clashes, which had sometimes been little more than raids for peasant hands or expeditions for plunder, now became a more general series of attacks all along the Lithuanian border.[2] Although the border was defined by both sides in 1494 when Casimir's heir married Ivan III's daughter, border warfare was soon resumed and in 1500 a general Russian attack resulted in the capture of Bryansk, Putivl' and Dorogobuzh. Toropets and its thirteen volosts were acknowledged as Moscow's in the settlement of 1503.[3]

The frontier situation meant that military servitors had to be settled in the area, especially along the new border with Lithuania. Fighting continued sporadically with Lithuania in the first half of the sixteenth century. The junior boyars settled here, however, were granted the lands of the former Lithuanian and Toropets boyars; the black peasant lands were not taken for this purpose.[4]

The source mainly used here to describe this uezd is a copy, made probably in the 1550s or 1560s, at any rate certainly before 1570, of a register of inquisition which has been dated 1540/41, though internal evidence suggests that the survey was not completed till 1543 at least.[5] Apart from the late-fifteenth–early-sixteenth-century registers from the Novgorod territories, this is the earliest surviving Russian register and contains a good deal of information on farming in the area. There are numerous references to an earlier inquisition; this was carried out by Prince Semen Kurbskii in the early sixteenth century.[6] In 1511 Semen Kurbskii was sent to Pskov to replace over-greedy royal representatives. He commanded a force against the

---

[1] Poboinin, 'Toropetskaya starina', 67.

[2] *Ibid.* 54.

[3] *PL*2, 224, 252. *SRIO*, 35.399. Solov'ev, *Istoriya Rossii*, III.114.

[4] Poboinin, 'Toropetskaya starina', 94.

[5] TsGADA, 'Boyarskie i gorodovye knigi. Toropets, kn. no. 1', published with a brief foreword by M. N. Tikhomirov and B. N. Florya in *AE za 1963*, 277–357. Subsequent notes in this chapter will give the folio number only for references to this register. The evidence for the survey having taken several years is that, for instance, on f. 166 (78) reference is made to grants on relaxed terms dating from 1543. The folio references to this register are complicated since there are two sequences, neither is in the correct order and there are misprints and omissions in the published version as well as a misleading explanation given by the editors.

[6] F. 204. Semen Kurbskii was a brother of Andrei Kurbskii's grandfather; he was related to Vasilii III and to princes of Smolensk and Yaroslavl'. Herberstein knew him as an elderly man.

Lithuanians in 1519 and died in 1527.[1] The nature of any particular register of inquisition is largely determined by the aims envisaged. The state was primarily concerned to know the number of units from which it could expect to obtain tax, and to increase that number. On the other hand, the tasks of the heritable estate holder were different and much more complex: he needed to know who lived where and how, and then who could pay and how much. Both aspects were equally important to him.[2] In Toropets, where there were extensive black peasant lands, the state itself played the part of the estate holder and required those more precise details of how each unit lived. This makes the document more useful than many other registers. Moreover, it antedates the reforms of the 1550s and so shows something of an earlier stage of Muscovy.

The 1540 register as it has survived is incomplete; about an eighth of the extant document deals with the town of Toropets, about a half with the lands held by service and the remainder with the 'black' (i.e. peasant) lands. The omissions appear to relate to the lands held by service.[3] There are, in addition, a number of minor defects in the document and, unfortunately, in its publication; nevertheless, the information provided is of great interest and there is enough of it to enable certain relationships to be seen and conclusions drawn about local conditions; in particular the account of black peasant lands, those held without any lord other than the sovereign, is unusually interesting.

The register records a total of 2,281 tax-liable persons on peasant lands.[4] In addition, 221 villeins and 2,053 dependent peasants are listed on the estates of about 70 landowners. This makes a total of 4,625 households. Assuming an average household size of 5 persons, this would give a total population of 23,125.[5] It is difficult to identify many of the settlements listed in the register, and even many of the lake names do not correspond with those found on modern large-scale maps (1:50,000 American Military Series). Estimates of the area, therefore, are somewhat tentative, but it appears to have been of the order of 2,000 square kilometres. This would mean an average population density of 11–12 persons per square kilometre. It has to be stressed that this average is derived from an incomplete register and the area taken is only that within which settlements have been located. A comparison of these figures

1 *DRV*, xx.18, 23. Poboinin, 'Toropetskaya starina', 137. Another Kurbskii, prince Mikhail Mikhailovich, was royal representative in Toropets in 1522. Rozhdestvenskii, *Sluzhiloe zemlevladenie*, 172, notes a lack of information on the Kurbskii family estates in the sixteenth century.

2 Grekov in: *Sb. statei . . . Platonovu*, 188.

3 For an extract from the register describing an estate held by service see Smith, *Enserfment*, 87–9.

4 Poboinin estimated a total of 2,233, but his figures are consistently lower than those arrived at by M. Pursglove and myself.

5 *Agrarnaya istoriya severo-zapada Rossii XV–XVI v.*, 32, 185, mentions average household sizes of 5 and 4.8 in the nearby southern parts of the Novgorod territory at this time.

with the information to be derived from the 1897 Russian census may be of interest.

	Area (km²)	Population	Density (persons per km²)
1540s	2,000	23,125	11.7
1897	6,000	97,840	16.5

Pervaya vseobshchaya perepis' naseleniya Rossiiskoi Imperii 1897g., vyp.I, 13. The area of Toropets uezd is given (with minor variations) in Efron Brokgauz, 66 (XXXIIIA). 641 and 54 (XXVIIA). 109. The inclusion of the urban population in the two calculations make little overall difference.

The rather high figure for population density in the 1540s is, thus, probably accounted for by an underestimation of the size of the uezd at that time, but it seems impossible on the basis of present evidence to make the necessary correction.

PEASANT LANDS

In most cases details of arable land and hayfield are not given for the black peasant lands; presumably because, from the point of view of the officers compiling the register, there was no need to interfere in well-established arrangements within the peasant communities; information on the number of tenements and adult (male) peasants gave an adequate assessment of wealth. We are fortunate, however, that there are exceptions, perhaps due to over-sights on the part of the clerks of the inquisition, more probably because the status of some lands were not immediately clear to Moscow officials used to different customs, and full details, as for estates held by service, were noted. Information relating to size of arable land and hayfield in the peasant lands is summarised in table 15 which also shows calculated average amounts per tenement and per adult male.

From this evidence it appears that, in terms of a one-man tenement, these peasant holdings varied in estimated wealth or size, arable land usually being noted in the range 5–15 chets (8–24 ha in all) and hayfield 0–37 ricks; the mean for arable was 7 chets (11.5 ha) and for hayfield 18 ricks. Overall, then, the average one-man peasant tenement in sixteenth-century Toropets, insofar as we are justified in generalising from this somewhat scanty evidence, fell within the limits of the model constructed in the first part of this work. If we take these estimates by the officers of inquisition as indicating real areas, the arable land averaged about 10 desyatinas (1 chet per field = $1\frac{1}{2}$ desyatinas in all) and 18 ricks would probably require about 2 desyatinas of hayfield. In part, no doubt, variations in size would occur as a result of particular locations. Clearances sometimes had smaller amounts of arable land, but sometimes reached the size of a normal fully operative tenement. Extra tillage and, in some cases, considerable extra hay occur in tenements with specialist functions; lines 1 and 2 were priests' tenements, and so not tax-liable; 3, 10 and 13 were held by post-horse men, 11, 22 and 29 were Rozderishin holdings (we will touch on these below); only Svyatitsa waste, line 18, does not appear to have had any such function. This hints at the benefits which might accrue through an increased labour force or local co-operation where this occurred.[1] It has to be remembered, of course, that each male listed in the register normally represents a family; the unlisted wife and older children would also work. Mowing and the pasturing of stock could be undertaken by women and children provided that there was someone to share the domestic chores. This information may possibly imply that such larger units had a more than proportional concern for livestock husbandry although, since we have no record of stock held,

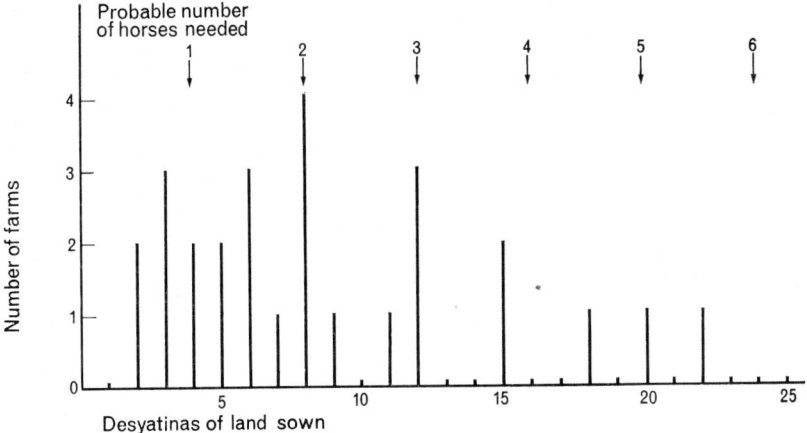

Fig. 4. Peasant farms by arable area.

it is impossible to show this. It is particularly noticeable that the variation in arable area is much less than that for hayfield. Two thirds of the holdings fall within the limits of what could be tilled with two horses (assuming one fallow field and two sown; see figure 4). Finally, for both arable and hayfield the variations in size per man are much smaller than those per tenement. Thus labour was evidently an important determinant of farm size.

We have no clear evidence about methods of cultivation on these holdings.

1 A recent collection of songs from Toropets, which was Musorgski's home area, contains one reference to communal work; the term *toloka* in the sense of neighbourly, unpaid help; Zemtsovskii, *Toropetskie pesni*, 24. Unfortunately, such songs cannot be dated, but the fact that they seem to be related to songs sung at the festival of Ivan Kupala hints at their early origin; *ibid.* 111.

TABLE 15. *Peasant farm size, Toropets uezd*

		Tenements	Males	Arable Land (chets)			Hayfield (ricks)			Notes	Source (folio)
				Total	per tenement	per man	Total	per tenement	per man		
1	Lyubuto volost	1	1	15	15	15	30	30	30	a	133(45)
2	Tura volost	1	1	12	12	12	20	20	20	a	144v(90v)
3	Tura post-horse men	1	1	8	8	8	60	60	60	b	127(39)–127v(39v)
4		2	4	12	6	3	40	20	10		127v(39v)–128(40)
5		2	2	10	5	5	30	15	15		127v(39v)–128(40)
6		2	4	5	3	1	20	10	5		127v(39v)–128(40)
7	Total	6	8	37	6	5	210	35	26	c	128(40)
8	Priluki	1	1	2	2	2	25	25	25	a	150v(62v)
9	Lobno desyatok	1	1	3	3	3	26	26	26	d	155(67)
10	Nebino, post-horse men	3	3	25	8	8	65	22	22		168v(80v)
11	Nebino	3	3	20	7	7	50	17	17		168v(80v)
12	Strugi	1	2	15	15	7	35	35	18	e	168v(80v)
13	Strugino	1	1	12	12	12	26	26	26		169(81)
14	E., M. & L. Rozderishin	1	2	22	22	11	0	0	0	f	169v(81v)
15	Vavulina	1	1	6	6	6	15	15	15		170(82)
16	Moshnitsa	1	3	20	20	7	55	55	18	g	170(82)
17	Dobsha	2	2	17	9	9	40	20	20		170v(82v)
18	Svyatitsa waste	(1)	(1)	12	12	12	30	30	30	h	170v(82v)
19	Empty clearance			9			12	12			170v(82v)
20	Tupitsyn			7			10	10		i	171(83)
21	Ratnoe			12						j	171(83)
22	V., Z., & G. Rozderishin, Gorodok	2	3	35	18	12	110	55	37	k	171(83)
23	Fed'ko Yakushov	2	2	22	11	11	40	20	20		171v(83v)
24	Golovan clearance	2	2	15	8	8	20	10	10		171v(83v)
25	Levko Glazov's clearance	1	1	3	3	3	5	5	5		171v(83v)
26	Gudinov's land, clearance	1	1	2	2	2	5	5	5		171v(83v)

Borovna	5	5	40	8	8	109	22	22	m	172(84)
30 Rodion's clearance	1	1	6	6	6	0	0	0		172(84)
31 S. & E. Rozderishin, Dudino	3	3	15	5	5	20	7	7		172(84)
32 Clearances	7	8	30	4	4	50	7	6		172v(84v)
33 Semen's waste			3						n	172v(84v)
34 Dankova perevara	5	4	20	4	5	30	6	8	o	183(95)–183v(95v)
Total	54	63	404			956			p	
Average		15	15	6	7	6	35	18	15	p
Standard Deviation		10	10	5	3	5	27	14	11	p

a Priest's tenement; non-taxable
b Non-taxable
c Total differs from addition of items
d Priest's son's tenement; non-taxable
e 'and in that hamlet the empty tenement place of Klimko Okatov the townsman, it had been an empty tenement at the [former] inquisition.'
f 'and they till beyond the Rudnitsa stream above lake Dobsha 5 chetverts of arable and mow hay on a site above lake Zbud and on the river Ovzha on the town side at the mouth, 120 ricks; they also mow a meadow on the river Zavzha, opposite Semen's meadow, 40 ricks.' Perhaps *Zavzha* should be read as meaning 'beyond the Ovzha'.
g 'And they till and mow their old settled site [*siden'e selishcho*], Moshnitsa, 8 chets of arable; and they mow a meadow above the Semnaya on Moshna side, 15 ricks; and they mow a meadow on Khilov and on the little pond, 25 ricks both; and they mow a water meadow [*navolok*] on the Toropitsa in Lyubovitsa held by Vasyuk the post-horse man, 40 ricks.'
h Formerly Svyatitsa hamlet with one tenement, now all overgrown.
i 'it became a clearance after the [former] inquisition, they till it from a distance.'
j Empty hamlet 'and there is a meadow to that hamlet on the Toropitsa...'
k Hamlet arose after the former inquisition 'and they mow the Govnya meadow on the Govnya on Startsova side by lake Toropets and they mow the Semenkovo site beyond lake Zbud and the Slizino site on the Kamenka and two meadows, Ploskush and Borisov, on the Studentsa and the Ogalovo site, and they till and mow them from a distance...'
l Arable all overgrown
m Hayfield includes a water-meadow along the Toropitsa and a bog.
n Semen also had in the remote land a quarter lot of arable, noted as 8 chets and 318 ricks of hay in nine locations.
o Church non-taxable tenements, including 1 held by a female host-maker
p Omits total and items which are incomplete (lines 7, 19-21, 27-8 and 33)
Unfortunately, many placenames have not been identified. Toropitsa may be an alternative for the river Toropa; lake Toropets is probably the lake now called Solomennoe.

There are, however, a number of indirect indications. The nearest we come to a hint of a three-field organisation on the peasant lands is an entry relating to two brothers who had a large tenement, one in which there were three other males and a widow. 'And from that hamlet Rodivonko and Ivanko, the Matveevs, till on two fields, the close is on the high bank of the streams of Trufanko and Romantse, children of Onan'ya Kuznetsov; and they till Trufanko and Onan'ya's land also in the other field, from the Tura road, the close is Rodka's'.[1] Here two 'fields' seem to be equated with a close and there is 'land' in 'the other' (or 'another') 'field'. In fact, the entry might be taken to refer to two rather than three fields, in the sense of discrete blocks of cultivated land, though probably with three courses in rotation.

A village of post-horse men contained two tenements of post-horse men and one peasant tenement. 'And of the same village they have ploughed up the tenements, the place of Gridka Olekseev...' There follows a list of such 'places', presumably empty tenement places, fourteen in all. The entry then continues: 'and they have let into the field the new unrecorded clearance of Ivanko Ofonasov as arable to the village'.[2] Here evidently the arable field has been extended by the incorporation of a new clearance; the 'places', on the other hand, formed part of the already existing cultivated land, but no indication is given of how this tilled area was arranged. The clearance and the field into which it was incorporated were perhaps compact units, but the fourteen 'places' may have been separate patches.

Another entry notes 'the Fedyaevo site on the field which at the former inquisition had been Fedyaevo hamlet'.[3] The hamlet with its tilled field, not fields, had become a site used for hay and pasture. Here, too, the field may have been a compact discrete unit of land by the former hamlet and it would have accomodated the main courses of the rotation.

There is other evidence which refers to separate pieces of tilled land. A peasant was 'to till from his hamlet Nekhai's strip of Rogov hamlet which is before his gates along the spruce wood towards Zabolot'e [lit. beyond the marsh], a chetvert of arable land'.[4] A large and numerous hamlet consisting of four tenements and nine adult males also had 'a more distant close, Chertezh, beyond the field by the Rozderishins' Roslov hamlet and Voikovo site and Milokhov bog on the river Kamenka'.[5] Taking these entries in conjunction with what has already been said about the layout of cultivated areas,[6] it seems that peasant lands usually, if not always, consisted of lands, places, strips and closes. Even when the term field was used of lands held directly by peasants it seems at times to have indicated a rotational course and not always a compact area.

This picture is what we might reasonably expect in a thinly occupied area

[1] F. 130v (32v.), reading an amended form of the third pair of personal names.
[2] F. 156(68).   [3] F. 142(54).   [4] F. 148v(60v).   [5] F. 164v(76v). On *chertezh* see p. 22.
[6] See Smith, *Forschungen*, 18.127.

of dense forest where the characteristic settlement consisted of one or two tenements. Even in the cases of the larger than average holdings cited, the tilled land, or at least the supplements detailed, comprised such small items. Thus, Poboinin's view that something like an infield–outfield system (*perelozhnaya sistema*) was then dominant appears to be correct.[1] Moreover, it seems possible to go a little further than this. The infield–outfield system in such an area might be worked with the help of the techniques of slash and burn. The occurrence of the term *chertezh*[2] in one of the quotations cited above may be significant in this respect. Fire would, in any event, be a likely means of clearing forest and it long remained in use in Russia for providing patches for flax cultivation. The Toropets area was well suited to flax cultivation; flax was regularly included in payments in kind in the 1540 register; women regularly worked the flax, producing cloth, fishing nets, etc.[3] The area was within range of Pskov, the probable centre of origin of cultivated flax. It therefore seems at least probable that the tillage of cultivated land was at least in part effected by means of slash and burn.

The register allows us to see something of the changes which took place as tenements declined and disappeared, moved from one spot to another or as new tenements and settlements arose and grew. This has already been mentioned in dealing with field systems and layout.[4] Now we must look at this material for evidence of both physical and social mobility. Were tenements or settlements disappearing faster than new ones were appearing? Did relocation lead to a diminution or increase in agricultural activity? What caused such changes and how were they effected? Questions such as these have to be asked even though the nature of our source in many cases prevents us from finding a clear answer.

It might be thought possible to conclude whether there was an increase or decrease in agricultural activity in the area by counting the number of tenements in new settlements and clearances, assuming that relocations in themselves indicated no change in the scale of activity, and finally, deducting the number of empty tenements. The net change arrived at in this way, however, would not be at all accurate. We can, of course, get a very crude indication of the *trend* in these peasant lands merely by pointing to the fact that 121 clearances are mentioned, but 92 wastes (there were 675 hamlets listed). When we look more closely, however, we see that figures such as these give a spurious impression of accuracy.

There are several reasons for this. The register is a relatively static document; it is true that the survey was carried out over perhaps three or four years, rather than simultaneously, but even though some changes doubtless took place in this period they are not likely seriously to influence the results.

---

[1] 'Toropetskaya starina', 95   [2] See above.   [3] Poboinin, 'Toropetskaya starina', 271.
[4] See Smith, *Forschungen,* 18.125–37, who uses evidence from estates held by service as well as the peasant lands dealt with in this section.

The clerks record changes from the situation at the time of an earlier inquisition, but that register does not appear to have survived, nor do we know whether they noted all changes. Effectively, therefore, we have a description of the area at about 1540 with some, possibly erratic, indications of change. It is, moreover, rarely possible to accept the terminology in itself as a reliable indicator, even though the clerks seem to have used it with great accuracy. The inquisition was carried out mainly for tax purposes and a concern with agricultural activity as such was not envisaged as being of prime importance. Most terms are ambiguous; changes took place leading both to an increase and to a diminution in activity. Headings themselves sometimes make this clear. One such reads: 'In Zbud perevara new hamlets and clearances which arose after the [former] inquisition along the sites which had been empty hamlets at the inquisition and along new places in Dobsha.'[1] Here empty hamlets had declined and become sites, used mainly for hay and pasture, and were now again growing as new hamlets and clearances; there was also growth in 'new places', or was this merely a relocation, not involving any net growth? This heading also illustrates another problem. If a hamlet became empty it might mean that the family had moved or died out; but if the empty hamlet was 'empty' only as regards tax liability and continued to be worked, at least for certain purposes, as a site, then we have to offset the desertion of the hamlet by the growth of the tenement which used the deserted site for its livestock. In fact, the terminology has to be looked at with great care since decline or growth is likely to affect only an element or some elements which make up the tenement or hamlet, but not necessarily the total range of activities. A man may die, but another hand grasps the sokha. If the sokha is abandoned there, someone may still continue to use the scythe and the axe. Moreover, the measurement of activity on a social scale may be quite other than that on the scale of the tenement or hamlet, the family or settlement. Individual families die out even though total population is increasing.

The sort of evidence which might be taken to indicate a decline in economic activity is when a hamlet became a waste or, more rarely, a site.[2] Sometimes the arable land of the former hamlet was 'all overgrown',[3] though no reason is given. Sometimes villages and hamlets 'became deserted due to the plague and the Lithuanian wars'.[4] This change from a hamlet which was at least one fully operational tenement to a waste or site may, perhaps, be taken as the most serious evidence of decline; but even here all is not what it may seem. For instance, the Doibuzh site formerly had 19 tenements.[5] Unfortunately we do not know what, if any, relationship existed between this site and the 'Doibuzh village above lake Verezhyun' which has now transferred to a new place along the end of lake Verezhyun' [6] and had ten live tenements. In the

---

[1] F. 166(78).

[2] Ff. 170v(82v), 182(94), 193v(105v), 206(118), 206v(118v), 207(119), 215(127).

[3] F. 170v(82v).   [4] F. 205v(47v); cp. f. 143v(55v) etc.   [5] F. 208(120).   [6] F. 198(110).

case of another large village in the same area, however, we are more fortunate. Bentsy, deserted, like Doibuzh, due to plague and war, had had twenty live tenements, and twenty names are listed.[1] However, in Startsova volost three hamlets with six tenements are noted as having transferred from Bentsy village.[2] In fact, of the twenty names listed seventeen are found, either in the same form or as patronymics, in the small group of hamlets in which the three transferred ones are recorded; it seems that most of the original tenants or their sons or brothers were to be found there. This suggests that, in this case, large settlements which were more susceptible to the incidence of disease and the disasters and exactions of war became fragmented into a number of smaller hamlets, the original locations becoming wastes or sites. Overall, therefore, there may have been relatively little decline in basic agricultural activity even when war or other disasters led to dispersal. But such disasters will, of course, have inhibited the growth of larger settlements and the possible development of more specialised and complex activities.

Formerly live hamlets might continue to be tilled although from a distance. For example, 'Sen'ka Filimonov tills and mows from a distance, from the artisan quarter of Toropets town, the Berezova hamlets and the Berezova site.'[3] In some cases, however, when a hamlet was empty its decline was vividly depicted; Zalogi, for instance, had four tenements, but 'the buildings in it have tumbled down'.[4] Sometimes smaller items than a hamlet became empty, but continued to be worked, though from a distance. Certain clearances in Nezhel' volost, for example, were empty, but tilled by the post-horse men.[5] Evidently we have here an unsuccessful attempt to expand, but the failure of the clearance to become a hamlet was mitigated by its continuing use by others.

If we look at the manner in which the wastes are registered we can see, again, that they are scarcely those 'abandoned lands' which Blum has taken as evidence of a decline, characterised by depopulation and abandoned holdings, in the period from the thirteenth into the fifteenth centuries.[6] Only two of the wastes are listed singly among other types of settlement.[7] All the rest occur in groups of between four and twenty wastes. The largest group, under the heading 'sites and wastes in Startseva volost which had been villages and hamlets with people at the inquisition, but became deserted due to plague and the Lithuanian wars', has already been referred to above; it included Bentsy. A group of nine wastes had, apparently, been empty at the previous inquisition, so does not represent abandonment in the immediate past.[8] Another

---

[1] Ff. 206v(118v)–207(119).   [2] F. 200(112).   [3] F. 161v(73v).   [4] F. 144v(56v).
[5] F. 158v(70v); cp. f. 171v(83v).
[6] *Lord and peasant*, 61. Blum recognises the sixteenth century as a period of growth, but has not, apparently, argued that there was any difference in the meaning of the term *pustosh'* in the two periods; *Ibid.* 121f. For an excellent discussion of desertion and its terminology see Goehrke, *Die Wüstungen*.
[7] Ff. 170v(82v), 179v(91v).   [8] F. 192(104).

group of nine, and two empty hamlets, are listed under the rubric 'Wastes and sites in Kudino and in Loshira and in Medenets and in Yazets which were hamlets at the inquisition'.[1] Here is a clear case of abandonment, but the inclusion of 'sites' in the heading, though none is listed, suggests that at least some of these places continued to be used as hayfield or pasture. A further group of nine, listed under 'wastes and sites in Strezhino', was made up of four which had been tenements (one of these was now tilled); two which had been empty tenements; one had been a hamlet and was now tilled; one had someone living on it and was now tilled and mowed; the last one had no details given.[2] This final group shows how dangerous it would be to take the number of wastes as a straightforward indicator of declining activity. Wastes may represent complete abandonment of a tenement or hamlet, but in many cases abandonment was incomplete; moreover, sometimes the abandonment had taken place before the last inquisition. A final point is that the grouping of the vast majority of wastes mentioned suggests that the officers of the inquisition regarded them as an item to be taken into account when evaluating the resources of the area. Whether they regarded them as on the debit or the credit side of the balance we do not know.

An item which is well represented in the peasant lands (it is scarcely found in the estates held by service) is the empty tenement place (*mesto dvorovoe pusto*). 316 empty tenement places are listed in peasant settlements, roughly 18 per cent of the number of tenements (14 per cent of the number of adult males listed).

While it is often impossible to be certain of relationships or to describe them precisely from the available evidence, nevertheless, as one would expect in such small and scattered communities, many people were related to one another; the extent to which such family connections existed seems to imply fairly stable communities. It is noteworthy that brothers by no means always share a tenement; there is little evidence here of an extended family. This register, however, records a situation at one point in time and it is, therefore, not a suitable source to use for extensive conclusions about the movement of the peasant population over time.

Let us now look at what was recorded about the (mainly arable) lands of empty tenements. Empty tenement places might arise when a man died. A two tenement hamlet above lake Zhyulebin had three empty tenement places listed, one 'of Grigorko Filin the reeve [*starosta*]'.[3] The hamlet itself was 'of Grigorii Filin the reeve', probably the original holder or founder of the hamlet, but his name is not among those of the adult males noted. Some of the other males listed could be brothers of the former holders of the other two tenement places. A similar linkage by name of at least four of the seven holders of tenements which had become empty tenement places is found in Starina

---

[1] Ff. 181v(93v), 182(94).   [2] F. 215v(127v).   [3] F. 200v(112v).

hamlet.[1] In this instance, however, it is added that 'those tenements had become deserted due to the Lithuanian wars after the inquisition'. Here, too, the original holder had evidently died, but the war had led the hamlet's inhabitants, or at least part of them, to set up a new hamlet called Novosel'e in which his son had a tenement.[2] In other parts of the same volost, Startseva, nine further tenement places are stated to have been empty due to the war with the Lithuanians.[3]

Such arable land, available when a tenement fell empty and probably the house was removed (this is presumably what distinguishes it from an empty tenement), sometimes continued to be tilled. The officers of the inquisition regarded 189 of the 316 tenement places recorded as still cultivable, in a few cases the land was actually tilled and brought back into full use. A nil entry, moreover, was coupled with the note in the totals for that area that 'they are to till, subject to their tenements, the tenement places from which they carried themselves off to the remote arable lands'. This suggests that cultivation of empty tenement places was regarded as a usual demand.[4] It should, of course, also be borne in mind that the register gives totals relating to only 272 of the 316 empty places; had totals for all areas been included the proportion considered cultivable would have been higher. Whether such cultivable arable land was in fact usually worked we do not know. We may well have here a conflict of interest between the officers who were concerned for tax purposes to ensure that arable land did not fall out of cultivation and peasants who, though anxious to avoid the arduous work of winning new arable land from the forest, might find themselves unable to till additional areas of arable. It is clear, however, that in a few instances empty tenement places were developing into fully taxable holdings and, at least to this extent, we must be cautious of accepting this item as evidence of nothing but decline.

The site (*selishche*) seems here to have been an abandoned or partly abandoned area formerly worked.[5] Sites were contrasted with 'new places' and were often a complement to empty tenement places; the new places provided potential arable land, the sites, hayfield, or sometimes both hayfield and arable land, rarely arable only. It does not seem possible to make a sharp distinction between sites and seats; both seem to refer to a former holding, but we have more information about the use of sites and this shows that in general they were used for hay. There is no clear indication as to whether this was a supplement in tenements which had more than the usual number of livestock or merely brought the tenement up to the norm. It seems more likely that the latter view is correct. We find a clearance with a site, though the use to which it was put is not stated.[6] Similarly, some newly established Strezhino

---

[1] Ff. 195v(107v)–196(108).   [2] The place names are significant here; *stary* old, *novy* new.
[3] Ff. 197v(109v)–198(110).   [4] F. 191v(103v).
[5] See Goehrke, *Die Wüstungen*, 35–43, for a discussion of the meaning of this term.
[6] F. 211v(123v).

hamlets 'mow hay from those hamlets on the old Strezhino site'.[1] Both entries happen to suggest that sites may have been used to maintain feed levels before the new tenement was fully established.

This brings us to relocation. One case has already been mentioned.[2] Doibuzh village 'transferred to a new place along the end of lake Verezhun''.[3] Another partial relocation, evidently, was involved in the matter of the hamlets of Starina and Novosel'e.[4] The old places of this hamlet were scattered 'among their distant lands and arable'. It seems that scattering, it is too much to speak of fragmentation for such small units, was commoner than straightforward relocation. Hamlets and peasants were 'now scattered among their distant hayfields'.[5] Another hamlet had 'now scattered along the patches beyond the fields'.[6] Pestno village was deserted due to the war 'and after the war the peasants scattered along their distant arable land'.[7] The taxable tenements in Chokhovichi (sc. Khochevitsy) village were 'now scattered haphazardly through the forest'.[8] In entries such as these we feel the presence of the forest with various patches of arable land and hayfield, many of them sufficiently small and remote to escape the attentions of marauders.

Some of the transfers of location seem obscure. 'Pesovets hamlet on the river Psovets transferred from Psovets hamlet...'; 'a hamlet in Kostritsa... transferred from Kostritsa'.[9] Even place names derived from a physical feature were evidently not very precisely located. Those derived from personal names may have moved with the man. There may well, therefore, be unnoted changes of location which we do not recognise. Finally, there was a waste which was noted under Startseva volost 'because it had gone off separately over the river Toropitsa'.[10]

Whether relocation involved a change in the amount or balance of activity seems impossible to judge. Probably the desertion of established tenements for more obscure ones deep in the forest involved at least a temporary decline in output; though in some cases this might be offset from the peasant viewpoint by the chance of at least a temporary avoidance of exactions. Last century Poboinin pointed out that the disasters of plague and war mainly affected settlements located on the road to Toropets from the south and he added that 'the volosts which suffered did not even lose their peasants; the latter only "transferred themselves, carried themselves off from their village to a new place, scattered among their distant arable, have now scattered in different directions through the forest" '.[11] The evidence clearly supports this view; but it is worthwhile adding that the small size and low level of technology of the peasant unit contributed to its viability when disaster struck.

---

[1] F. 213v(125v).   [2] See p. 168 above.   [3] Cp. f. 164(76) for another case.
[4] Ff. 195v(107v)–196(108).   [5] F. 175(87): similar formulae on ff. 178(90) and 179(91).
[6] F. 164(76). See also Smith, *Forschungen,* 18.125–33, for similar evidence from estates held by service.
[7] F. 196v(108v).   [8] F. 192v(104v).   [9] Ff. 202v(114v), 204(116).   [10] F. 208(120).
[11] 'Toropetskaya starina', 95.

TABLE 16. *Old and New Settlements*

Folio	Hamlets	Clearances	Tenements	Empty tenements	Empty tenement places	Adult males	Tenements per settlement	Males per tenement
			Old Settlements					
131 (43)	63(61)	—	113(112)	6	17	212(207)	1.8	1.9(1.8)
143 (55)	90(92)	—	207	27(30)	44(39)	315(316)	2.3	1.5
126 (38)	58(60)	—	162(166)	—	—	295(300)	2.8	1.8
157v(69v)	21★	—	72	6(4)	71(58)	88(86)	3.4	1.2
167v(79v)	39	—	81(80)	—	22	117(116)	2.1	1.4(1.5)
175 (87)	11	—	15(18)	1	7	21(27)	1.4(1.6)	1.4(1.5)
181 (93)	43(44)	—	92	14	11	147(151)	2.1	1.6
191v(103v)	62(63)	—	82(79)	—	—	154(152)	1.3	1.9
204v(117v)	79†	—	189(89)‡	5(6)	66(67)	308	2.4	1.6
214v(126v)	33	—	41(40)	2	—	53(54)	1.2	1.3(1.4)
Total	499(503)	—	1054	61(63)	238(221)	1710(1717)	2.1	1.6
			New Settlements					
131 (43)	13(16)	1	(24)	—	1	36(44)	(1.5)	(1.8)
143 (55)	6	17	25	—	—	36(37)	4.2	1.4(1.5)
126 (38)	4	5	9	—	—	19	2.3	2.1
157v(69v)	3	16	25(26)	—	—	32(33)	8.3(8.7)	1.3
167v(79v)	16	5	23(24)	—	6	34	1.4(1.5)	1.5(1.4)
181 (93)	16(23)	—	34(35)	—	—	58(60)	2.1(1.5)	1.7
191v(103v)	25	4	31(34)	—	—	64	1.2(1.4)	2.1(1.9)
204v(117v)	5	24	34(33)	—	—	56(55)	6.8(6.6)	1.6(1.7)
214v(126v)	46	7	58(59)	(1)	—	71(69)	1.3	1.2
Total	134(144)	79	239(269)	(1)	7	406(415)	1.8(1.9)	1.7(1.5)

Note: figures shown in brackets are those totals given in the source which differ from totals obtained by addition

★ Including 5 villages

† Including 6 villages

‡ Clearly a clerical error for 189 and therefore disregarded.

The peasant tenement with direct control of family labour had at least some advantages over the lords' estates dependent on slave or serf labour.

The evidence for growth seems a little less ambiguous than that for decline. The register contains ten partial totals (plus those for 'old Trinity church hamlets' and Tura post-horse men) in the section on peasant lands, and these give us the information on old and new settlements shown in table 16.

Most of these entries occur under a heading which reads, 'New hamlets and clearances which arose after the inquisition'. Certain features are regular; no clearances are noted for old hamlets, while new settlements frequently have clearances. Empty tenements are usually only found in old settlements and empty tenement places are rarely found in new settlements. New settlements, then, are characterised by the growth of clearances into hamlets. Old settlements do not have clearances, but have had time for some tenements to fall empty, though, as we have seen, this does not always mean a complete cessation of activity. Finally, on average, old settlements are slightly larger than new ones at least in terms of numbers of tenements, but the range in terms of tenements per settlement is much greater in the case of new settlements since some are very much above average size in this respect. Both old and new settlements appear to be similar as regards the average number of adult males per tenement; this evidently reflects the fairly stable size of the basic farm unit in terms of labour.

Hamlets were sometimes recorded as being 'of' someone. The first hamlet recorded on the peasant lands, for example, was 'Pabezho hamlet above the river Toropitska of Ivashko Sushko; in a tenement Fomka Ivanov and the empty tenement place of Fed'ko Ermolin...'[1] The hamlet is 'of' Ivashko, but the name of the man in the tenement is not his. This suggests a distinction between the holding of and the occupancy of the hamlet. In some other entries the names coincide. Probably, therefore, such entries indicate the man who first established the tenement; he might later move, but retain ownership although not occupation. Another possible interpretation is that these entries are the names of those who were recorded at the previous inquisition. Such men are here termed 'original holders'; this avoids deciding whether they were first colonisers or simply those in possession at the previous survey.

A high proportion of such original holders were in tenements in clearances and in new settlements in general (the formula is usually: 'in a tenement x himself'). Thus, about 169 of 267 new tenements listed had original holders shown (sometimes, though rarely, two or three names were recorded for one tenement). There were 115 of these original holders in tenements and a further 64 names appear to be those of a brother or son. The proportion of original holders or their relatives in tenements would, of course, probably be even higher if we had fuller information. Table 17 summarises this information for new hamlets and for clearances only.

[1] F. 120 (32).

TABLE 17. *New hamlets and clearances, and original holders*

Folio	New hamlets	Clear-ances	Tenements with original holders		Total tenements		Total of tenements known to contain			
	a	b	a	b	a	b	holder		relative(s)	
							a	b	a	b
128v(40v)	14	1	13	1	18	1	4	1	8	1
142(54)	4	8	4	8	4	8	2	5	3	3
142v(54v)	2	9	2	9	3	10	1	8	—	1
125v(37v)	4	5	3	3	4	5	2	3	3	—
156v(68v)	3	16	3	15	8	17	1	14	2	6
159v(71v)	2	6	3	6	3	6	2	6	—	2
166(78)	14	3	11	—	16	3	6	—	3	1
167(79)	2	2	—	3	2	2	—	3	—	—
179v(91v)	16	6	19	6	27	7	17	5	10	1
189(101)	18	2	12	1	18	3	5	2	5	—
190v(102v)	5	—	2	—	6	—	—	—	1	—
191(103)	1	2	1	2	1	2	—	—	—	2
191(103)	1	—	1	—	1	—	1	—	—	—
203(116)	5	18	4	16	6	20	3	14	2	5
204(117)	1	6	1	3	2	6	—	1	1	2
210v(122v)	17	6	8	4	19	7	3	4	1	—
212(124)	24	—	3	—	24	—	1	—	1	—
214(126)	4	1	1	1	7	1	—	1	—	—
Totals	137	91	91	78	169	98	48	67	40	24

While old settlements with 1054 tenements on the peasant lands had original holders shown in less than 50 per cent of cases, the new hamlets and clearances had original holders mentioned in over 60 per cent of the total number. Moreover, there is a noticeable difference between hamlets and clearances. Tenements in hamlets have original holders shown in about 28 per cent of their number, but in clearances the figure rises to 68 per cent; relatives mentioned show a much less marked trend: hamlets 23 per cent, clearances 24 per cent. This, of course, is not conclusive evidence that the 'of' entries refer to founders or colonists, but it shows that there was a closer correlation between this category of person and clearances than between them and hamlets.

Only occasionally is clear evidence found for ownership being separated from occupancy. In a village and hamlets (which are not separately recorded) above lake Gryadets one tenement was held by Vasyuk Olekseev, his son Stepanko and Eremko Esipov; they also had three empty tenement places.[1] Among the new hamlets and clearances 'which arose after the inquisition'

[1] Ff. 159(71)–159v(71v).

was 'a hamlet above lake Gryadets of Vasyuk Olekseev the post-horse man; in a tenement his crop-sharer Onanko Olferov; in a tenement Vaska Kuzmin'.[1] Vasyuk Olekseev also had a tenement in Nebino hamlet 'above Lake Toropets opposite the boatmen, where the post-horse men live'.[2] 'Vasyuk the post-horse man' also had a meadow in Lyubovitsa which was mowed by some members of the Rozderishin family.[3] It seems unlikely that there was more than one post-horse man of the same name though it is difficult to see why one man lived in two tenements. Possibly we have here an instance of double counting due to the slow compilation of this inquisition, but it seems more likely that Vasyuk was a prosperous peasant with more than one tenement. Clearly this post-horse man with at least one tenement well provided with labour (three males), and with possible additional income from the empty tenement places, was a man of some substance. He had a two-tenement hamlet and a crop-sharer dependent on him.

The type of settlement which is most closely associated with growth is the clearance. Clearances were frequently recorded scattered among the hamlets (unlike wastes), but there are a few instances when clusters of clearances are recorded.[4] Two Rozderishin brothers, Vokhromei and Zakharka, had five clearances, one of which was empty while one had its arable land all over-grown. The three 'live' clearances had four tenements, and both the original owners mentioned were living in tenements.[5] Another Rozderishin, Semen, had a group of four clearances with seven tenements, one of which had two men in it; the two original holders clearly mentioned were both in tenements.[6] Another cluster of clearances was recorded in Strezhino perevara.[7] The heading refers to 'hamlets and clearances of the Trinity [monastery] which are in Toropets within the town'. Nine clearances are listed; eight of them have a tenement occupied by an original holder and the ninth was occupied by a blacksmith. In addition, there were five Trinity clearances in the same area, not specified as being within the town; four of these had a tenement occupied by the original holder, the fifth was occupied apparently by the son of the original holder.

The evidence we have for change in the black peasant lands, then, suggests that internal colonisation and growth was probably greater than is indicated by the figures for the number of clearances as against the much smaller number of wastes. It looks as if relocation did not in every instance lead to a great diminution of agricultural activity, though this is hard to demonstrate conclusively. The causes of this colonisation are more difficult to establish and we shall have to return to this problem later.

The changes which have been considered so far relate only to the arable

[1] Ff. 159v(71v)–160(72).
[2] F 168v(80v). This was one of the tenements the size of which is given in table 15, line 10.
[3] F. 170v(82v).   [4] Ff. 142v(54v), 158v(70v).   [5] F. 171v(83v).
[6] Ff. 172(84)–172v(84v).   [7] F. 216(128).

land and the hayfield of the tenement. Perhaps due to the very abundance of forest at this time, evidence on gathering is not so well represented in the register. Twenty-five entries refer to forest bees; 56 marks used to indicate claims to bee-trees are listed.[1] Most of the entries are in Startseva volost, though it seems unlikely that other districts were so deforested or otherwise unsuitable for bees that no bee-trees were to be found. Startseva volost was assessed at three rubles for 15 puds of honey, and Nezhel' at one for 5.[2] In fact, the evidence of obligations on the estates held by service suggests that honey was normally available on peasant holdings. The form in which the entries are recorded implies that bee-marks were regarded as of the settlement; the marks are listed after the details of the settlement, including any empty tenement places, have been given. It is possible that bee-marks were found in most cases where there was a slightly higher than usual number of men available; this is based on the evidence we have for bee-marks in single tenement settlements, most of which have more than one male listed.

More than 50 lakes and the river Dvina had fisheries on them. In most cases details of the fish taken and the means used are given. The commonest fish were pike (mentioned at 48 locations), roach (48) and perch (for which two terms were used, *okun* and *ostrets*) (49); bream (34) and carp (22) were fairly widespread, and there were fewer occurrences of tench (5) and pike-perch (1). Pike was the characteristic fish of the area. Payments for fishing rights were calculated according to the number of barrels of pike; even today there is a lake Shchuchina in the area (lit. pike flesh; cp. *shchuka*, 'pike'). Fishing was with nets, with trap seines, another type of seine (*kerevoda*) and with traps (*neroty*).[3] Sometimes fishing took place both in winter and in summer; this was always at fisheries where all four means for taking fish were used. In some cases seines were only used in winter, when labour would be available for them and they could be used in conjunction with weirs where the rapid flow prevented ice from forming. The fisheries were sometimes exploited jointly, being shared with lords or rich peasants. For instance, some were run jointly with the Chikhachovs and the Golenishchovs.[4] How such joint operations were undertaken the register does not tell us. Moreover, it is impossible, unfortunately, to estimate the catch. Payments recorded are at the rate of half a ruble for a barrel of pike, but the actual amount taken is not stated.

On the basis of this evidence, then, it seems difficult to assess the part which gathering played in the peasant economy of the area. Some, possibly fairly restricted, areas had forest bees worth taxing; and fishing, sometimes on a fairly large scale, seems to have been widespread. The petty income of the

---

[1] Ff. 151(63)–151v(63v), 153(65), 155v(67v), 156v(68v), 193(105), 194(106), 196(108), 197(109)–198(110), 199(111)–201(113), 202(114)–202v(114v).

[2] Ff. 158v(70v), 206(118).   [3] See pp. 64–5 above on these terms.

[4] Ff. 168(80), 191v(103v)–192(104). Further information on these families is given below.

lords, however, shows that game was available in the area and it is difficult to believe that peasants did not take advantage of this fact. Fungi and berries are not mentioned, but again this is probably due to their abundance and the difficulty of transporting such produce without damage.[1]

If we look at the evidence for other activities we are struck by their paucity. Only half a dozen or so mills are recorded. These were, as was usual in Russia, water-mills. Klimko Demidov had a mill by the hamlet of Vzoto 'on the river Vzoto and it grinds in spring and in autumn'.[2] In the same perevara, Zbutsk, were two hamlets, 'and they had by both hamlets a churn mill (*mel'nitsa mutovka*) and it grinds in spring and in autumn'.[3] Others were in Toropets perevara, Nezhel', Lyubuta, and in Tura volosts.[4] One in Tura and one in Toropets worked in spring and autumn and had two sets of stones. The last mill noted in the peasant lands was in half of Startseva volost, in Terebezhova hamlet, 'above lake Terebezho of Luka Ivashkov: in a tenement Sovostko, son of Petr Borilin, and his son Nezhdanko, and that hamlet has as an appurtenance, a churn mill which grinds in spring and in autumn'.[5] These mills evidently were relatively few and far between; it may be significant that Klimko Demidov was in a hamlet that was among those listed as 'of townsmen of Toropets', while one of the two hamlets by the second mill had had a townsman as its original holder. The vast majority of peasants presumably relied on hand mills to grind their flour, as they continued to do for centuries.

The section of the register relating to the peasant lands has eight references to blacksmiths. Two of the smiths were evidently dead at the time of the inquisition; in one case the man was the original holder of an empty hamlet now tilled from a distance, and in another the smith had held what had become an empty tenement place.[6] Another was recorded in a clearance within the town of Toropets.[7] There thus appear to have been five blacksmiths in the peasant countryside of Toropets uezd, plus the one already mentioned in a non-taxable clerical tenement. One had a hamlet and was in a tenement of above average size (four males).[8] Another shared a tenement with his brother.[9] The other three were in settlements which were above average in size, or had been so. One was in Doibuzh village (10 tenements), another in a six tenement hamlet and the third in a hamlet which now had only three tenements, but also five empty tenement places.[10] This rather small number of smiths for such a large area reflects the low population density and also the comparatively small part played by metal in peasant life. Poboinin states that iron was not

---

[1] Obozov *i dr.*, *Pobochnye pol'zovaniya*, 56 and 104, shows the importance of cowberries and cranberries in the Pskov, Novgorod and Velikie Luki areas in the 1960s.

[2] F. 162v(72v).   [3] F. 165(77).

[4] Ff. 122(34)–122v(34v), 128(40), 129(41), 138v(50v), 141(53), 152(64).

[5] F. 193v(105v).   [6] Ff. 186(98), 199v(111v).   [7] F. 216(128).   [8] F. 188v(100v).

[9] F. 136v(48v).   [10] Ff. 198(110), 200v(112v), 201v(113v).

worked in the area, but this is probably an overstatement.[1] The larger than average size of settlement in which smiths were found indicates the importance of population density as an aid to specialisation.

A few other trades are mentioned; a carpenter tilled a waste from the town, so was not a true countryman, but there was a master tailor who shared a tenement with one other man and another was the original holder of a hamlet; and there are references, in the form of personal names, to a locksmith and a drying-kiln man (*osetnik*).[2] There were also ten churches, with thirty-four officials (including ten women host-makers), one very small monastery with two persons in it and some minor officials in the countryside.

There were thus about twice as many priests as blacksmiths recorded in the countryside, and many more clerical officials than other craftsmen. It is noticeable that more than half of this small number of churches were new. This is probably connected with the reestablishment of Orthodoxy when the area had been recovered from the Poles and Lithuanians in the early sixteenth century; under their rule Catholicism had been forcibly introduced.[3] The statement that one of the churches had been established by peasants may, therefore, in some measure reflect the feelings of the local peasant population towards Catholicism.

There are very few references to administrative officials. One reeve (*starosta*) was recorded in a normal tenement, another in a two-man tenement in a hamlet which probably belonged to a townsman in Zbuts perevara.[4] Another also in Zbuts, was in a tenement which had been removed to a new place.[5] A fourth was the original holder of a hamlet, but was probably dead, since there was also a reference to his empty tenement place.[6] There is also a single reference to a hundred-man.[7] The administrative organisation of the area, however, would seem to imply many more officials than this. There appear to have been six volosts mentioned within the uezd: Nezhel', Toropets, Lyubuta, Tura, Dankova and Startseva.[8] There were thirteen volosts at the time of the area's incorporation in the Moscow state.[9] Each volost was sub-divided into a number of perevaras, though it is not always clear to which volost a perevara belonged. Thirteen are mentioned: Porech'e, Toropets, Zbuts, Lashira, Kudino, Medino, Yazvets, Dankova, Serezha, Zimtsy, Solovskaya, Khochevitsy and Strezhino.[10] There are also references to a desyatok (*lit.* a ten); presumably, therefore, under the hundred-man there

---

[1] 'Toropetskaya starina', 79.
[2] Ff. 179v(91v), 133v(45v), 196v(108v), 210v(122v), 185v(97v).
[3] Solov'ev, *Istoriya Rossii*, III.110–114 and 225, citing *AZR*, II, no. 22, pp. 24–6; cp. *PSRL*, XXIV.216.
[4] Ff. 153v(65v), 163(75).   [5] F. 164(76).   [6] F. 200v(112v).   [7] F. 192(104).
[8] Ff. 156v(68v), 159(71), 183(95), 192v(104v).
[9] Poboinin, 'Toropetskaya starina', 93; ff. 133(45), 144v(56v).
[10] Ff. 119v(31v), 121v(33v), 162(74), 175v(87v), 183(95), 187(99), 194(106), 199v(111v), 208(120).

might be ten other officials.[1] Such ten-men are mentioned rarely.[2] In three cases hamlets are specified as 'not subject to any perevara'.[3] One of these was hamlets of townsmen in Toropets volost by lake Toropets; another was hamlets in Strugi. Strugi was one of thirteen district names which appear in headings but which do not clearly seem to have been within any perevara. The perevara, then, seems to have been an administrative unit of mainly peasant holdings for some specific purpose, possibly the performance of certain obligations or the exercise of certain rights. The suggestion that perevara peasants were those, around the halts of officials on circuit, who had originally to deliver beer does not seem convincing.[4] It was, however, not a unit into which the entire area of peasant land was divided. It thus looks as if the officers of the inquisition had only limited concern for the administrative organisation within the peasant areas; there seems to be no reference to all the volost heads, nor to the perevara officials.

Certain categories of peasant are represented in the register. The ordinary peasant is not normally designated as such, but other categories are sometimes indicated. The commonest of these is the labourer (*bobyl'*), a man who worked for someone else, though he might have a tenement of his own.[5] Thirty-four labourers are recorded on the peasant lands in the register. Sometimes the labourer is found in a tenement by himself; occasionally it was without arable.[6] Sometimes he was in a tenement with peasants; two brothers, for instance, had a labourer in their tenement and his holding 'had been an empty place at the inquisition'.[7] Sometimes the entry is more suggestive of the significance of this term. In a hamlet 'in Sel'tso' (the name indicates a manorial settlement; moreover, Izvoilovo hamlet had one of the few artisans recorded) there was a labourer of the original holder of the hamlet or of the two tenants' father in one of the two tenements.[8] This hamlet mowed half of their 'old seat'; the other half was mowed by a cossack (i.e. a wage labourer, the only one recorded in the peasant lands) and his son, who were in a tenement in a new hamlet.[9] Two other labourers were in tenements in new hamlets; one in a tenement with the original holder and his brother, the other was the labourer of three brothers who had their own tenement, the third tenement of the clearance being occupied by their father and his brother who was the original holder of the settlement.[10] There are several other cases where labourers were in tenements of three to five men.[11] In Dankova volost there was a group of exceptionally large tenements; the largest had 7 men, including

[1] Ff. 157v(69v), 158(70).   [2] Ff. 150v(62v), 151(63), 153(65).
[3] Ff. 154v(66v), 160v(72v), 173v(85v).   [4] Poboinin, 'Toropetskaya starina', 98.
[5] Recent work on the labourers has mainly dealt with those found on estates, not on peasant lands. See Sakharov, *Russkaya derevnya XVIIv.*, 157–71 and *VI* (1965), 9.51–67; Shapiro, *Ist. SSSR* (1960), 3.49–66; Gorfunkel', *TrLOII*, v.640–7; Koretskii, *Zakreposhchenie*, 161–82.
[6] Ff. 137(49), 194v(106v).   [7] F. 188(100).   [8] Ff. 185(97)–185v(97v).   [9] F. 189(101).
[10] F. 203(115).   [11] Ff. 122v(34v), 129(41)–129v(41v), 141(53), 147(59), etc.

a labourer, and a widow recorded in it.[1] There were three brothers, their father and uncle, their labourer; the seventh man had the same name as both the labourer and one of the elder brothers, so this may possibly be a clerical error. Finally, there was a small group of labourers who were probably linked with the town in various ways. In Strezhino perevara there was a hamlet of the Toropets man Fed'ko Popryatov. He seems to have been the head of a family with a number of tenements (see figure 5). Fed'ko's old hamlet had a crop-sharer, a dependent of Zik, in a tenement.[2] Fed'ko and his

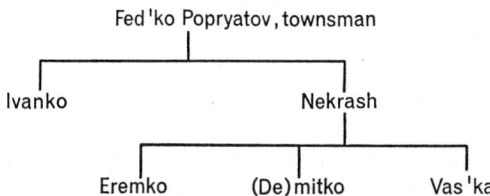

Fig. 5. The family of Fed'ko Popryatov.

children also had a site two thirds of which were now mowed by Nekrash, Demitko and Vas'ka (Eremko had a tenement of his own); the other third was mowed by Zik, his brother Osif and relatives.[3] Later in the register there is an entry for Zik's clearance; this shows a single tenement and notes the third of the site mowed, but the man in the tenement is Oksenko; there is no mention of Zik.[4] Immediately preceding this entry is one relating to a new hamlet of Nekrash and Vas'ka.[5] Nekrash also had a bit of arable land 'along the Strezhino road which goes to Toropets town' by another new hamlet.[6] Ivanko had formerly had one tenement of three in a hamlet in Strezhino perevara; this was now a four-tenement hamlet with no identifiable Popryatovs, but with a labourer in the fourth tenement. Ivanko himself was now in a tenement 'of Kondrat Gridov, a Toropets man', and mowed one quarter of an unidentified site.[7] In the same immediate area there was another tenement which had been empty at the inquisition, but was now occupied by a peasant and his labourer.[8] Finally, also in Strezhino perevara, under a heading of 'hamlets and clearances which are inside Toropets town', a six-tenement hamlet had two tenements held by labourers, one being further described as a new arrival.[9]

Thus, labourers are here associated with tenements which were above average in terms of labour force, engaged in opening up new land, or possibly working larger than usual holdings, and sometimes linked with other enterprising peasant families; there are also some indications that their masters had links with the town. They do not appear to be of inferior social status to

[1] F. 184(96). [2] F. 208(120), 208v(120v).
[3] F. 208v(120v). [4] F. 211(123). [5] F. 210v(122v). [6] F. 211v(123v).
[7] F 208v(120v). [8] F. 209(121). [9] F. 216(128).

peasants, whatever their economic position, since it is rare for them to be recorded by forename only.[1]

In one case, as we have seen, the labourer was also a new arrival. Only a total of twelve new arrivals are recorded. One was in a new hamlet where he shared a tenement with a peasant; another had a tenement in a three-tenement hamlet; two had evidently married into peasant families and now held tenements.[2] One held a tenement with a peasant in it on a clearance named after him.[3] This meagre evidence indicates that new arrivals were less involved in entrepreneurial activity than the labourers. They were also differentiated from them in that they were recorded by forename only; the only exceptions were those who were sons or sons-in-law of established peasants. This evidence suggests that the starting of clearances and new farms in the area was carried out largely by the new adults of a relatively stable local peasant population.

Few men accepted into peasant families are mentioned, perhaps four, possibly only one. Makarko Fedor's son shared a tenement with a peasant and was designated *vlazen'*.[4] In other cases, the term may have been used as a surname.[5] The term for a hired man (*kazak*) is only found once on the peasant lands and his lowly status is stressed by his being referred to only as 'Fedorko's hired man'.[6]

One entry relating to dependents is unique. 'Gorodok hamlet arose after the inquisition: in a tenement Vokhromei and Zakharka, children of Gornostai [Rozderishin]; in a tenement their peasant Oleshka Mikiforov...'[7] Here the term peasant clearly indicates dependence and seems analogous to the usage of 'people' to indicate villeins, the servile dependents of lords. In fact, this latter term is used in one instance of dependents of another Rozderishin. A hamlet called Chertezh had one tenement which was occupied by two 'people of Efimko Rozderishin'.[8] Pigasei, son of Filip Rozderishin, lived in a tenement in Dobsha hamlet; the second tenement there was occupied by his 'fellow' (*chelovek* – a term of disdain), Borisko.[9] Borovna hamlet, held by Rodivon, son of Efimko Rozderishin, had four tenements, one occupied by Rodivon, another by Davydko Yakhov and the other two by his fellows Istomka and Istomka Onishkin.[10] Similarly, in Dudino hamlet held by Semen, Rodivon's brother, one tenement was occupied by Semen and his children, the other two by his fellows D'yak and Chyurilko. These 'fellows', who usually have but one name and no patronymic, are analogous to the lords' villeins, servile dependents settled on the land and of small account, not deserving to have their progenitor noted. It might, in fact, be thought that the Rozderishins were not peasants, but lords. Certainly they have between them considerable holdings (see table 18). The editors of the Toropets register

---

[1] F. 141(53). [2] Ff. 129(41), 166(78), 201v(113v), 198(110).
[3] F. 216v(128v). [4] F. 122v(34v). [5] Ff. 146(58)–146v(58v).
[6] F. 189(101); see p. 180 above. [7] F. 171(83).
[8] F. 166v(78v). On the place-name see p. 22 above. [9] F. 170v(82v). [10] F. 172(84).

regard the section describing the Rozderishin holdings as being out of sequence and baldly state that 'a fragment of the description of lands held by service (the lands of the landowners Razderishin – ff. 170(82) – 173v (85v)) has been included in the description of the black lands'.[1]

The Rozderishin family, as far as it can be reconstructed from the information given in the register, is shown in figure 6. In reality the family was larger than this; no females are recorded and therefore all marriages of females into other families are unknown. Nevertheless, this is a large family. We have here at least eighteen males (twelve of whom were alive at the time of the inquisition); the largest of the families of lords holding by service, the Golenishchovs, had ten members, including two women, recorded.[2] Unfortunately, we lack adequate information on peasant family size. The structure of the family itself does nothing to indicate whether it was a peasant or a lord's family. The entries relating to their holdings, however, seem to indicate an answer.

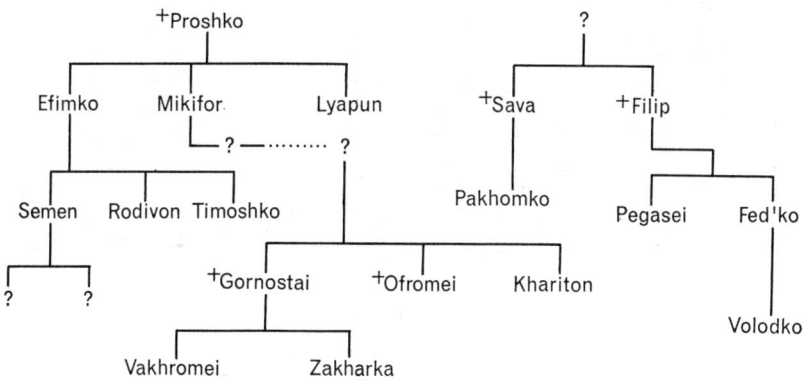

Fig. 6. The Rozderishin family.

The Toropets register, after the description of the peasant lands, has a heading: 'Villages and hamlets held by service by junior boyars in Toropets uezd'.[3] Each entry is in the following form: 'An estate held by service by so-and-so' (*pomest'e za*) or at least in the form of one which mentions holding by service (*pomest'e*). The Rozderishin entries are not in this form. They omit the term for a service estate (*pomest'e*) and usually read simply 'held by so-and-so' (*za*). This is not the formula used of other peasant lands which are listed under placenames: 'in such-and-such a village' or area. This might, therefore, be taken to indicate that the Rozderishin lands were not peasant; but it does distinguish their entries from those of lords holding by service. Moreover, in the case of entries relating to those holding by service the description of the

---

[1] *AE za 1963*, 278.

[2] See p. 191. Other members of the Rozderishin family are mentioned, but their relationship to those shown is obscure; f. 2v.   [3] F. 217(129).

TABLE 18. *Entries relating to Rozderishin holdings*

Heading	Settlement	Males in Tenements*	Notes	Folio
1 In Zbutsk	Chertezh hamlet	*Efimko* Proshkov Rozderishin *Timosh* Efimov Rozderishin	a	163v (75v)
2 New hamlets and clearances in Zbutsk	Chertezh hamlet	Leonidko and Ivanko, people of Efimko Rozderishin		166v (78v)
3 (In Zbutsk perevara)	Sychok waste			168 (80)
4 In Zbutsk perevara	Vavulina hamlet	*Mikifor, Lyapun*	b	169v (81v)
Hamlet held by Efim, Mikifor and Lyapun	Vavulina hamlet	Kudin'ko Ivanov		170 (82)
5 Held by Pegas and Fed'ko and their cousin Pakhomko	Moshnitsa hamlet	*Fed'ko,* his son *Volod'kya* and *Pakhomko*	c	170v (82v)
	Dobsha hamlet	*Pigasei* his fellow Borisko Fed'ko Isakov, Klimko Sergeev		
	Pustyn'ka hamlet		overgrown	
	Svyatitsa waste			
	Stepanko Okhobnev's clearance		empty	
	Tupitsyn clearance		d	171 (83)
	Ratnoe hamlet		empty f	
6 Held by Vokhromei and Zakharka	Gorodok hamlet	*Vokhromei, Zakharka* their peasant Oleshka Mikiforev	e	171v (83v)
	Bobrovka hamlet	Grishka Lasin Grid'ka Tolstoi Ondreev		
	Golovan's clearance	Golovan himself Loginko		
	Clearance of Levka	Levka himself		
	Glazov			
	Clearance of Gudinov land	Demidko Fed'ko		

			empty & overgrown	
7 Held by Rodivon	Clearance of Ivanko Kholmit		empty	172 (84)
	Borovna hamlet	*Rodivon* himself		
		Davydko Yakhov		
		his fellow Istomka		
		his fellow Istomka Onishkin		
		Davydko Semenov	f	
8 Held by Semen	Clearance of Rodivon Goryainkov			
	Dudino	*Semen Rozderishin* himself and children		
		Semen's fellow D'yak		
		his fellow Chyurilko		
He also has clearances:	of Kondrat Ermolin	Kondrat Ermolin himself		
	of Grid'ka Terekhov	Grid'ka himself, Petrushka Borisov		
		Gridka Vasilev		
		Fed'ka Klyapik		172v (84v)
		Filipko Yagoda		
	Dudkin clearance	Onisimko Lukin		
	Kolenidkov clearance	Goryainko		
			g	

* Each line indicates a tenement. Names of members of the Rozderishin family are in italics.
a Efim Prokof'ev was granted this waste with relaxation of obligations for 10 years from 1543 'and he is to plough up the arable in those ten years and establish tenements and summon people, and after this relaxation he is liable to his share with the volost peasants'.
b Hay not present on holding, but elsewhere, also an additional parcel of arable land.
c Additional arable and hayfield elsewhere, including a meadow of Vasyuk the post-horse man.
d Tilled from a distance.
e Additional arable and hayfield elsewhere.
f Additional hayfield elsewhere.
g Semen, Ofromei and Gornostai had additional arable and hayfield elsewhere; some of the hayfield was in place of some Semen had given to the Trinity church at Pustyn'ka in Kudinets perevara.

estate is followed by items detailing totals of settlements, persons, arable land and hayfield, as well as by items specifying the income derived from the estate and its value in terms of vyts (the *vyt'* was a tax unit). The Rozderishin lands lack all such entries. The fact that no income is listed for them and they are not recorded in terms of vyts seems crucial, since all entries relating to estates held by service in this register have such entries, while peasant lands do not. It seems, therefore, that we here have a fortunately surviving example of a family of peasant landed proprietors, substantial peasants akin to the English yeomen.

In fact, there is direct evidence of this in the register. In the section describing the town of Toropets, in the area 'about the church of St George' was a tenement 'of the *semtsy*, Semen and Rodivon, children of Efim Rozderishin, and of their nephews, Ofromei and Khariton Rozderishin'.[1] The term *semtsy* is probably a Pskov dialect form of *zemtsy*; the *zemtsy* were 'small-scale non-serving holders of heritable tenure'.[2] This type of tenure akin to allod in western Europe, seems to have been in decline at this time; it was characteristic especially of the Pskov and Novgorod area, and the area of free peasant tenure in the northern territories, and it may have been largely abolished by the government in the third quarter of the sixteenth century. Poboinin suggests that it might formerly have been commoner in the Toropets lands, but he regards it as a form of land-holding resulting from the distribution of lands to townsmen on advantageous terms in order to open up unworked land.[3] The evidence does not seem to support him in this view. The grant made to Efimko Rozderishin, for example, (item 1 in table 18) expressly stated that after the term of relaxed obligations, during which he was expected to establish viable tenements, he was then to be liable to his share along with the volost peasants. The general impression from the available information in this register is that this very large peasant family held, for some generations, several heritable tenements, including one in the town, and had dependents working for them.

In this area on the western borders of European Russia, settlements in dense forests in the mid sixteenth century were very small in size (see table 19). Clearance of forest for cultivation was heavy work, justifying relaxation of obligations for periods of ten years in some cases. We have seen that the closest correlation in the size of family farms in the area was with the number of adult males. This supports the view that labour was the crucial factor of production, due no doubt to the relatively simple and readily available means of production; capital investment was unimportant as long as techniques

---

[1] F. 2v.

[2] *PRP*, IV.141, 612. Cp. Novgorod *svoezemtsy*, on whom see Pomyalovskii, *ZhMNP* (1904), 7.94–135, esp. 120f.

[3] 'Toropetskaya starina', 62–3. For a general survey of this form of land-holding see Klyuchevskii, *Sochineniya*, VIII.207–18.

remained unaided by more complex and expensive implements and land was abundant. The labour of clearing land, however, was such that men were reluctant to abandon it; if some disaster prevented all aspects of the farm unit's work continuing, some aspects might be foregone, but others would be continued wherever possible. Even where a tenement or settlement was forced to move, as many as possible of the old activities were retained at the former site. A combination of regular tillage of patches of cleared land with both clearance and cultivation by means of fire resulted in the farm relying to a considerable extent on the resources of the forest. This was needed not only to supplement the demand for arable land, and as a possible refuge in case of disaster; it was needed especially for hay, in order to maintain the livestock without which cultivation could not take place.

The normal peasant unit was the household with at least one adult male. Family and local ties were probably of considerable importance; the usage of personal names in different ways for different social groups indicates this. But the male head of the household might well act despotically and 'in the extreme case hostile clashes between family members were resolved by the ease with which families could split up'.[1] It is sometimes possible to trace apparent relatives in tenements within a fairly restricted locality; but the nature of the register does not allow us to see the causes or means of achieving the implicit family divisions in any detail. The numerous patches of worked land, however, clearly sometimes offered opportunities for new family units to establish themselves away from the original home. There were few peasant newcomers to the area. The established local peasant was evidently the main force in internal colonisation.

Larger peasant families seem to have been somewhat above average in terms of status and economic situation. The Rozderishins were not only the largest family for which we have evidence, but also clearly a kin group which stood at the top of the peasant society of the area. Peasant society, even when, as here, not dependent on any immediate overlord, was differentiated. At the lowest levels were fellows and hired men; new arrivals were also evidently low in status, but might advance to the level of the labourer or the rank and file peasant by marriage. Post-horse men, and some others who had functional links with the town, were sometimes in an advantageous situation. The general impression, however, and it can be no more than that due to the limitations of the sources, is that while there was some social movement especially at lower levels, the higher levels of peasant society were fairly stable. The Rozderishins seem to have been in the area for four generations by the mid sixteenth century, going back, perhaps, to the first half of the fifteenth century. In 1596 there were members of the family who had become junior boyars, that is servitors, with the smallest allocation of land in return for service, 100 chets. In 1606 there were more than ten such members of the

---

[1] Poboinin, 'Toropetskaya starina', 318.

family recorded with allocations in the range 100–250 chets (in the Toropets area at that time junior boyars were granted from 100 to 600 chets).[1] Although by this time they had ceased to be peasants, they remained in a social situation which was little higher than that their forebears had had as *zemtsy*.[2]

LANDS HELD BY SERVICE

If we compare the information on settlement size in terms of number of tenements on lands held by service and on black lands, we can see that the differences are not very remarkable (table 19). The main differences appear to be rather in the types of settlement than in their size. Villages and manorial settlements, as is only to be expected, were commoner on estates than on the black lands. It is more surprising that hamlets are both more numerous and a higher proportion of settlements on the estates; this seems mainly to be because clearances are very much more common on the the black lands, and account for almost 15 per cent of all their settlements. Thus, estates seem to have settlements which are either villages or hamlets, the former often having an establishment of the lord and villeins in it, the latter being somewhat larger than equivalent settlements on the black lands. The black lands had fewer villages, but very many more clearances.

Small size of settlement was usual both on estates held by service and on peasant black lands. Nevertheless, there are some differences in the distribution of settlement size between the two types. A somewhat greater proportion of settlements on black lands are of very small size than of those on estates held by service. Less than 10 per cent of all settlements had five or more tenements each.

Nearly fifty allocations of land held by service are listed in the register, but the junior-boyar families holding by service were fewer than this since several members of one family might have allocations. Various members of the class of servitors exchanged parts of estates between one another; such exchanges were noted by the officers of the inquisition. In order to illustrate some of the complexities of tenure let us take two of the largest families recorded, the Chikhachevs and the Golenishchovs. Some details of their estates are given in table 20.

There is little that can be said about the Chikhachovs as a family. The five members of the family holding by service were all sons of Ondrei Chikhachov; one of the brothers, Fedor, had two sons, Danil and Mikhail. No other family members appear to be mentioned. Fedor had formerly lived at a place which at the time of the survey was a site (*selishche*) of

[1] Poboinin, 'Toropetskaya starina', 63.
[2] Cp. Klyuchevskii, *Sochineniya*, 8.218, who stressed that *zemtsy* differed from junior boyars 'though they stood no lower than they did'.

TABLE 19. *Settlement size, Toropets uezd*
(Number of settlements)

	total	empty	*with stated number of tenements*														
			1	2	3	4	5	6	7	8	9	10	11	12	13	15	19
A. Lands held by service																	
Villages & manorial settlements	24		1	1	2	2	4	6	3	1							
Hamlets	857	53	361	208	116	59	29	20	6	3	1		2		1	1	1
Clearances	19		17	2													
Totals	900	53	379	211	118	61	33	26	9	4	1		2		1	1	1
Percentages*			44.8	24.9	13.9	7.2	3.9	3.1	1.1	0.5	0.1		0.2		0.1	0.1	0.1
B. Black lands																	
Villages	12			1	1	1	1		1	1		2		1		1	
Hamlets	675†	17	385	139	60	38	19	8	2	2	1	1	1	1			
Clearances	121		103	16	2												
Totals	808†	17	488	156	63	39	20	8	3	3	1	3	1	2		1	
Percentages*			61.9	19.8	8.0	5.0	2.5	1.0	0.4	0.4	0.1	0.4	0.1	0.3		0.1	
TOTALS A + B	1708†	70	867	367	181	100	53	34	12	7	2	3	3	2	1	2	1
Percentages*			53.0	22.4	11.1	6.1	3.3	2.1	0.7	0.4	0.1	0.2	0.2	0.1	0.1	0.1	0.1

* Percentages have been calculated excluding those empty and for which no number of tenements is stated.

† includes 3 for which no figure is given.

TABLE 20. Estates of two families holding by service

folio		manorial settlements	hamlets	tenements	villeins	peasants	arable land (chets)	hay
	Chikhachevs							
17(134)	Fedor & children	3(0)	13(2)	21+2	3(0)	25(1)	320(49)	2096
54(172)	Fedor & children (suppl.)		4	19		27(26)	71	241(240)
20(137)	Mikita	2	17	49+1	7	66	432(438)	1650(1645)
21v(138v)	Mikita (suppl.)		3	10		12	45	140
255	Ofonasi	1	26	59(52)+1	6	52	260(263)	1382(1367)
256v	Ofonasi (suppl.)	1	5	18+1	(3)	(15) ⎱(19)	97	294
260v	Petr & Elizarei	1	34(31)	64+2	12(11)	55(58)	455(466)	1220
	TOTAL	8	102	240+7	31	252	1680	7023
	Golenishchovs							
35(153)	Matfei & Ivan	2(0)	34(36)	60+2	11	82	439(441)	1419(1609)
35(153)	Matfei & Ivan (suppl.)			4	4		9	20(10)
61v(179v)	Stepan		27(26)	45(43)	(6)	(51) ⎱(58)	215(188)	671(691)
63v(181v)	Kostantin & Vasili	1	12	25+1	4(5)	34(27)	145(148)	533
65(183)	Ivan & mother		6	21+1		24(23)	113(103)	135
71v(189v)	Ivan & mother (suppl.)	2	76	117(115)+3	16	113	616(619)	1990(1980)
	TOTAL	5	155	272+7	41	304	1537	4768

Notes: Figures shown in brackets are those totals given in the source which differ from totals obtained by addition. The description of Fedor Chikhachev's estate is incomplete or there is a lacuna at the start of the entry.

Ondrei Koshnev.[1] Fedor's supplementary grant, intended to bring his allocation up to standard, was $6\frac{1}{2}$ vyts, half of an estate originally held by Laptev Il'in which had then been granted to Stepan Zeleny.[2] Stepan Zeleny's allocation, in its turn, was made up to 13 vyts by an additional grant of $6\frac{1}{2}$ vyts.[3] Mikita's allotment was in Kazarino volost and in Zales'e (lit. beyond the forest) and details are given under five placename headings. His supplementary grant came from the estate of Vasili Zeleny, a brother of Stepan Zeleny.[4] Ofonasi's lands were in the same volost and are also listed under five placenames, but he had also exchanged land, with three hamlets on it, with Ivan, son of Ivan Baida or Zeleny.[5] The last two brothers had no supplementary grant, nor do they seem to have been involved in any exchanges. Thus, the Chikhachovs seem to have held land in proximity to one another and to have exchanged land in some cases with members of the Baida or Zeleny family who were their neighbours. This is clear from an entry on Baida's estate:

And Baida has a boundary with Ofonasi, son of Ondrei Chikhachov and with his brethren by his old estate and with Ofonasi's supplementary grant and the Zagorodno land from prince Oleksandr's Vanbol'sk land down the Serezha to the prince's Sel'ski land; on the right-hand land, forest, pastures, meadows and appurtenances of Baida, and on the left, land, forest, pastures, meadows and appurtenances of Afonasi Chikhachev and his brethren.[6]

The Golenishchovs are a little more interesting as a family. The members mentioned are shown in figure 7. The brothers Matfei and Ivan jointly have a grant but they lived in separate hamlets; their estate, like that of the Chikhachovs, was in Kazarino. Stepan, Kostyantin and Vasili, and also Ivan and Anna, his mother, had their estates in Serezha perevara.[7] Kostyantin and his brother Vasili lived in one house, as did Nastasya, widow of Timofei, and her son Vasili.

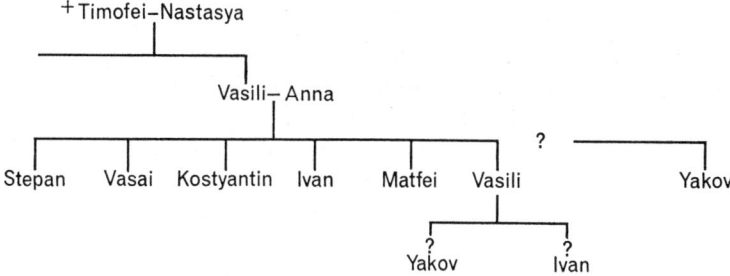

Fig. 7. The Golenishchov family.

[1] F. 16v(133v).   [2] F. 53v(171v).   [3] F. 56(174).   [4] Ff. 21(138), 55(173).
[5] Ff. 254, 255; 261–3.   [6] Ff. 264v–265.
[7] Ff. 12, 15v(149v), 32(150), 59(177), 62v(180v)–64v(182v).

In both families, as in others in this area, rights in forest and in the lakes for fishing were shared with some other member or members of the family, and also with peasants. In some cases the register shows concern for a standard allocation to these junior boyars in terms of vyts. For example, Kostyantin and Vasili Golenishchov held 15 vyts, and, the clerk added, 'no supplementary grant has been given, but they are to be given 19 vyts'.[1] Thus, it seems that around 1540, in Toropets at least, the system of centrally allocated holdings in exchange for service which became fully developed in the second half of the sixteenth century was still growing. Moreover, the officers of the inquisition do not seem to have been particularly rigorous in attempting to grant a standard norm to each tenant holding by service.

Each servitor lived on his estate and any settlement where he had an establishment was known as *sel'tso* (translated here as 'manorial settlement').[2] Either in the tenant's own establishment, or in separate tenements were his 'people' (*lyudi*) or villeins who are normally distinguished from peasants. In the two large families of tenants holding by service we have just looked at, such villeins amounted to about 12 per cent of the number of peasants on these estates; the average on estates held by service was 8.9 per cent. However, on estates the term 'people' is also used, at least in some cases, to indicate dependent peasants. Sometimes it seems that servile dependants on estates held by service were used particularly when land was being extensively worked or opened up. If, for example, we look at the estate held by Elizarei and Petr Chikhachev, which had an exceptionally high proportion of villeins in its labour force (about one villein for every five peasants) we find much evidence of such activity. Petr Chikhachev himself lived in an establishment in a village with a church situated above Lake Toropets. Apart from his own dwelling, and that of the priest, there were two tenements occupied by his villeins 'and the Dubrovka hamlet above Lake Toropets has been let into the field of that same manorial settlement, and another Dubrovka hamlet, and from that same manorial settlement they till and mow the Bolalobova site, 50 chets of arable per field and 306 ricks of hay'.[3] An empty hamlet above the lake was tilled from a distance by Petr's villeins, as was the empty hamlet of Yuri Ivanov.[4] Elizarei lived in a hamlet which had in it his own dwelling and the tenement of one of his villeins with 12 chets of arable and 70 ricks of hay. A separate hamlet had one tenement with a villein of Elizarei in it with 2 chets of arable and 5 ricks of hay. 'And Elizarei also set up a clearance, an out-settlement (*vystavok*) from Parfenko Olyukhin's hamlet; in a tenement

[1] F. 64v(182v).

[2] This can cause difficulties since in the totals the manorial settlement is sometimes included in the number for a specific type of settlement (e.g. village or hamlet), thus altering the total.

[3] Ff. 257–257v. The average amount of arable for all Chikhachev tenements was just over 7 chets, and the number of ricks of hay was 29 (see table 20).

[4] Ff. 258–258v.

Ivolko Olyukh Svin'in without arable.'[1] The two brothers also had two sites and a waste which, although overgrown with forest, produced 150 ricks of hay.[2]

The great bulk of the estates of lords holding by service, however, consisted of peasant tenements. On average an estate had nearly 40 tenements, the average settlement consisting of two or three tenements. There were remarkably few clearances, but wastes and sites (which provided hay and, presumably, other fodder and products of gathering activities) amounted to about a quarter the number of tenements. Thus estates held by service were not, in general, particularly active in expanding by internal colonisation. The number of newly established settlements on such estates was 19, compared with a total of 900 of all types.

The servitors holding estates derived their income by exactions from the peasants on the estates; the exactions were in money and in produce. The register, however, lists three types of 'income' and, additionally, specifies other assets of the estate (normally forest and fisheries, occasionally items such as clay for pottery). Income in money is self-explanatory. Income in grain included, apart from raw grain, processed grain in the forms of buckwheat meal and of malt, and also the item 'peas and hemp' (presumably the oily hemp seed was intended here, not the fibre); all these were thus items of food. Flax, too, was listed more frequently under grain income than under petty income, probably since it, too, was here as an oil seed. Petty income included the produce of livestock husbandry (meat, hides, fleeces, butter and cheese) and of gathering (hay, game and honey).

The items of produce exacted from the peasants on the estates are not always the same, nor are they always recorded in the same order. It seems very difficult to establish whether, in fact, a series of possible patterns of exaction was in use. Flax, for example, may occur in a different position according to whether it is regarded as a food crop or as petty income (presumably as fibre). Moreover, certain items are sometimes omitted. Salt is only once listed; buckwheat is omitted 19 times, hay 4 times, hares and grouse 3 times, hens and hops twice, peas and hemp, and malt once each. Some of these omissions may be oversights, of course, but the frequency with which buckwheat is omitted suggests that it may not have been grown everywhere. The overwhelming majority of items, however, seem to have been exacted on each estate.

The rates at which exactions of various items were made seem to have been proportionately related to one another. Usually, for every 16 chets of oats, 8 of rye were to be delivered, 4 of malt, 2 of wheat and of hops, and 1 of buckwheat meal and of peas and hemp.[3] For every sheep, side of meat, hare, grouse, cheese, sheepskin, fleece and load of hay, two hens had to be handed

---

[1] Ff. 260–260v. Parfenko Olyukhin and the occupant were presumably related.

[2] Ff. 260v.   [3] The pairing of one unit of rye and two of oats was known as *yuft'*.

over. It appears to be the butter figure which may have formed the basis for the actual measurements, since this figure multiplied by ten most frequently gives the figure for sheep etc. (and from this all the others may usually be derived with maximum accuracy).[1] This curious phenomenon may perhaps be due to the fact that it would be much easier to measure a *grivenka* (one pound) of butter than a chet of rye; moreover, the livestock resources might well be a fairly accurate indicator of the farm's wealth.[2]

In a formal sense the amounts exacted were determined according to the number of vyts on the estate.[3]

Unfortunately, the available evidence does not enable us to see what determined the number of vyts for each estate. While the amounts of money, of grain and of petty income appear in most cases to be interrelated in broad terms, the number of vyts does not appear to bear any close relationship to the size of the estate as indicated by the number of settlements, or of tenements (though this figure is usually the closest) or of males (peasants and villeins). Veselovskii pointed out that the vyt tax unit varied in content between over 9 and 15 chets of arable land, while the number of vyts in the sokha-unit ranged between $29\frac{1}{3}$ and 72.[4] 'Equally, there is no correspondence between allocations in the sokha-units and vyts, on the one hand, and the number of tenements and people in them, on the other. Vyts and sokha-units in this register differ individually.' Thus, we have here a stage in the development of the state administration and taxation system which had not yet reached the more uniform condition achieved after the reforms of the mid sixteenth century. It is thus impossible with any degree of certainty to assess the burden of exactions of various sorts on the peasant holdings on the estate. Nevertheless, the general impression is that the amounts per tenement were quite small, on average less than one-and-a-quarter chets of rye, for example.

LINKS WITH TOROPETS TOWN

In Toropets volost the 'hamlets of townsmen of Toropets near Toropets town around lake Toropets which are not subject to any perevara' have fifteen names of original holders.[5] None of these is found in the names noted

[1] I am grateful to Michael Pursglove for calling my attention to this.
[2] The existence of terms such as Icelandic *smérlinga*, (colloquial *smérrentur*) and Danish *smörgåld* suggest that perhaps these speculations are not as far-fetched as might at first sight appear. I am indebted to Jakob Benediktsson for the information that in medieval Iceland butter (*smjör*) had a fixed price; as a means of payment, a *fjórðungr* (10 Icelandic pounds or about 5 kg) was valued at 20 *álnir* (originally ells of cloth) or 1/6 the value of a cow. In normal use butter was used to pay for hired cattle and sheep; the payment for one *hundrað* (one cow or six ewes) was one *fjórðungr* of butter. This rent was called *leigur* or *smjörleigur*. If we assume a fat content of 2.5%, a yield of about 43–5 gallons would probably be needed to produce a *fjórðungr* of butter.
[3] On f. 15 this is made quite clear.   [4] *Soshnoe pis'mo*, II.341–2.   [5] F. 160v(72v).

in Toropets town. In fact, however, the first name mentioned, Golash Esipov, or Osipov, is found in a tenement in Strezhino. In Ogafonov *zaselitsa* (a term of uncertain meaning in this context), Strezhino perevara, in Ogafonova hamlet of Osip Ontsyforov, Golash Osipov and his brother, Fed'ko, held a tenement and also mowed a site, the seat of Mikhal Osipov and his brother Matveiko. There are other hints that this area had a number of more than usually substantial peasants. The relations are complicated and the data are unclear, but it seems that the families of Fed'ko Popryatov, a townsman, and two brothers, Zik and Osif Prokoshev, held several tenements, some newly established, and some worked from the town; they supplemented these tenements by working a number of other items of land; Zik also had a crop-sharer.[1] There thus appears to be a link with town-dwellers, though the details are obscure. The families concerned perhaps lived in the area near the market and the main street; several names found here may be related to those in the countryside.[2] Moreover, a certain Ivanko Goloshov, possibly a son of Golash Osipov, had two tenements in this part of the town.[3] A Vas'ka Golashov had another tenement nearby.[4] The same district had tenements held by a trader in wax, a blacksmith, a saddle-maker and a quorn-maker, all trades likely to have links with the countryside.[5] Finally, another Golashov, Onisimka, had a hamlet in Ovzha; it had one tenement and in it Kostya Ortemov, a townsman 'who additionally mowed a site'.[6] The same Kostya Ortemov had the next hamlet which he tilled from the town; he, too, was found in a tenement, near the market, which he shared with his brother.[7]

Ovzha was included in Zbuttsk perevara where there were a number of 'hamlets of townsmen of Toropets', though, again, no other names have been found in the town.[8] In Kamenka, in the same perevara, another townsman had a hamlet and shared a churn mill with the next hamlet.[9]

There appear to be other connections between Zbuttsk perevara and the town. In a district of Toropets town by St George where a number of families of lords such as the Chikhachevs as well as members of the Rozderishins, who were substantial peasants, lived, there was a tenement 'of Shchur Karpov and Kutya Pershin and Trofimko Nefed'ev and Grisha Ondreev and Fed'ko Kalikhin from Porech'e perevara...'[10] Shchur Karpov also had a tenement in another part of the town, where he had a hut (*yzba*) evidently with a dependent in it.[11] Next to his tenement was another one held by Kutya Pershin. Shchur Karpov was granted in Zbuttsk perevara empty tenement places with a remission of various obligations for ten years. After this term, he was to be liable to impositions on a par with the other members of the volost.[12] In the same area Vasyuk Yakimov was granted the same privileges;

---

[1] Ff. 208(120)–211(123). See also pp. 181f. above.   [2] Ff. 107v(19v)–108v(20v).
[3] Ff. 109(21), 110v(22v).   [4] F. 108v(20v).   [5] Ff. 109(21)–110v(22v).
[6] Ff. 163(75)–163v(75v).   [7] F. 108(20).   [8] F. 162(74).   [9] F. 165(77).
[10] Ff. 2–2v.   [11] F. 115v(27v).   [12] Ff. 169(81)–169v(81v).

this man was probably the Vas'ka Yakimov who had a tenement near the grain stores.[1]

The tenement next to his in the town was held by Vas'ko Sopronov. The Sopronovs also appear to have been a family with holdings both in the town and in the countryside. Fedor Sopronov, together with at least four other relatives, men of the artisan quarter, had a tenement in the town. Another Sopronov, Dorofeiko, was one of six artisan quarter peasants who had a tenement in the same district.[2] Vas'ko Sopronov had a tenement on a street between the grain market and the Transfiguration of the Saviour.[3] Fed'ka Sopronov was in a hut of the tenement of another man in another district of the town; he held a full tenement in Modentsa hamlet in Lashira. In Seredka, Toropets perevara, Sopronka (probably the originator of this family) and his brother Karpik, 'children of Demid Okulov' had a tenement.[4] In another street in Toropets Ivan Sofronov, a clerk, had a tenement; in this street there was also a member of the Rozderishin family and a post-horse man as well as the tenement of Ustimko, 'son of Sofron'.[5]

The general impression, unfortunately the data does not allow us to be more specific than this, is that there were a number of individuals involved in activities, either individually or through their family, in both town and countryside. Even if we leave out of account the activities of lordly families, such as the Chikhachevs and the Golenishchovs, as well as the intermediate categories represented by the Rozderishins, there are still probably a few dozen peasants with holdings in both town and countryside. These links, however, appear to be somewhat narrowly circumscribed. The areas involved are: Porech'e, Buttsk, Ukleino, Startseva and Lashira; all are close to Toropets town itself.[6] It also seems probable that at peasant level two main groups were involved in these links. First, those involved in various trades and artisan activities relying on raw materials derived from farming and gathering, and second, the post-horse men who appear in several instances to have been somewhat wealthier than the average peasant.

---

[1] F. 112v(24v).   [2] Ff. 9–9v.   [3] F. 112v(24v).   [4] F. 122v(34v).
[5] Ff. 115(27)–115v(27v).   [6] Ff. 2, 2v, 5v, 6v, 8, 8v, 9, 10v.

Plate 1. Sowing and reaping, 16th century.

Plate 2. Sokha–harrow, *c.* 1600.

Plate 3. Sokha–harrow, *c.* 1600.

Plate 4. Slash and burn, from a life of St Antony of Siya,
16th century.

Plate 5. Slash and burn from a life of St Antony of Siya,
16th century.

Plate 6. Sowing and harrowing, *c.* 1600.

Plate 7. (*a*) Mill on high timber platform, 17th century (?).

7. (*b*) Modern mill in Yur'ev raion, Ivanovo oblast.

Plate 8. Seine fishing, four-timber fence, 16th century.

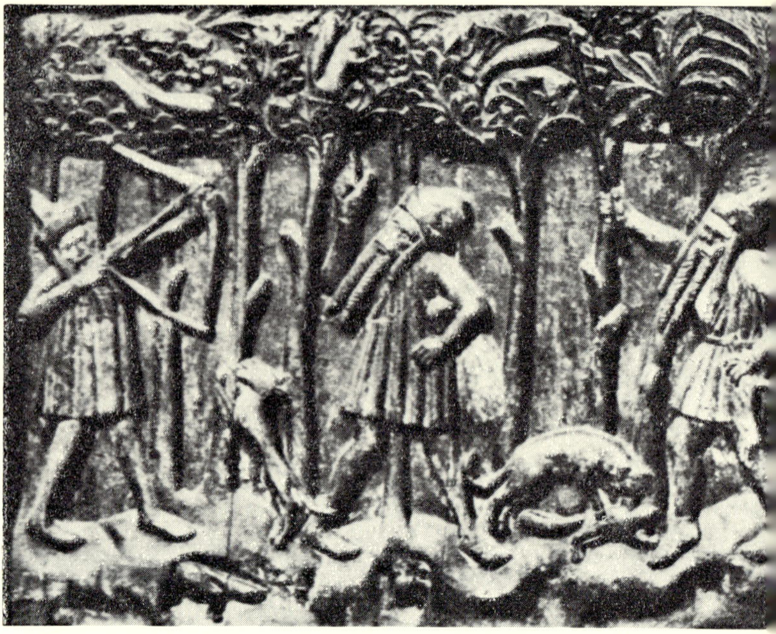

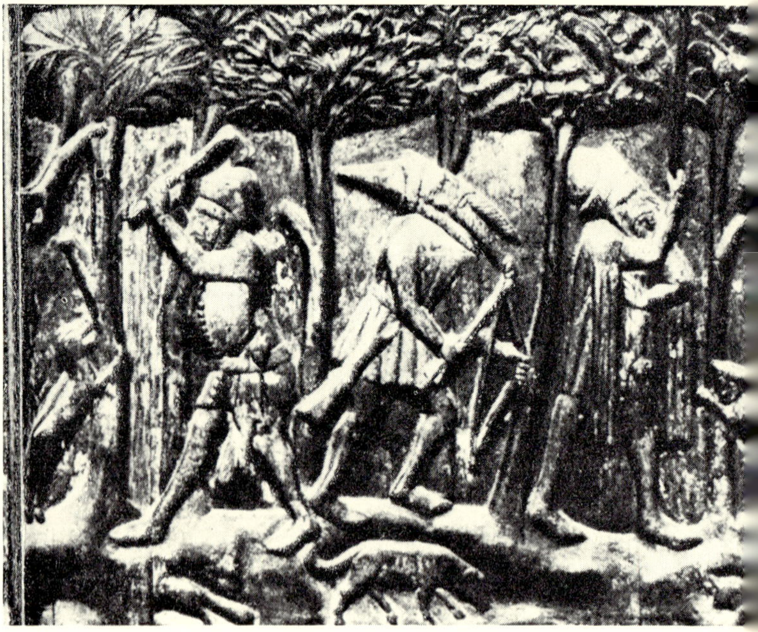

Plate 9. Squirrel hunting and the fur trade. Wood
carving in a Stralsund merchants' hall, 14th century.

Plates 10 and 11 show plans from the middle or second half of the seventeenth century. They reflect the continued existence of central villages, surrounded by cleared fields, usually three, and by forest dotted with wastes or clearances, sometimes with 'land' nearby.

Plate 10. This plan shows a village, Sofarino, with three ill-defined fields around it. Forest encloses this complex except at two points where other settlements impinge on it; within the forest a number of wastes are indicated by circles.

Gerasim Semenovich Dokhturov, who held the village and hamlet shown on the right hand edge of the plan (18, 19) was an official in various state administrative offices between 1650 and 1678.

1 Sofarino village; 2 Field of 35 desyatinas; 3 Field of 27 desyatinas; 4 Tenement which was held by heritable estate holders; 5 38 sazhens; 6 34 sazhens; 7 Pond on the river Talitsa; 8 Meadow; 9 Rakhmantsovo village; 10 Rakhmantsovo field; 11 From the Talitsa stream to Rakhmantsovo field; 12 Land and forest of Davyd Derabin; 13 Grigorieva hamlet of Davyd Derabin; 14 Kleiniki hamlet of the metropolitan's estate; 15 Kolomino waste of the metropolitan's estate; 16 Land of Kolomino waste; 17 Land of Garasim Dokhturov; 18 Svarobino hamlet of the clerk of the council, Garasim Dohkturov; 19 Kozhukhovo village of the clerk of the council, Garasim Dokhturov; 20 Land of the metropolitan's estate, of the two Volkovo wastes; 21 Big Volkovo waste; 22 Little Volkovo waste.

Plate 11. This plan of Mikhailovskoe village, Zvenigorod uezd, to the west of Moscow, shows the central settlement, the churches dominating the boyar dwelling, and numerous cultivated patches in the surrounding forest; in some parts the clearances have resulted in the apparent elimination of the forest. I. D. Miloslavskii was a boyar from 1648 to 1668. B. I. Troekurov was a chamberlain from 1653 to 1673. The map therefore relates to the period 1653–68. (I am most grateful to R. O. Crummey for supplying me with information on these two noblemen.)

1 Land of the boyar Il'ya Danilovich Miloslavskii, of Anninskoe village; 2 Land of Elisei Verigin, of the Palaumova waste; 3 Land of Iakov Kakoshkin of Old hamlet; 4 Land of Mikita Berazhnikov; 5 Land of Fedor, son of Mark Pozdeev, of Pokrovskoe village; 6 Land of Saveli Danilov; 7 Land of the chamberlain, prince Boris Ivanovich Troekurov, of Nikolskoe village; 8 Land of Onofreevo monastery; 9 Wastes of that monastery; 10 Road to Mikhailovskoe village from Onofreevo monastery; 11 Mikhailovskoe field; 12 Church of the Emperor Constantine and his mother Helen; 13 Church of the Archangel Michael; 14 Church of St Nicholas the Wonderworker; 15 Boyar tenement.

15
16
17
105
156
2 Поле
18
21
20
2 Поле
4
5
6
19
1
22
7
14
3 поле
8
13
10
9
11
12

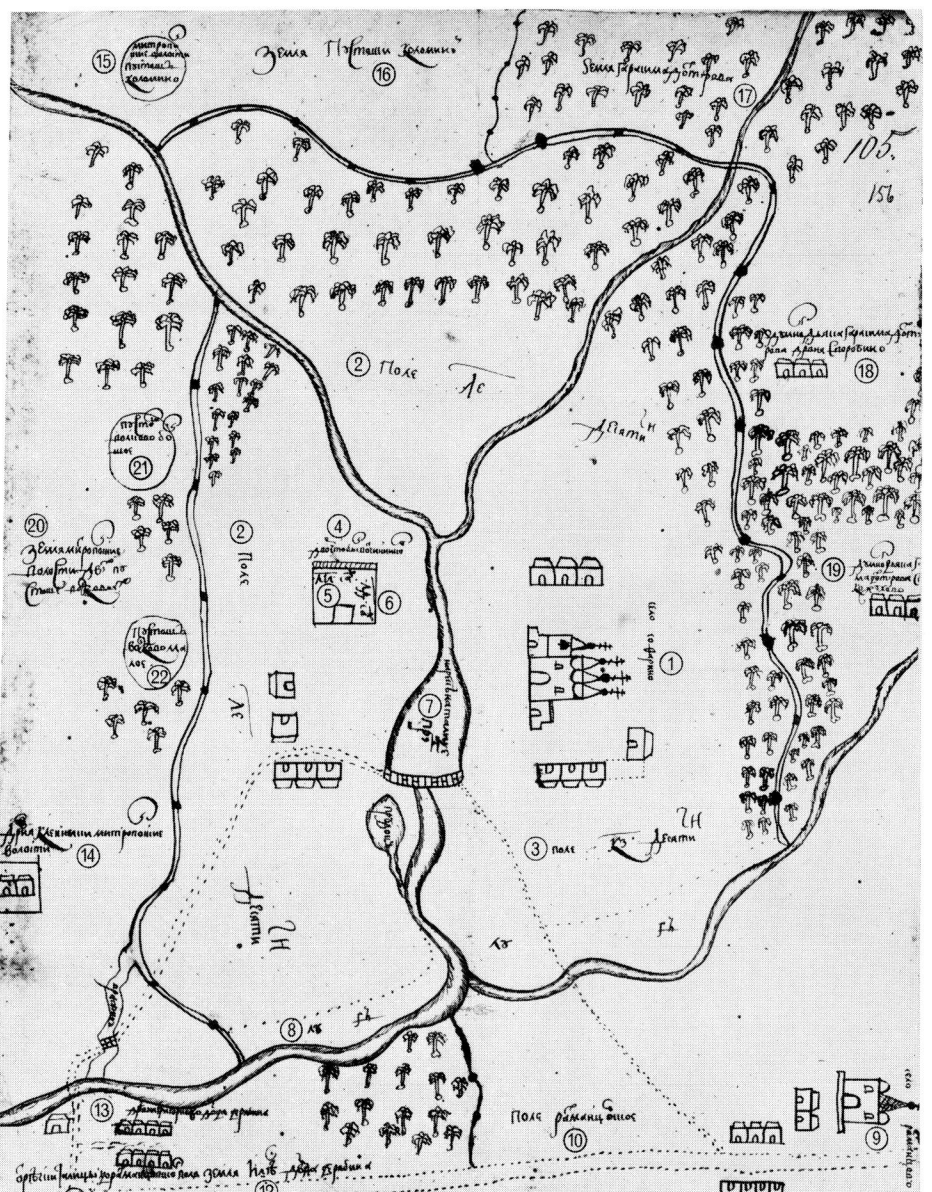

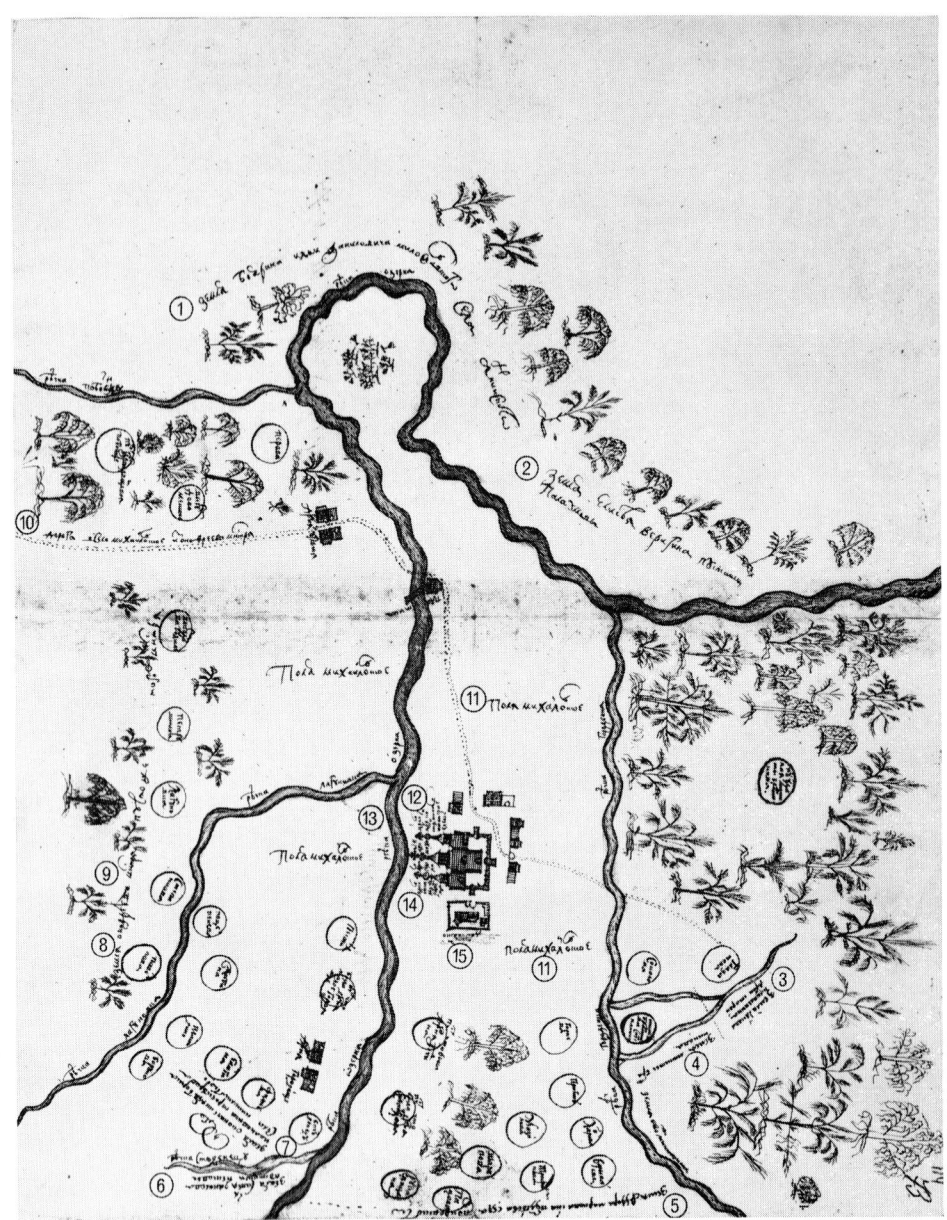

II

9

Kazan'

The Mongol invasions of the thirteenth century perhaps called a halt to the advance of East Slav venturers along the mid and upper Volga; Nizhnii Novgorod had been founded in 1221, but expansion and colonisation seems subsequently to have been retarded. In the second half of the fourteenth century, however, when there is much other evidence of a general Russian recovery, Vyatka was seized – from about 1375 it was a Russian base in the centre of a rich fur-getting area. Russian settlements arose south of Nizhnii Novgorod in the Sura basin and contact was established with the Chuvash (though they were not yet distinguished as such by the Russians). The Sura long remained a frontier or contact zone for Russians and Chuvash.

In the mid and late fifteenth century Russian princes were sometimes in alliance with, but often in conflict with, the Kazan' khanate which was formed in 1445; the population of the khanate was mainly Tatar, but included Chuvash, a Turkic people, and some other groups (Mari, Udmurts) of Finnish stock. These complicated and obscure relations are associated with the dynastic struggles both of the Russian princes themselves and, on the other hand, of the Tatars during the decline of the Golden Horde. These relations were further complicated by the existence of the Crimean khanate, and behind it the Turkish sultan, in the south and by Polish and Lithuanian efforts to use the khanates as allies against Moscow. In 1439 raiding Kazan' Tatars had reached Moscow itself; they captured the Moscow Grand Prince, Vasilii, at Suzdal' five years later. The advance of Asia on Europe in the fifteenth century, however, was much more successful and enduring in the south. Constantinople fell to the Turks in 1453. In the forests of European Russia the attacks from the east, though destructive, and costly in terms of the defensive effort that had to be made against them, resulted in no lasting successes in this period. By the early sixteenth century, the initiative had clearly passed to the Moscow state. Russian forts were established in the Volga region: in 1523 Vasil'sursk, later known as Sura, at the mouth of the river of that name; in 1551 Sviyazhsk at the mouth of the Sviyaga. Sviyazhsk was intended, but failed, to rival Kazan' in dominating the mid-Volga area. In 1548 an attack was mounted by the Russians against Kazan', but was largely unsuccessful, as was another intervention in 1550. The capture of Kazan' in 1552 was of crucial importance and led rapidly to Russian domination of the

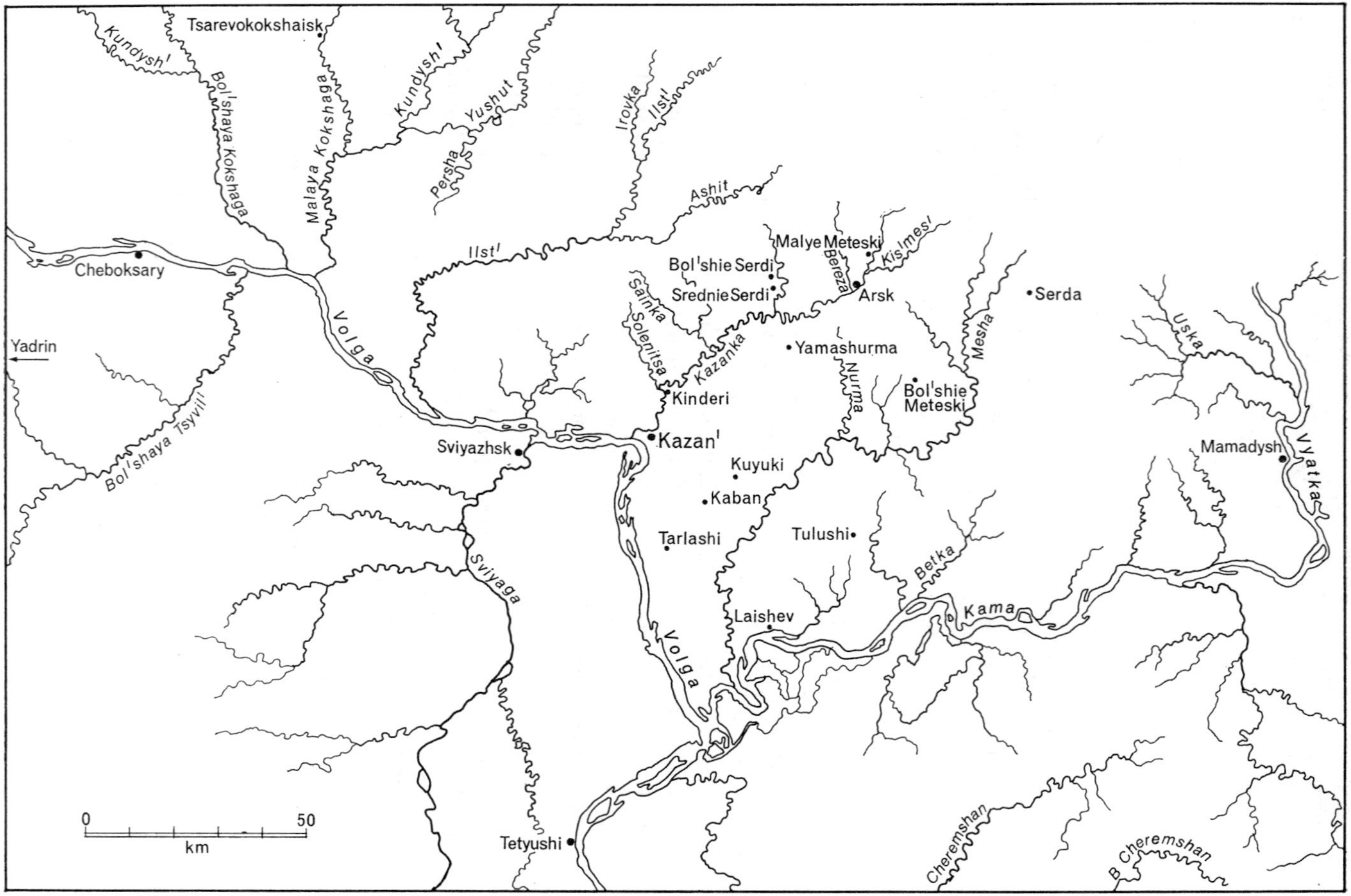
Tsarevokokshaisk
Kundysh'
Bol'shaya Kokshaga
Malaya Kokshaga
Kundysh'
Yushut
Persha
Irovka
Ilst'
Ashit
Ilst'
Cheboksary
Yadrin
Volga
Bol'shaya Tsyvil'
Sviyazhsk
Sviyaga
Tetyushi
0
50
km
Malye Meteski
Kis'mes'
Bereza
Bol'shie Serdi
Srednie Serdi
Arsk
Sainka
Solenitsa
Kazanka
Kinderi
Kazan'
Yamashurma
Nurma
Bol'shie Meteski
Mesha
Serda
Kuyuki
Kaban
Tarlashi
Tulushi
Laishev
Volga
Betka
Kama
Uska
Mamadysh
Vyatka
Cheremshan
B Cheremshan

Map 4. Kazan' uezd.

entire Volga route; Astrakhan, at the mouth of the Volga, was captured in 1556. The capture of Kazan' should be regarded as a major turning point in European history. Within a century Russian settlements extended across Siberia and there were Russian colonies on the Pacific coast.

The Kazan' area was a frontier zone on the borders of forest and steppe. Extensive mixed woods, mainly oak and birch, were interspersed with blocks of open steppe. In the late sixteenth century, after the conquest of Kazan', the Russian surveys give information which shows widespread forest 'scattered around the fields' and 'scattered along the area beyond the fields'.[1] Apart from the very numerous references to oak and birch woods, there are many to lime trees[2] and to pines.[3] Elm (*vyaz*) and wych elm (*ilm*) seem to have been somewhat less common and perhaps more locally restricted.[4] Willow was apparently almost as common as elms.[5] Aspen, maple, spruce, alder and silver fir were fairly uncommon.[6] The black poplar was also found.[7] This evidence on tree species is taken from a register of boundaries to estates and may not truly reflect actual proportions of undisturbed forest, but it shows the importance of oak, birch, lime and pine, and the mixed nature of the forest.[8]

The area is divided by the major rivers, the Volga (here mostly about 10 km wide) and the Kama. The south-western part of the area lying on the right, the high, bank of the Volga was known as Tau yagi (the Hill Side). It slopes towards the north-east, descends sharply to the Volga, and is intersected by the Sviyaga; the whole part, though less to the south and west of the Sviyaga, is much intersected by gully erosion which was already noticeable in the sixteenth century. This suggests that the area may have long been tilled; certainly there is some evidence for a relative abundance of grain and livestock at the time of the Russian attacks on Kazan' in the mid sixteenth century.[9] In order to be near water and hay from the large water-meadows most settlements seem to have been on lower ground, usually the second river terraces.

The south-eastern part, the Trans-Kama, between the low, left bank of the Volga and the Kama is mainly steppe. The western section of the Trans-Kama is a broad plain intersected by a few river valleys and with little gully erosion. The eastern section is higher; the Bugul'ma plateau lies in the south-east of

---

[1] TsGADA, fond 1209, kn. 152, folios 114v, 115v; 116 etc. Further unspecified folio references in this section are to this source.

[2] Ff. 220, 220v, 225, 226, 228, 228v, 230, 231, 232v, 233, 235, 238v, 239, 240, 243v, 247v, 248, 248v, 250, 252v.   [3] Ff. 219v, 223v, 224, 229, 230v, 231, 231v, 232, 234v.

[4] Ff. 225v, 230, 233, 234, 235, 236, 237, 237v, 240, 243v, 244, 250v, 233v.

[5] Ff. 223v, 224, 228v, 229, 232, 232v, 234, 242v, 248v, 249v.

[6] Ff. 231, 232, 233v, 234v, 238, 248v, 254v (aspen); 230, 233v, 234, 243v (maple); 229v, 234v, 235, 244, 244v (spruce); 229, 251v, 255v (alder); 221 (silver fir).

[7] Ff. 242v, 248, 251.

[8] Oak is commonly referred to when forest is specified: see pp. 207, 209, 212 below.

[9] Fennell, *Kurbsky's History*, 31, 33.

this part. There are deeper river valleys and steep slopes, but the river network is less dense than elsewhere, so watersheds are more extensive, sometimes 20–30 km across. In the north, along the Kama and towards its junction with the Volga, there were black-earth soils, and such fertile soils were also found throughout the southern part of the region.

The area north of the Volga and the Kama, and as far east as the Vyatka, was called Kazan' Arti; it is somewhat disected by the rivers Ashit, Kazanka, and Mesha; the steep slopes come together in the north-east and descend sharply to the Vyatka. The watersheds here are small. Kazan Arti was evidently well populated and cultivated; black-earth type soils extended along the Kazanka and Mesha. The rivers had good water meadows and were rich in water fowl, especially geese; both rivers and lakes had many fishes. The forest here was coniferous, mainly spruce, with pines found only on glacial sands. This was the core of the Kazan' khanate.[1]

Climate differs appreciably from the other regions examined. The maximum monthly average temperature (18–19°) is higher than elsewhere, but winters are as cold as in the north, reaching $-16°$ in the coldest month. Although the frost-free period (191–5 days) approaches that of Moscow (218–51 days), spring and autumn are both shorter than in the other regions dealt with here. The climate is thus appreciably more continental than in our other examples; moreover, evaporation (420–50 mm) approximates to the amount of precipitation (450–500 mm) and there is a consequent threat of summer drought.

Kazan' itself lay on the Volga trade route; it handled northern furs, eastern silks, and local fish, grain and hides, as well as spices and gem stones. At the time of the Russian attack in 1552 there were said to be 5,000 foreign merchants in the town.[2] Agriculture had long been practised here by all the ethnic groups of the region, but the Tatars also exacted dues in furs (*yasak*) and other produce from the Chuvash, Mari, Mordva and Udmurt. Farming, however, differed in certain respects from what we have seen in Moscow and Toropets; livestock husbandry was much more important. Gathering, too, was well represented, especially, in some parts, hunting and fowling; gathering bee products was also widespread.

The main primary sources for the history of farming in the area are the surveys and registers of various types compiled by Russian officials after the conquest of 1552. Unfortunately no complete account of these sources seems to be available, but what is known suggests that there were at least two periods of considerable activity of this type. First, a number of surveys and registers of inquisition were compiled in the 1560s. These included registers of lands held by service, registers of mills and boundary surveys. The main source used

---

[1] Further information on the physical environment may be found in Semenov, *Geografichesko-statisticheskii slovar'*, II.415; Semenov, *Rossiya*, VI.12–15; Vorob'ev, *Kazanskie tatary*, 43–51.   [2] *PSRL*, XIX.130.

for this chapter is one such unpublished register compiled in 1565–8 by
N. V. Borisov and D. A. Kikin.[1] This has survived in a late-seventeenth- or
early-eighteenth-century copy of the original; it contains references to an
earlier register compiled by Semen Normatskii in 1563.[2] In 1565–8 a
register of inquisition was also compiled for Sviyazhsk and its uezd.[3] By this
time a certain uniformity in such administrative matters was already noticeable. A second period of activity seems to have occurred in the early seventeenth century. A register of 1602–3, compiled by Ivan Boltin, deals with the
estates held by service and the settlements of the Tatars, Chuvash and other
non-Russians of the area.[4] Between 1616 and 1619 a series of other surveys of
different types were carried out.[5] It is difficult to see reasons for the dates of
these two bouts of activity; one might expect surveys to be made immediately after the conquest; the first seems somewhat late for this. The second
round is perhaps in part to be associated with measures to re-establish order
after the disorders of the early seventeenth century. Nevertheless, it needs a
fuller survey of these registers to account for the particular dates of activity.

In the mid Volga there may well have been a continuity of agriculture
practised by the non-Slav peoples between the periods before and after the
Mongol conquests of the thirteenth century.[6] Documentary, archaeological
and linguistic evidence shows that farming was well established, particularly
in the Chuvash area, between the Sura and the Volga, and, beyond the Volga,
in Kazan Arti.[7] Mukhamed'yarov has argued convincingly that the Kazan'
khanate had a three-field system of sorts. But he stressed that he was not
asserting 'that a regular three-field system with a planned sequence of winter-
sown, spring-sown and fallow and all three fields equal was completely
dominant in the Kazan' khanate'.[8] Indeed, he argues for a variety of farming
systems being present: the three-field arrangement may have involved an
irregular fallowing system with both winter and spring sown crops; some
lands in the Chuvash area were worked from a distance and involved tempor-

---

[1] TsGADA, fond 1209, kn. 152. I am most grateful to the Director of TsGADA for
enabling me to obtain a microfilm copy and to E. A. Dudzinskaya for her help in this.
The section of this book relating to the town of Kazan', was published in *Materialy po
istorii Tatarskoi ASSR. Pistsovye knigi g. Kazani 1565–68 i 1646.* K. I. Nevostruev
published excerpts from the section dealing with the uezd, in *Spisok s pistsovykh knig
po gor. Kazani s uezdom*, 62–81; this is of little value.

[2] The Central State Archive of Ancient Documents (TsGADA) also holds a late-
eighteenth-century copy of kn. 152, listed as kn. 643.

[3] Published by A. Yablokov, as *Spisok s pistsovoi mezhevoi knigi goroda Sviyazhska i uezda;
pis'ma i mezhevaniya N. V. Borisova i D. A. Kikina (1565–67).* The TsGADA copies are
numbered fond 1209, kn. 485 and kn. 432.

[4] TsGADA, fond 1209, kn. 642. See Mukhamed'yarov, *Izv. Kazanskogo filiala AN SSSR,
seriya gumanitarnykh nauk* (1957), 2.191–6.

[5] TsGADA, fond 1209, kn. 153. See Mukhamed'yarov, *MISKh*, III.116–17, 120.

[6] Mukhamed'yarov, *MISKh*, III.95.

[7] *Ibid.* 89f.   [8] *Ibid.* 117.

ary abandonment; but slash and burn, in his view, was restricted to the northern forests of the Udmurts.[1] This last opinion is probably not completely accurate; but he is surely correct in pointing out that slash and burn might be an independent system or combined with two- and three-field systems. At any rate, both cultivation and the system of exploitation seem to have been at a level by the sixteenth century which meant that the Russian conquest introduced modifications but not radical innovations.[2] Shares from the Kazan' area probably dating from the thirteenth to sixteenth centuries lend support to this view.[3] Present available evidence, however, seems insufficient to elaborate a detailed account of pre-Russian farming in the area; moreover, our concern is primarily with Russian peasant farming. It is, therefore, to the evidence of the Russian surveys of the 1560s that we now turn.

The main immediate modification introduced by the Russian conquest was, of course, the change of ownership. This was more than a formal change. It was not merely that Tatar lords were displaced (the khan himself, the four great noble families called *karachis*, the princes (*beks*) and their younger sons (*murzas*) with their mounted servitors (*ulans*); in addition, Tatar institutions were replaced by those of the Russians, in particular the Orthodox church and especially the archbishop and the monasteries. Tatar lords were replaced by Moscow gentry; there was no Russian heritable tenure in sixteenth-century Kazan', land was held in return for service. Russian settlers in the main began to till around Kazan', the Tatar population moved off into the hinterland.[4] One entry in the 1565–8 register refers back to an allotment of land six years after the conquest, in 1558, to a group of peasants performing guard duties against incursions by Nogai and Crimean Tatars; apart from this the register gives no information on peasant tenure.[5] There were no black peasant lands, and, as the area was newly conquered, it was entirely at the disposal of the conqueror and his servitors.

At the grassroots level the movement of Tatars and Chuvash from their old holdings seems to have been both 'voluntary', i.e. probably due to force of circumstances, and the result of more deliberate Russian policy. One section of the 1565–8 register, for example, deals with 'empty and escheated lands which were held by no-one and old Tatar and Chuvash lands intended henceforth to be distributed for service'.[6] These included a waste $1\frac{1}{2}$ to 2 verstas ($1\frac{1}{2}$–2 km) inside the forest from a hamlet; 'the Aishi Tatars and Chuvash held

---

[1] *Ibid.*

[2] For a differing view see Tikhomirov, *Rossiya v XVI stoletii*, 490. The 1523 yarlyk of khan Sahib-Girei gives a picture of the pre-Russian system of exploitation; *Vestnik nauchnogo obshchestva tatarovedeniya*, 1–2, 1925, 29–37, gives the Tatar text and a Russian translation; the translation is also given in *Istoriya Tatarii*, 101–2.

[3] Items 4222, 4223, 6052–5 in the Zausailov collection in the National Museum of Finland, Helsinki.

[4] *PSRL*, xx.583. Khudyakov, *Materialy po istorii Tatarskoi ASSR*, xviii.   [5] F. 192.

[6] F. 129v.

that waste without documents and mowed hay on it, and in 70 (i.e. 1562/3), prior to the distribution of lands for service, the Aishi Tatars and Chuvash were ordered to mow the waste and after that to hand it over for distribution'.[1] In other cases, the reasons for moving are not so explicitly stated, though clearly enough implied. At another waste 'formerly held by the Chuvash Meretek and his fellows, those Chuvash quit that waste and dispersed to live in Tatar and Chuvash villages and hamlets'.[2] As with the dispersals noted in Toropets, such movement sometimes involved, apparently, a reduction in the size of settlement. Chuvash quit a village called Big Termerlik (or Temerlik) 'and settled on that lot on the Shusha at Temerlik Minor, downstream towards the Kama'.[3] From old Cheremsha 'the new, little Cheremsha hamlet was transferred to the third field beyond the river'.[4] Abandoned Chuvash lots of land were taken to make up Russian estates.[5] At the same time there appears to have been some concern for native property rights; for example, a waste was recorded as shared by two Russians and 'if that waste is not Tatar or Chuvash, they are to set up tenements on that waste, to till the arable, to mow the hay and cut the forest'.[6] In the sixteenth century the forced conversion of the Muslim to Orthodoxy had not reached the scale it did in later times, and the Russians, while imposing restrictions on the native peoples did not pursue anything akin to a policy of Russification. Many Tatars, indeed, not only gave military service to the Russians but even sometimes intermarried with them.

In general, however, the Russian register naturally gives less information about Tatars and Chuvash than about its main concern, the distribution of land and other resources to tenants holding by service. The general attitude towards the native population seems to be reflected in brief entries such as this, without further detail: 'on the third field of that waste, beyond the Mesha, live Chuvash'.[7] In other cases Tatars and Chuvash live 'in place of peasants'.[8] In such cases these native peoples were treated as dependent, enserfed peasants.

While, in broad terms, there was a similarity between pre- and post-conquest situations, there were also some changes. Perhaps the most important of these for farming was the further development of a three-field system. Sometimes it seems that the phrase 'in three fields' was used in a notational fashion only; for instance, land 'which is noted in the registers of Semen Narmatskii as being held by Stepan in three fields of open steppe'.[9] In some cases, an amount of arable land was noted and it was stated that it was made up of equal amounts in the three fields: '60 chets [33 ha] in three fields... twenty a field'.[10] Similar examples show strikingly regular amounts, all being divisible by five, and this may well indicate a conventional and approximate usage, even if not a mere notational device.[11] Similarly, references to arable

[1] F. 130v.   [2] F. 131; cp. ff. 121–121v, 132.   [3] F. 131v.   [4] F. 162v.   [5] F. 144v.
[6] F. 127.   [7] F. 194.   [8] Ff. 109v, 211v, 212, 214v.   [9] F. 109.   [10] F. 171.
[11] Ff. 159v, 171v, 179v, 206.

land and tillable forest 'by the three fields' or perhaps 'attached to the three fields' (k trem polyam) might be taken as only a general, rather than a literally accurate, statement.[1]

Other references, however, show that a three-field organisation or layout undoubtedly existed. Bulgak hamlet, for example, was located on the Brysa stream and had tillable forest along both banks; its third field was beyond the stream together with a glade in the forest.[2] There are other instances of third fields being situated beyond a river.[3] Such evidence seems to indicate specific locations and to justify the belief that there were actually three fields present. This is confirmed by additional material. There are a number of references to two fields.[4] In some cases the entries are such as to make it plain that there were fields in separate locations, two in one place and the third elsewhere.[5] Two fields of Iski Yurt waste were distributed to three villages and a later entry notes that 'on the third field of that waste, beyond the Mesha, Chuvash live'.[6] Tillable forest is recorded 'behind two fields of Sakurkadysh' village; but then 129 desyatinas [140 ha] of tillable and untillable forest are recorded 'for three fields, 43 desyatinas [47 ha] a field'.[7] This entry suggests that the forest may have been in one or in two blocks, but this reality was recorded in the standard notation of 'three fields'. At Kuyuk manorial settlement there was forest both 'around two fields...and to the third Kuyuk field which is between Bulgak Onuchin's Tugasheva hamlet, held by service, and Kuyuk'.[8] A boundary was established to two fields of the tsar's wastes at Malye Chepchugi; 'and on the left, the land and forest of Chepchugi village, the third incorporated field of Little Chepchugi'.[9] Thus, it seems that, quite apart from the conventional notation in terms of three fields used by the officers of the inquisition, a three-field layout existed and there were sometimes newly incorporated areas.[10]

Fields had ends; a boundary, for instance, ran 'along the end of Tarlashevo forest to the lime tree and the oak with marks on them, and from the oak along the end of the field'.[11] Land for a church was set aside 'along the end of Bulgak hamlet's field and along the end of the Tatar hamlet Balakhchei's field'.[12] There were empty lands along the end of Temereino field.[13] Other references show that these ends of the arable fields were often remote or forest. Among the abandoned lands is mentioned a glade 'along the end of the field, beyond the gulley, by the dense forest'.[14] There was a site (selishche) 'beyond the stream, along the end of the field...beyond the gulley, behind the Elani village field'.[15] The remoteness of these areas, no doubt together with their intermittent use, sometimes gave rise to differences of opinion.

---

[1] Ff. 144v, 155, 174v, 204, 205, 205v, 210, 213, 213v, 214.   [2] F. 105v.
[3] Ff. 224v, 225, 225v.   [4] For example ff. 146v, 155v, 159, 165v, 172v, 175v.
[5] F. 157v, for example.   [6] Ff. 193, 194.
[7] Ff. 209v, 210. In this instance the figures do not appear to be rounded and so may well be genuine.   [8] Ff. 213–213v.   [9] Ff. 220v, 221.   [10] Cp. pp. 208–9 below.
[11] F. 240; cp. a similar entry, f. 219v.   [12] F. 106.   [13] F. 108v.   [14] F. 130.   [15] F. 161.

There are no Chyulmen lands mentioned anywhere apart from that; and the glade which is abandoned land of the Chyulmen field beyond the forest, along the end of the field of the Chuvash hamlet, Dolgaya, that glade has not been recorded as Chyulmen's, nor has it been measured, because the old-established Chuvash claimed that small glade as of their Dolgaya hamlet.[1]

The immediately preceding entry in the register relates to Solmen hamlet (presumably a mistake for Chyulmen) and notes a patch beyond the fields (*zapolitsa*) by the steep gulley along the Cherpa.[2] Such patches beyond the field (*zapole* is the commoner term) are not frequent in the 1565–8 register, but seem often to be associated with forest glades, natural clearances that is, as well as with man-made ones, both of which were evidently cultivated. All entries cited below come from the section of the register which records the boundaries of estates. Glades occur in the following cases. 'By the swamp to the patch beyond the fields of Kruglaya Polyana [lit. Round Glade] hamlet to the rivulet that flows from the swamp beyond the Kruglaya patch beyond the fields...and between the patch...Kruglaya Polyana hamlet'.[3] 'And from the pine along the patch beyond the fields of the Koval' glade.'[4] 'And to the left the (arable) land, hayfields and forest of Yagodnaya Polyana [Berry Glade] village held by service, along the patch beyond the fields, along the Kazan', downstream to the mouth.'[5]

There are more references to glades, natural clearances in the forest, which were used both for taking hay and as arable land. In one case a waste was noted with a glade.[6] In general they were, as would be expected, 'in the forest', 'beyond the forest', and 'by dense forest'.[7] Often, too, they are associated with long fallows; one settlement had 20 chets (11 ha) of long fallows 'including glades'.[8] On the Shalsha 65 chets (35 ha) of good quality arable land were recorded, 'and $40\frac{1}{2}$ chets [22 ha] of long fallows and clean-burnt patches, including the long fallow which is beyond the forest and the Memeri Glade and that which is beyond the river Yashchurna, above Memeri glade as far as the dense forest'.[9] The use of fire is noteworthy. Three lots in another hamlet were recorded as 45 chets (25 ha) of good arable land and 60 chets [33 ha] of long fallows 'together with the glades'.[10]

From such entries it is not possible to see the size of these glades; they may have been quite small.[11] The nature of the area, however, with blocks of steppe vegetation intermingled with forest meant that there may also have been large natural clearances. On the Iya three brothers had a hamlet and behind their glade beyond the forest, beyond the gulley, is a tilled glade...and on it 60 chets [33 ha] of arable land...and behind the same forest glade at the top of the gulley which is above the middle glade...along both sides of the gulley and on the glade are 15 chets [8 ha] of arable land attached to the fallow field.[12]

[1] F. 113v.   [2] F. 113.   [3] F. 235v.   [4] F. 238v.   [5] Ff. 236v–237.   [6] F. 194v.
[7] Ff. 111v, 113v, 155, 130.   [8] F. 123.   [9] F. 155.   [10] F. 165.v   [11] Cp. ff. 173v, 245.
[12] F. 171: as has been noted above, these figures may have been rounded.

Even more significantly, at another location there was an 'empty' glade (that is one which was no longer tax-liable, probably because it was not being worked) by dense forest 'and on the glade 80 chets [44 ha] of arable land in three fields'.[1] Glades, then, were by no means always poor, small clearings in this area, but they seem often to have been associated with some sort of out-field system, and, sometimes, with the use of fire.

Other entries speak of human clearance. 'Along the boundary, along the patch beyond the fields of Kuzma's tilled patch site' (*selishcha Kuzminskogo zaimishcha*).[2] Another entry refers to the same site: 'from the first stream which flows from the forest beyond the site's patch, beyond the fields into the Kazan' '.[3] The term here translated as tilled patch (*zaimishche*) several times occurs as a placename.[4] Its meaning is shown, first, by a heading which refers to the tsar's holdings in order of diminishing importance, his 'villages, hamlets, clearances, wastes and patches'.[5] The tilled patch, therefore, was akin to the clearances and wastes. Second, we have the placename Smoili's Patch which is described as a clearance (*pochinok*); it is listed as one of the clearances which 'arose anew on the distant land of Tarlashi village'.[6] These settlements formed part of the estate of the archbishop and were all of considerable size in terms of tenements; Smoili's Patch had eleven tenements, one of which was held by a Chuvash, Smoili Iseinov, possibly the original holder of the place. Total arable land, however, was estimated as only 28 chets [15 ha]; this valuation may reflect a decline in the economic state of the settlement as a result of its being taken over by the Russian Church. It is possible that, as in many other cases, the native population fled. Thus we have some limited evidence for the clearance and use of natural and of man-made patches in the forest surrounding the regularly cultivated fields.

Further, such evidence is afforded by a term meaning worked land (*rozpash'*). A boundary, for instance, passed over the river Noksa by a ford and then went 'directly by the newly worked lands'.[7] Such places might usually be expected to be of fairly small size. Yapsheika hamlet had, 'on the shore by the Kama, in Yapsheika forest in Ilman, newly worked land; and on the newly worked land 5 chets of good quality'.[8] If this were a real measure of total arable land, it would be enough to maintain one tenement, assuming 5 chets ($2\frac{1}{2}$ ha) in each of three fields. Also on the Kama, by the town of Laishev, beyond Laishev fort, 150 Russian prisoners (i.e. Russians captured in earlier times by the Tatars) were granted holdings; these were made up of 510 chets of good quality in each of three fields on newly worked land and 90 similar chets of 'remote arable land of their newly worked land, the Meltiska [sc. Myateska] waste'.[9] This works out at a total of 6 desyatinas (12 chets, $6\frac{1}{2}$ ha)

---

[1] F. 130. [2] F. 234.
[3] F. 237v. The hamlet of Kuzma's Patch is also mentioned on ff. 173v, 180.
[4] Ff. 156, 203, 209 etc. [5] F. 219. [6] F. 203. [7] F. 230.
[8] F. 150. Other references to this patch are at ff. 115v, 116v. [9] F. 128v.

a man in all three fields, which was a normal holding in this area.[1] In Novya (New) Serdi hamlet four peasant tenements held 40 chets (22 ha) of newly worked land, approaching a normal size.[2] It thus looks as if the term *rozpash'* at times survived and continued in use even when such clearances had established themselves to the satisfaction of the officers of inquisition, as fully fledged holdings; it is immaterial from this point of view whether these were real amounts of arable land, or an estimation of the wealth of the holdings for tax purposes.

There are also a few other references to remote arable lands. By the river Noksa a boundary passed 'from the birch tree across the remote arable lands'.[3] In other cases it is made clear that such lands were long fallows; this suggests that they were part of an infield–outfield system. Such long fallow remote arable lands were found around Likhoe swamp or lake.[4] At Tulush hamlet there were also 'arable lands under long fallow in the forest at a distance along the road that they ride to Devlezerevo'.[5]

Evidence for wastes is also not very frequent. They are mentioned in only one heading which, among other matters, refers to 'wastes, empty surpluses of long fallow lands, those overgrown, and tillable oak woods, which remained after the distribution to the new Kazan' servitors; and what estates held by service and wastes were allotted by the former allocations of estates in 74 and 75 (i.e. 1566–8); how much in any estate on surplus land and on wastes is long fallow and overgrown with tillable oak woods, and hay and forest; and which wastes were not held by service; that is recorded as separate items'.[6] Despite this impressive rubric, the contents of the section are not very illuminating. Four rich or large wastes are mentioned, estimated at 120, 208, 240 and 190 chets (65, 113, 130, 104 ha) of arable land, as well as long fallows.[7] The register of boundaries has a reference to a waste in terms of a three-field layout.[8] Other wastes, Buchyunbi and Yaltyki, both overgrown with oak woods, were recorded at 30 desyatinas (33 ha) each per field, that is 60 chets.[9] Even a waste which was tilled from a distance, Dertuli Apcheik, was noted as having 10 chets ($5\frac{1}{2}$ ha) of arable land, as well as 45 chets (25 ha) of long fallows and 4 desyatinas ($4\frac{1}{2}$ ha) a field of tillable oak woods.[10] Again, as with other items of land beyond the main three-field layout, it seems that wastes here were often quite substantial units, at least in terms of tax-liability.

In general, therefore, we seem to have evidence on these estates held by service of some three-field layout combined with various areas of outfield, some possibly of considerable size, in the forest. Only occasionally, however, do we find further details of layout. The existence of strips of tilled land is mentioned, but only in two entries.[11] In places the extension of cultivation was

---

[1] See p. 216 below.   [2] F. 165.   [3] F. 232.   [4] Ff. 114 and 114v.   [5] F. 172.
[6] F. 190v. See appendix 5.   [7] Ff. 192–3.   [8] F. 213: see above.   [9] Ff. 152v, 153.
[10] F. 128.   [11] Both on f. 248.

bringing units of worked land into contact. By Novotsarevo hamlet the remote oakwoods (not specified as arable land, but evidently worked) of this new settlement extended 'from the boundary of the Transfiguration monastery's Engildeevo clearance, from the gulley up the Kutcha, along the gulley to Tsarevo hamlet's field'.[1] As the result of an exchange of lands with the archbishop, there was a deficit of 70 chets (38 ha); it was not made good 'because the arable lands of other hamlets of the archbishop's villages had not come next to Sakurkadysh manorial settlement'.[2] There was some concern, therefore, for convenience of layout, probably especially among the larger and better organised estates such as those of the archbishop.

Some of the outfield areas were incorporated into the arable (*pripushcheno v pashnyu*). This, again, shows concern for layout; but this process seemed, in the Moscow area, to be associated with the development of the three-field system. In Kazan', too, it may be associated with an attempt to enlarge the basic cultivated area to effect better control of the labour force on estates; it may also be one of the innovations introduced as a result of the Russian conquest. This last aspect may make it of some importance; at the same time, however, it was not, apparently, widespread.

A total relating to the estates of 155 tenants mentions the incorporation of three wastes into the arable.[3] Unfortunately, parts of the register are missing and earlier entries mention only the incorporation of one site (evidently taken as a waste in the total) and one waste.[4] Both were located on the river Brysa. There are two other references to the incorporation of wastes.[5] In one of these cases the settlement incorporating the waste was named Kaimar Mamysh tilled patch and the tenant had 'started to set up a bit of a tenement'.[6] Evidently not all incorporations were initially regarded as large-scale operations; the holding was noted as having 2 chets (1 ha) of good quality arable land. At Novaya Kruglaya Polyana (New Round Glade) 'which had been transferred from the other side of the Kazan', from old Kruglaya Polyana to Babikovo waste' a tilled patch was incorporated.[7] At Novaya Malaya Cheremsha hamlet (New Little Cheremsha), transferred from old Cheremsha across the river to the third field, the hamlet itself, i.e. presumably its economic activities in some sense, was incorporated into the third field of the original settlement.[8] This entry seems to indicate the establishment of a small unit, hived off from the mother settlement with an estimated 5 chets (3 ha) of good quality arable land, and its subsequent inclusion in the mother settlement's expanded third field. Here the third field seems to indicate a general location or administrative area. A contrary case was the inclusion of the third field of a waste into the arable of the large settlement, Bolshie (Big) Chepchyugi village.[9] This village

---

[1] F. 119v. The clearance is also mentioned on ff. 245 and 245v.   [2] F. 210.   [3] F. 119v.
[4] Ff. 104v, 106.   [5] Ff. 182v (184v also refers to the same one), 156.
[6] F. 156; *pochal dvorishko stavit'*.   [7] F. 209, also ff. 218v and 234.   [8] F. 162v.
[9] F. 145v.

had 36 tenements, apart from 3 empty ones; it had formerly been a crown village 'without the new incorporated lands'. Here we seem to have an example of the regularisation of lands on a very large unit.

Thus, while we have evidence of concern for layout in the Kazan' area and of the incorporation of outlying land into the central fields, reorganisations of this sort do not appear to be confined to large units. All such measures, however, probably related to Russian, as distinct from native, attempts to control the labour force more closely. They may also suggest some change of stress from livestock husbandry to increased arable farming.

One further point needs to be made about the arable land. The view that slash and burn was confined to the northern part of the Kazan' area occupied by Udmurts has been called in question.[1] The register of 1565–8, in fact, has some relevant material. It mentions a hamlet called Podseki; this is the plural form of a term which means slash and burn, or an area used for such a system of farming.[2] From this hamlet the holders had the right 'to ride into that forest for every sort of useful thing' (*po vsyakoe ugod'e*).[3] Burnt areas (*vygari*) and burnt patches (*gari*) are also mentioned in other settlements associated with dense forest. The Transfiguration monastery's hamlet of Davydko Borisov (evidently the original holder of this settlement) had a boundary which, in part, ran 'by the track to the left by the old cutting in the forest... and by the burnt areas straight to the elm on the old cutting'.[4] The surrounding forest had several bee-trees in it and this may indicate that it was relatively undisturbed by human activities, even though it was quite close to Kazan'.[5] Sharymash village had an estimated 80 desyatinas (87 ha) of tillable oak woods and cleared burnt areas attached to a Chuvash hamlet.[6] Big New Shalshi hamlet had long fallows and cleared burnt areas beyond the forest.[7] There were clean-burnt patches on the Shalsha.[8] In determining the boundaries of Kruglaya Polyana hamlet and Tsaritsyn village several burnt patches are mentioned in the dense forest.[9] There is, thus, some limited evidence for the use of fire as a means of forest clearance or cultivation.

The 1565–8 register has interesting, though not entirely clear, information on mills. The officers of the inquisition 'recorded in Kazan' uezd mills liable to rent held by men of the Kazan' artisan quarter and by peasants...along the rivers, streams and sources'.[10] They noted such details as the amounts paid to the Treasury, any remissions, when mills started working and details of the mill sites, including the sources of materials for repair. The entries which follow this heading, when specified, are all large-wheeled mills, and include some held by the archbishop along the Noksa. Apart from the large-wheeled mill (*melnitsa bolshoe koleso*) i.e. one with horizontally mounted wheel, the churn mill (*mutovka*) is mentioned, but there are two main terms used: *melenka mutovka* and *melnitsa mutovka*. The available material does not allow

---

[1] P. 202 above.   [2] Ff. 143v, 144, 191, 191v, 230.   [3] F. 189v.   [4] F. 244.
[5] F. 244.   [6] F. 169.   [7] F. 169.   [8] F. 155.   [9] F. 235.   [10] F. 196.

any clear distinction to be made between these two, but it seems best to assume initially that there was some difference. At one village, Big Shigazda, three examples of a *melenka* and one *melnitsa* are noted.[1] This suggests the two were differentiated in some way. It may be significant that, although all the names of the places where the mills were found are non-Russian, the term *melenka* seems to be more closely linked with the non-Russians: in three out of the eleven occurances of this type of mill Mordva and Chuvash are specifically mentioned in connection with them.[2] In none of the 10 cases where the term *melnitsa mutovka* is used is there any specific evidence of their being associated with non-Russians.[3] Possibly, therefore, the more complicated large-wheeled mills were usually Russian and belonged especially to the larger lords or men with capital; any peasant engaged in milling on this scale had links with substantial men of the Kazan' artisan quarter. The churn mills were simpler structures held in the 1560s by Russians, though the term *melenka mutovka* (or *kolotovka*) may have been used for some particular type of non-Russian churn mill. These last were probably simpler than the *melnitsa mutovka*, one of which is noted as having two sets of stones.[4] The output of a churn mill (*melnitsa mutovka*) is given as half a chet a day, i.e. approximately 30 kg of rye or 20 kg of oats.[5]

In the uezd four millers are mentioned; three of these were peasants, but evidently men of substance (two of them acted as guarantors for the third who had a large-wheeled mill), the fourth was a monk.[6] They are all linked with the Transfiguration monastery. There was a quern maker (*zhernovnik*) in a tenement in the large village of Shigaleevo (32 live tenements).[7] There are only two references to dealers in processed grain products in the countryside. One is to a bread man (*khlebnik*), the other to a roll man (*kalachnik*); both appear in the register as guarantors for mill owners and are not themselves countrymen, but from the artisan quarter of Kazan'.[8] This reinforces the impression that when milling became a specialist activity, no longer accommodated within the normal round of activities of the peasant household or hamlet, it formed a link outside the peasant world and acted as one of the channels of communication with the wider world through trade.

If we turn from arable farming and its products to look at the livestock sector, we find much less evidence. As is usual in such registers, there are no details of livestock on holdings; moreover, the fact that the 1565–8 register has no detailed note of obligations means that even this indirect source of informa-

[1] F. 148v.   [2] Ff. 114, 155v, 169.
[3] Occurrences of the *melenka* are mentioned on ff. 105v, 114, 122 (also 154), 148v, 154v, 155, 155v, 167 (also 167v), 169; of the *melnitsa* on ff. 110v, 128v, 148v, 178v, 187. There is, in addition, one use of the term *melenka kolotovka*, a churn mill found on land formerly held by native converts to Christianity: f. 175v (see also ff. 247v, 248).
[4] F. 110v. A guess is that the two terms distinguished churn mills of the type set directly in the stream bed and those for which the water was channelled and so controllable.
[5] F. 178v.   [6] Ff. 191, 196v, 211, 211v, 213v.   [7] F. 185v.   [8] Ff. 198, 198v, 200.

tion on livestock is wanting.[1] That livestock were important is shown by the concern for hay. Moreover, this parkland zone with interspersed blocks of steppe vegetation had for centuries been inhabited by Tatars and others; they were famous horsemen and the horse provided them with important items of diet. Tatars still esteem horsemeat above all other flesh; fermented mare's milk (*kumys*) is an important drink.

Hay was taken wherever it could be found. There are frequent references to hay along the rivers and streams by settlements. For example, hay was taken 'along both sides of the Bima, round the fields, and along one side of the Mesha'.[2] Entries such as 'along the river, between the arable lands and round the fields' or 'along the oak woods, gullies, between the arable lands and along the water meadows' are characteristic.[3] Smaller sites were also used; one such, a promontory jutting into a river, provided only 50 ricks.[4] Apart from these sources of hay in the immediate vicinity of the settlement and its fields, more distant sites were utilised. Remote glades and meadows are listed, such sites sometimes contributing considerable amounts of hay, especially along the great rivers, the Kama and Volga.[5] There were 30 (probably 'drag') ricks to a desyatina.[6] This implies a yield of around 1,120 kg to the hectare.[7] The quit-rent to be paid for hayfield was a denga per rick, plus a customary due of the same amount for each transaction.[8] Some care was taken to ensure that unsuitable plants and unmowable areas did not count in the record. For instance, 'sedge, swampland, pits and brome grass' were not included.[9] A hayfield of 550 ricks held by the protopope of the Cathedral Church of the Annunciation was noted, 'without mowing 5 desyatinas [5½ ha] of sedge and brome grass'.[10]

Hayfields of all sorts were measured, just as the arable land was, though doubtless estimation was commoner. For example, although some of the archbishop's meadows were, allegedly, measured 'and measured out, in all, 50 desyatinas [55 ha] of clean mowing land', 15 desyatinas (16 ha) of sedge and brome grass in different places were estimated (*po smete*) and not included.[11] Apart from the large hayfields of the archbishop and the monasteries, the town of Kazan' itself had livestock pastures.[12] The importance of these sources of feed for the livestock throughout the winter appears from a note relating to some of the Transfiguration monastery's meadows; these were 'all beyond the Ichki Kazan'; and in summer and in autumn prior to the setting in of the winter road it is not possible to cart the hay to the monastery'.[13] The impor-

---

1 Information on the income of lords from obligations had not been included in the 1563 register either; f. 134v.
2 F. 110v. 3 Ff. 121v, 114. 4 F. 116v.
5 Ff. 106v, 128v, 203 refer to amounts of 1000 and 1500 ricks.
6 F. 206v: *po 30 kopen volokov*; see also f. 215v. 7 See chapter 2 above. 8 F. 199v.
9 Ff. 215v, 242. There are many similar entries, especially for sedge and brome grass.
10 F. 218. 11 F. 206v. 12 Ff. 191v, 236, 242. 13 F. 216.

tance of horses for the archbishop and, by implication, for other great lords, is indicated by the mention of a farrier, a peasant of the archbishop.[1] Unfortunately, there seems to be no evidence in this register of livestock at lower social levels. One leatherworker is listed, but the two butchers mentioned were both men from the artisan quarter.[2] The constant presence of hayfields, however, shows that livestock were an integral part of all holdings.

Forest, similarly, is recorded everywhere: 'tillable and untillable forest scattered around the fields', 'around the fields and scattered along the areas beyond the fields'.[3] Typical entries such as these record amounts in two ways: first, forest was measured in desyatinas; secondly, particularly when dense or remote forest was concerned, a rough estimate was given in terms of the notional length and breadth of the area. Thus, there are entries noting 'tillable and untillable forest estimated in various places, a versta in length and in breadth, from all three fields, a versta'.[4] In such contexts the expression 'from the fields into the forest' is common; it seems to indicate forest still uncultivated, or at least sufficiently little used, for an approximation in terms of distance rather than area to be adequate for the purposes of the register. In some places remote forest is specified.[5] A small clearance with two peasant tenements had '15 desyatinas [16 ha] of tillable oak woods to the three fields' and also 'tillable forest to the clearance around the fields'.[6]

Oak woods are commonly mentioned, sometimes they were 'remote'.[7] Birch woods and pine woods or woods with pine dominant are referred to, but not frequently.[8] The categorisation of tillable or untillable is virtually universal, but forest is also frequently specified as dense (*ramen'e*) or black (*cherny*). Sometimes both the latter terms occur together. For instance, one hamlet had an 'estimated three verstas [3 km] in length and one versta [1 km] in breadth of tillable dense forest and untillable black forest around the fields and along both sides of the river Mesha'.[9] It may be that black forest was considered untillable. A heading also refers to 'tillable dense forest and untillable black forest'.[10] At Kuyuk manorial settlement there was tillable and untillable forest 'around two fields', while the third field had dense forest.[11]

Forest was utilised for all sorts of extractive activities. First, timber was taken for firewood and for buildings. Podsekina or Podseki hamlet had ascribed to it some remote forest beyond Kilderle hamlet to which the inhabitants were to ride 'for every good thing, because there is no useful or building timber by Podseki hamlet'.[12] Perhaps this was because the forest had not had time to regenerate itself from the burning implied by the name. Timber, of course, was used not only for the construction of houses, but for

---

[1] F. 197; he is mentioned in a list of guarantors for the rent of a mill. At f. 208v he is shown in a tenement, but his trade is not quoted.   [2] Ff. 191v, 196v, 197v.
[3] Ff. 114v, 116.   [4] F. 115v.   [5] F. 191v, for example.   [6] F. 205v.   [7] F. 119v.
[8] Ff. 219v, 168, 203, 220, 235v, 236.   [9] F. 136: cp. f. 219v.   [10] F. 135.   [11] F. 213v.
[12] F. 191v.

many other structures. Entries in the register of mills usually include the phrase 'timber to repair the mill is to be cut near the mill, apart from the bee-trees'.[1] In one case the phrase reads 'timber for every sort of store for the mill'.[2] Three carpenters are listed in tenements, but there seems to be no other reference to woodworkers.[3] The pits frequently mentioned in describing boundaries were evidently sometimes the result of charcoal production.[4]

The forest was also important as a source of wealth in the form of bees, fish and beavers, as well as various plants which are not mentioned at all in the registers. Bees and beavers were of especial importance for the non-Russians and were expressly excluded from the allocations to Russian service tenants in the 1565–8 register, 'because the (bee-tree) appurtenances in the forests and the beaver runs in the rivers are all Tatar and Chuvash in Kazan' uezd'.[5] Some Russian prisoners had, however, held such resources from before the capture of Kazan' and they were to continue to pay obligations in kind (*yasak*) 'according to the old registers'. Tatars and Chuvash also might meet their obligations in honey instead of money, the amount of honey being estimated according to the market price.[6] The bees seem to have nested almost as often in oak trees as in pines.[7] Most of the evidence for wild bees comes from the archbishop's estate with its largely Tatar and Chuvash population. The bee woods by the three Kaban settlements are frequently referred to.[8] A heading in the section of the register detailing the archbishop's holdings refers not only to arable lands, but also to bee-tree areas and fisheries.[9]

The concern of church institutions with fisheries is fairly well demonstrated in the register. The archbishop had fisheries

in the Volga, on its islands and small lakes, wherever there is a fishery; from the mouth of the Kazan' along both sides of the river Volga, apart from the hayfields, along the Kazan' side and along the Sviyaga side to the Kama, including the spring-time flood areas around the town, wherever the spring water lies; and the fishery from the mouth of the Kazan' and Cherny Pesok [Black Sand] along the Chertysh head; and the winter and summer fishery in the Kazan', wherever he wishes to build a weir, and lake Tsarevo [the Khan's lake] together with other lakes in the meadows around lake Tsarevo.[10]

The archbishop was also granted some noble fish from the tsar's Kama weir, 'from the winter fishing: 15 salmon, 10 stellate sturgeon, 5 sturgeon, in all 30 fishes'.[11] Other fisheries of the archbishop were located in Lake Kezemetevo,

---

[1] Ff. 197, 197v.   [2] F. 200.

[3] Ff. 142, 150, 218. Another carpenter mentioned was a guarantor from the artisan quarter; f. 196v.

[4] F. 233.   [5] F. 135v. Cp. *Spisok s pistsovoi i mezhevoi knigi g. Sviyazhska i uezda*, 65.

[6] F. 203v, 205, 207v etc.

[7] Oaks: ff. 221, 223, 224, 228v, 233, 239, 243v, 248v, 250, 253, Pines: ff. 219v, 221, 231v, 236v, 238, 238v, 239, 244, 248, 253v.

[8] Ff. 208v, 233, 234, 238, 240v, 243.   [9] F. 208.

[10] F. 207v. A weir on the Betka is mentioned on f. 253v.   [11] Ff. 207v, 208.

along the river Sumka and elsewhere.[1] The archbishop petitioned for a place 'for a tenement to receive his fishermen and for pasture for the horses on which they bring the seines; and for guards, because, he claims, they poach fish in those lakes across the Volga and there is nowhere to guard against that from a distance'.[2] It was in the same area of these lakes by the Volga that the archbishop was allowed 2, 3 or 4 sazhens (4–8½ m) around each lake and also had fish nurseries.

Monasteries, too, had their fisheries. The Transfiguration monastery had fisheries at the mouth of the Kazan', on the Volga, at Krasnoe Lake, Tevlevo Vody, and the Kabans.[3] No money rents were paid for these fisheries on the grounds that they provided for the needs of the monastery's refectory, presumably implying that the catch was not traded.[4] The Ilantov monastery had fisheries in the same area along the Kazan' and the Volga.[5]

Information on the fisheries of lay lords is scantier. All those mentioned were on lakes.[6] There seem also to have been special spots where fish were stunned (*glushitsy*), presumably by blows on the ice.[7] In some cases the benefits to be derived from fisheries, in terms of the money rents paid for them, were notionally converted into arable land and so counted as part of the allocation to lords holding by service. A fishery at Lake Shanby, for instance, had paid a quit-rent of 5 rubles a year and was, therefore, taken as the equivalent of 50 chets (27 ha) of arable land; the rate of 10 chets per ruble seems to have been a usual valuation of arable land.

It thus seems that the source fails adequately to describe the resources for gathering, as it does for livestock. Certainly fish and fowl (there is no mention of the latter in the 1565–8 register) were of greater importance than these entries suggest. The explanation of this is probably that the officers of the inquisition were focussing their attention on the needs of the Russian authorities in the newly won area. The need of the church for fish to be consumed on so-called fast days was of considerable importance, but the consumption of fish and fowl at lower social levels was of little concern or interest to them. That these traditional items of diet continued to be gathered is implied by the fact that Tatars and Chuvash who lived as peasants on the archbishop's estate had to pay in kind and money for honey and fish.[8] It seems improbable that Russian peasants living in the same environment totally ignored such resources as lay at hand.

The register is also defective in a number of other ways. It is not simply that there are gaps, as well as errors in the transcription. The copyist was aware of at least some of the defects. 'But where after this item in the register something was written there are no sheets; they have rotted.'[9] Another entry noted that seven sheets were missing.[10] The officers of the inquisition had themselves

---

[1] Ff. 210v, 211, 240v, 241.   [2] F. 241v.   [3] Ff. 215v–216.
[4] F. 216.   [5] F. 218.   [6] Ff. 139, 139v, 141, 141v, 173v.   [7] Ff. 173v, 254.
[8] F. 212.   [9] F. 107. Similar entries are on ff. 110, 172v, 201v, 204.   [10] F. 223v.

attempted to correct errors arising in earlier registers, amending a double entry and trying to establish the true holder of certain lakes.[1] They had also carried out enquiries in order to distinguish between prisoners' and converts' holdings in a large settlement.[2]

In such cases they seem, too, actually to have surveyed the land, and not to have treated this as a mere book entry. At least we have the lively entry, relating to land to be divided between two lords, that a boundary was not established 'because snow fell and the ground was frozen, and it was impossible to measure out their arable land equally'.[3]

From the point of view of our concerns, the defects of the 1565-8 register lie not in the gaps and the minor errors introduced in the process of copying, but rather in the fact that certain major aspects of the economic life of the area are not dealt with. First, details of the income of the lords holding by service were not included. The rubric relating to grants made in 1565-7 refers expressly to the fact that in the 1563 register of Semen Narmatskii 'no income is noted according to his record in those court villages and hamlets, other than from the arable land; the peasants of live vyts in the villages and hamlets tilled for the Sovereign Tsar and Grand Prince desyatinas a vyt, a desyatina of rye and a desyatina of the spring sown crop'.[4] The new officers of the inquisition themselves did not note any income for the lords from the peasants 'other than that the peasants are to till arable land for the lords, a desyatina [1 ha] a vyt, as it was for the Sovereign'.[5] The lack of this information means that we have even less material relating to the livestock and gathering sectors than we had for Moscow and Toropets; in the registers for those areas material on income was included and the items of petty income (in kind) helped for these sectors. Secondly, the Kazan' register, focused on the allocation of lands to tenants holding by service, pays little heed to peasants. It gives the impression, however, that diminishing attention is paid in sequence to Russians, prisoners (i.e. those formerly captured by the Tatars), converts, Tatars, Chuvash and Mordva. Nonetheless, it seems possible to say something more about the nature of these estates and of the peasant units on which they were based.

If we classify various types of estate by the different composition of the labour force and calculate average amounts of arable land per tenement and per person, we can make some deductions and indulge in some speculations.[6] It appears odd that there should be a group of lords with no peasants or people (i.e. with neither serfs nor slaves); moreover, among lords holding before the 1565-8 inquisition (I) this was the largest category, and among those holding as a result of the new allocation (VI), it was the second largest. The existence of this category, however, may be the result of defects in our source. At any rate the very high standard deviation for these data (column D) and

---

[1] Ff. 175v–176, 241. [2] F. 175. [3] F. 184.
[4] F. 134v. This standard obligation would be a *yuft'*. [5] F. 135.
[6] See appendix 4. Available data is too scanty for further statistical evaluation.

for lords with slaves only (column C) suggests that only the data on lords with both peasants and people (A) or with peasants only (B) are sufficiently reliable to be considered anything like real. Even then, the reality is of a particular type; the data derived are more bunched, in terms of both tenements and persons, when lords are included, so this reality was apparently a conventional one. The material need not be taken to imply that lords actually worked the land with their own hands alongside their labour force, though that is a possible interpretation. It is more likely that a strong evaluative, normative element in the attitude of the officers of inquisition was reinforced by a real, though widely varying, requirement of 7–13 chets per field of good arable per head for adequate maintenance, i.e. 10–19 desyatinas (11–20 ha) of total arable area. The amount of land per working man was probably about 10–17 chets ($5\frac{1}{2}$–9 ha), but with amounts fluctuating widely about this range.

Only for lords holding from new allocations (VI) do we have a full series according to the classification in appendix 4. This category is, thus, of particular value. The average amounts indicated for the large category of such lords with peasants only (VI B), both per tenement and per man, fit well with the figures indicated above. There were about 10 or 11 chets per man and slightly more per tenement, i.e. a total arable area of about 15 desyatinas (16 ha). The proportion of arable land to total area granted was, of course, small.

The estates not taken up (VII), though we have data for only a small number, were too poor or too small at around 7 chets per working man to maintain a lord.

The ecclesiastical estates (VIII and IX) also had small to miniscule amounts of arable land per working man. Two possible explanations are that this is an under-valuation of church estates or, more probably, that these estates, based on the holdings of the former Moslem church, had an unduly large Tatar and Chuvash population; in any event, not all the labour force on such estates would be directly engaged in working the land and some who should be excluded from this category may in fact be included due to the nature of our source.

The indications then, are that viable estates were evaluated by the officers of inquisition at about 15 desyatinas (16 ha) of arable land per working man; and that on estates held by lords before 1565 as well as after it, in many cases, this was the average size of a dependent peasant holding. In this area, of course, there were no free peasant lands.

The links between the countryside and the town of Kazan' are not as clear as could be wished. In part this is due to the nature of the information available, but in part it reflects the partial maintenance, at first, of such important lords as the archbishop and the monasteries by grants of produce from the tsar. The archbishop, for instance, had ninety peasant households in the late sixteenth century; almost thirty of these were engaged directly in the provision or processing of food. There were ten fishermen and two seiners, four butchers and two oil-men, for instance.[1] The archbishop's 'peasants' 'put up buildings and fencing at the archbishop's palace, cart firewood and water to the palace, make kvas and till the gardens beyond the town, work at the archbishop's mill and do every other sort of work; and they said that a quit-rent was taken for the archbishop's treasury from the peasants of the free settlement in place of work'.[2] The thirty-nine peasant tenements of the Transfiguration monastery had similar duties; 'they cart firewood and water to the kitchens, make kvas, build stables for the cow-yard beyond the town, till the garden and do every sort of monastery work'.[3] The last item in this case, however, probably includes general fieldwork. The gardens (*ogorody*) sometimes amounted to areas of 5 or 10 desyatinas (5½–11 ha), though individually they were probably quite small. One of half a desyatina (½ ha) is mentioned; this was a newly cultivated patch. 'And Petrushka has 10 beds [*gryady*] added to his garden'; probably raised beds used for vegetable growing.[4] One group of gardens held by three brothers had connections with an onion seller.[5]

The archbishop was maintained by a large grant of grains from the state Treasury: 1,300 chets of rye, 500 of rye flour, 300 of wheat, 100 of wheat flour, 1,000 of oats, 60 puds of butter and, later, 100 puds of honey. By the seventeenth century, however, this grant of produce was being replaced by grants of lands and this resulted in increased involvement with the countryside around the town (shown in table 21). At the same time the number of monasteries and the size of their estates also increased.

TABLE 21. *Holdings of the archbishop of Kazan'*

	1565–8	1603	1623
Villages	4	4	7
Hamlets	4	15	20
Peasant tenements	106	406	695
Arable (desyatinas)	2,775*	11,613	26,589
Meadow (ricks)	10,000	20,730	22,860

Source: Khudyakov, *Materialy po istorii Tatarskoi ASSR*, xix–xx.

* This information does not agree with that given on ff. 211v–12, where total good arable for the archbishop's holdings is given as 1,068¾ desyatinas, plus 1,168¼ desyatinas of long fallows and 358 of overgrown and tillable oak woods. Khudyakov also makes the astonishing error of adding and then multiplying the versta measurements of forest (see p. 212); his figures are thus meaningless and so are not included here.

The market had twenty merchants, mostly from Pskov and Vologda; there were also links with Kostroma. There were 376 stalls, 102 benches, 86 shelves

[1] *Materialy po istorii Tatarskoi ASSR*, 42–3.
[2] *Ibid.* 43. There is no explicit general obligation to till desyatinas.
[3] *Ibid.* 44.   [4] *Ibid.* 47.   [5] *Ibid.* 48.

and 35 tents.[1] About half the rows of stalls, to judge by their names, dealt in foodstuffs: fish, groats, wheaten rolls, meat, milk. Preserves, pies, sweetmeats and fools were sold.[2] Only ten of the stall-holders, however, were specified as peasants: 4 in Iron Row, 2 in Groats Row, 2 in Fish Row, 1 in Silver Row and 1 in Kostroma Row.[3] Evidently in this newly colonised area where there were no black peasant lands 'peasant' often merely indicated a dependant and not necessarily a land worker; so there is a lack of evidence for any considerable marketing on a regular basis by peasants. Moreover, much of the foodstuff needed to maintain the town's population was either traded by men of the artisan quarter (not farmers) or supplied direct to dependants without passing through the market. Kazan' was more important for its long distance trade, along the Volga and to Central Asia but this took time to recover from the interruption of the Russian conquest.[4]

Apart from the important position of the archbishop and monasteries from the time of the conquest, boyar land-holding grew, especially during the period of the Oprichnina when a number of nobles were despatched to Kazan'.[5] Nevertheless, holdings of Tatar military servitors also remained in the seventeenth century and were probably dominant in the countryside beyond the immediate influence of Kazan' itself. Such holdings were quite small; often, about 50 chets (27 ha) a field.[6] In fact, 'the subordination of the Meadow Side [i.e. the low-lying area east of the Volga] to the Moscow government only took place at the end of the sixteenth century when the Tsarevo Kokshaisk, Sanchursk and Yaransk forts were built there'.[7] Even then control of this hinterland was evidently not sufficient to prevent Tatars, Chuvash, Mordva, Mari and Udmurts from taking part in risings against the Russians in the early seventeenth century. Thus, we have a picture of the Russian conquest displacing Tatar rule from the town and its immediate vicinity, but only gradually able to exert effective control of the more remote parts of the hinterland where native rule long remained. Nevertheless, Kazan' was the crucial staging-post for the Russian advance on the Urals and across Siberia. Within a hundred years of the capture of Kazan' there was Russian settlement on the shores of the Pacific.

[1] *Ibid.* 70–1.   [2] *Ibid.* 58.   [3] *Ibid.* 61, 63–5.

[4] *AAE*, I, no. 329, 389–90; *AI*, I, no. 184, 346; no. 193; Fekhner, *Torgovlya*, 56–7.

[5] *Razryadnaya kniga*, 213, 214(7073); 241(7079); 256(7083).

[6] Tikhomirov, *Rossiya v XVI stoletii*, 496, citing Mel'nikov, *Akty istoricheskie i yuridicheskie*, I.7.

[7] Tikhomirov, *Rossiya v XVI stoletii*, 506.

PART III

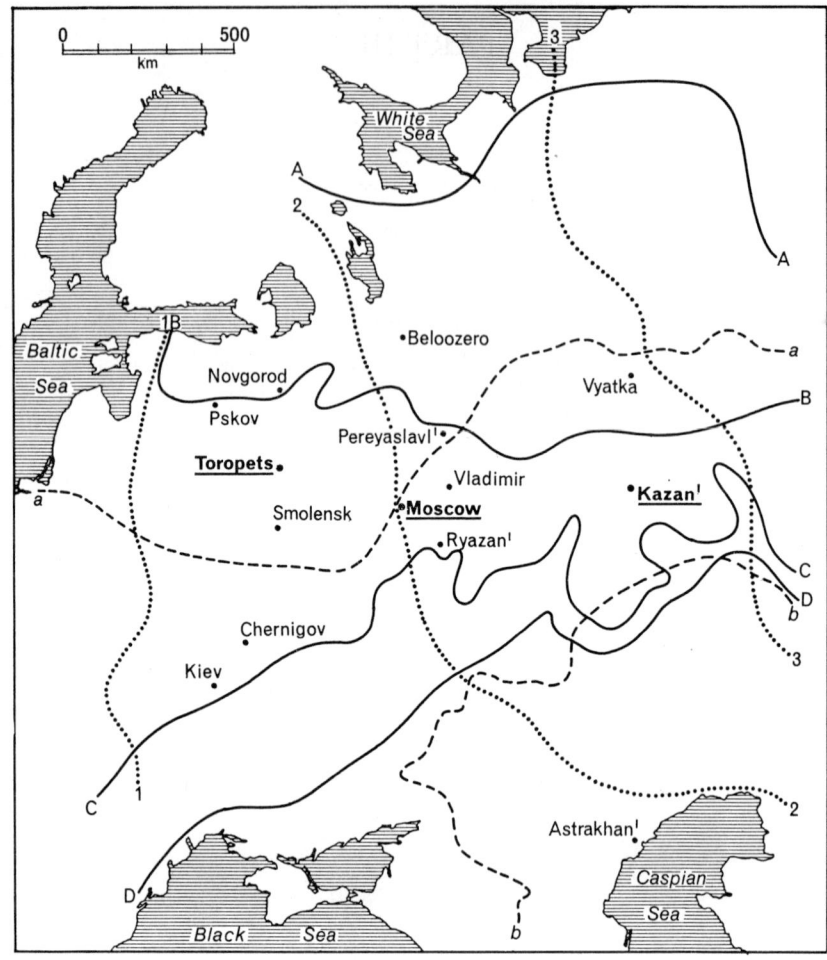

Map 5. European Russia: climate and vegetation.

Vegetation: A, northern limit of rye cultivation; B, southern limit of coniferous forest; C, southern limit of mixed and deciduous forest; D, southern limit of forest steppe.

Moisture: *a*, southern limit of surplus moisture; *b*, northern limit of area liable to moisture deficiency.

Winters	*mean temperature in coldest month*
east of 1 moderately mild	− 5°C to −10°C
2 moderately cold	−10°C to −15°C
3 cold	−15°C to −20°C

10

Peasant farming and the state

Conditions of soil and climate in the three areas examined differ appreciably (map 5 and table 22), the climate becoming increasingly difficult for farming as one moves from west to east. There were also differences other than those of physical environment. Moscow is centrally located in Russia in Europe. Toropets, however, was historically on the fringe of the Polish-Lithuanian state; its Russian dialect, for example, is basically North Russian, but with Belorussian features. Kazan' was only colonised by Russians after the military conquest in 1552; it remains a basically Tatar area. The Slav areas had a grain-based diet, the Tatar area had a livestock-based diet.

The differing areas were linked by the interrelated processes of colonisation by Russian peasants in the post-Mongol period and during the emergence of the Moscow state. We have to distinguish various types of colonisation. There was internal colonisation, that process of settlement within the vast forest area occupied by the East Slavs which led to axe meeting axe, to settlements and estates in some places coming to have contiguous bounds. Often this was the natural creep of peasant land settlement associated with an early marriage pattern; this, together with partible inheritance at the level of the household production unit, encouraged the farming of available land in closes with a simple, cheap and readily accessible technology. This was the more positive side of the somewhat gloomy view propounded by Lyubavskii, for example. 'Settlement throughout an extensive, wild country long condemned the Russian people to a primitive extractive and agricultural economy, long deprived them of live and close contacts through economic and cultural exchange, long paralysed the state centre, its development, energy and authority.'[1] The different stress is partly a difference of viewpoint: Lyubavskii is looking at the process in terms of state concerns; peasant flight and resistance seems to demonstrate that most peasants preferred life without the state. Their struggle with nature was hard, at times brutal, but they often evidently felt it was not as hard as the exactions and injustices imposed on them by the state.[2]

The natural creep of peasant land settlement could continue only until axe

---

[1] Lyubavskii, *Istoricheskaya geografiya*, 196.
[2] The right to opt out of liability to state taxation remained a live peasant memory, at least in the north, throughout the seventeenth century; Veselovskii, *Soshnoe pis'mo*, II.144–5.

TABLE 22. *Summary of soil and climatic data*

	Moscow	Toropets	Kazan'
	Southern taiga sub-zone of turfy podzol soils		Deciduous forest zone of grey forest soils
Av. temp. (t) warmest month (°C)	16.5–18.5	17.5–18.5	18.0–19.0
coldest	−2−−8	−8−−14	−13−−16
Days above 0°C	218–251	198–228	191–195
10°C	117–150	115–145	119–123
15°C	55–58	55–85	71–75
No. of spring days 5–15°C	54	53	43
No. of autumn days 15–5°C	57	50	44
No. of frost-free days	141–174	120–150	111–115
$\Sigma t > 10°C$	1600–2200	1600–2000	1800–1900
of which, after harvest of early spring-sown grains	0–600	0–400	200–300
Precipitation (mm)	550–700	500–600	450–500
Estimated evaporation (mm)	350–450	450–500	420–450
Snow cover (cm, av. max.)	20–60	40–70	55–70

Source: Pochvenno-geograficheskoe raionirovanie SSSR, 382–3, 390–1, 398.

met with axe. Deliberate colonisation by lords, often involving seizure of peasant lands, and the extension of the state frontier then became more characteristic. But the acquisition of new, sparsely settled areas, such as much of the Volga, the steppe zone and, of course, Siberia, opened up possibilities for a new phase of that first type of peasant land settlement. This occurred from the mid sixteenth century. It coincided with and contributed to the diminution of agricultural activity in the core of Russia, a diminution resulting from the growth of a developed system of tenures held by service, the Oprichnina, the strains imposed by the exhausting war against Livonia (1558–83) and, later, at the end of the sixteenth and the early seventeenth centuries by the complex events of the Time of Troubles.

Internal colonisation resulting in a pattern of small, probably non-nucleated settlements, characteristically hamlets of one to four tenements, certainly continued at least into the sixteenth century. Only in areas such as Opol'e, where forest cover was absent, were large farms important before this. The tenement was crucial for this aspect of colonisation and farming; indeed, hamlets sometimes had only one tenement. Manors and estates were in the main composed of a number of such settlements; even villages were of small size, except in the Moscow uezd, and, as regards farming, were not always sharply differentiated from their subordinate hamlets. Even where, at such nodal points, the three-field layout was still developing and being extended by the incorporation of outlying sites in the sixteenth century, production took place on both central demesne arable land and on outlying peasant arable land in a variety of forms.

The claims of the Grand Princes, later of the tsars, in the fifteenth to sixteenth centuries had been to rights over land; but these rights varied and included an amalgam of administrative and economic elements. Similarly, the extension of the claims of ecclesiastical and lay lords often regarded as the growth of large-scale land-owning, meant the assertion of rights (with certain exceptions) to administer justice over, and exact obligations from, those on the land. The direct working of the land, however, was less often involved. Moreover, at least on the black lands, deals in land by peasants who sold it or disposed of it by gift or bequest, were commonplace alongside the superior rights of the nobility and monarchs.[1]

The peasant tenement, or average hamlet of a few such units, exploited the land with a simple, though by no means primitive, technology: sokha, scythe and axe were cheap (if they had to be bought) and also easily maintained. The combination of farming on closes, clearances and patches of all sorts in the forest and of gathering provided more than subsistence, though it stressed use rather than exchange values. Largely self-sufficient though they continued to be, such units always had some links with the market; and these links grew and developed as state demands, especially those at least nominally in money, increased during the sixteenth and seventeenth centuries. The impact on the work unit, however, often took the form of an increase in labour rent. Military service also meant experience outside the isolation of the hamlet.

Both colonisation and, we assume, population pressure, on the one hand, and demands for military servitors and their attendants by the emerging tsardom, on the other, led to a number of consequences in the late fifteenth and sixteenth centuries.

First, estate boundaries had evidently come into contact with one another at least in the central area. The law code of 1497 (paragraph 61) mentions common fences between both villages and hamlets.[2] Even when bounds were contiguous, however, there was still much land available for cultivation; arable land (in variable units, if measured at all) was rarely in contact with other arable land, especially at peasant level. The isolated close, not the open field, here remained the characteristic unit. The vague indication of use enshrined in the fourteenth–fifteenth-century three-fold formula 'wherever the sokha, scythe and axe have gone' slowly gave way not only to an indication of boundaries, but also to new concepts of property and associated rights which

---

[1] Veselovskii, *Soshnoe pis'mo*, II.97–8; Makovskii, *Pervaya krest'yanskaya voina*, 125f., especially 132, 135; his *Razvitie*, 313–439 deals with peasant land ownership; see also Cherepnin, in Novosel'tsev, Pashuto & Cherepnin, *Puti razvitiya feodalizma*, 210f.

[2] *Sudebniki XV–XVI vekov*, 27, 99–100. The commentary stresses the growth of land-holding encroaching on common rights.

were alien to the old system based on use and usage. There were conflicts over land rights between the two systems, as well as between lords within the new system.[1]

Secondly, grants of privilege in the fourteenth to fifteenth centuries did not have the formula, found in sixteenth century documents, that peasants pay tribute and certain other main taxes 'according to the registers' or 'according to what the officer of the inquisition records'. They had to pay what they could (po sile, literally, 'according to their strength').[2] Kashtanov goes on to argue that the officers of the inquisition had no right to enter old-established lands, unless a court case arose, even in the early sixteenth century. Evidently this distinction between old lands and the new ones liable to inquisition arose in connection with the development of inquisitions in the second half of the fifteenth century. The great boyar and monastic landowners saw the inquisitions of the Grand Prince as an encroachment on their rights; the latter was evidently only able to effect such inquisitions piecemeal. We know that extensive inquisitions were carried out after the wars of the second quarter of the fifteenth century, in the 1480s and 1490s, but the function of the first officers of the inquisition was evidently largely limited to the establishment of boundaries and decisions about land use.[3] Their records were thus allocation record books rather than true registers of inquisition, and included neither details of tenements and the peasants in them, nor the number of sokha-units in the estate. Intermediate registers included the number of sokha-units, but not details of tenements, as in later cadastral records. The hundred books (sotnitsy), though the hundreds themselves were of ancient origin, illustrate this.[4] The delimitation of the estates of the boyars and monasteries thereby helped to establish what was crown and state land and thus to make clearer the basis of the Grand Prince's power. The lords came to be as it were a lower rung of the state administration. Peasants were evidently no longer able, as in the fourteenth century, to claim a right to a say in the disposition of the land off which they lived; in the late fifteenth century they were being reduced to a mere labour force of the estate.[5] Use values, moreover, were being supplanted by a new stress on exchange values which violated the autarky of the household unit.

Thirdly, as the distribution, in exchange for military service, of court lands and the lands of black peasants became a developed system around the middle of the sixteenth century, the agricultural basis of the national economy had to support more men, officers and their attendants at a higher standard. Moreover, the increased size and extent of the organisation of the military forces

[1] See appendix 1.
[2] Kashtanov unpublished MS. For Moscow uezd examples of the latter formula occur in *ASEI*, I, nos. 291, 305, 401.
[3] See *AGR*, 1.70; Shumakov, *Obzor*, II. no. 116 (1482).
[4] See Shumakov, *Sotnitsy*, I.122–4, no. I–II; 128, no. V; 134–6, no. VIII.
[5] Smith, *Enserfment*, 15–16.

(and of the colonisation of the eastern regions and Siberia from the late sixteenth century) meant increased overheads. The number of officials grew.
Greater quantities of grain were collected, processed, transported, stored and
distributed to the army; and at all stages in this extended chain of operations
losses would occur: from spoilage, insect and animal pests, human peculation.
This pressure for increased agricultural productivity, or at least for a greater
quantity of delivered produce, occurred at a time when agriculture was
devastated in many parts of the country by the Oprichnina and warfare.
What emergent peasant entrepreneurial activity existed was interrupted; the
late sixteenth century saw a great development of villeinage. It was lordly,
mainly ecclesiastical, entrepreneurial activity which, in so far as there was
any, became characteristic.

There was no significant widespread technological change in implement
types at this time, probably because of shortage of metal (required for warfare)
and the absence of a general peasant demand for improved implements or
metal implement parts on any scale. The increased monetisation meant, in the
main, an increased state demand for money, but only a limited amount of
money was in peasant hands. Peasants in areas such as Moscow, having a
developing market, will have had more money, as well as greater incentives
to market produce, than those in remoter areas such as Toropets. Perhaps this
suggests an explanation for the several finds of virtually identical pairs of
sokha irons in Moscow;[1] this was linked with increased population densities,
more regular demand for agricultural produce and changes from cultivation
on closes to field farming. The adoption of some three-field organisation
probably contributed to more regular production, but not necessarily to
higher yields.[2] It must be noted, though, that rising market prices for produce
did not necessarily stimulate production. The peasant production unit was
not market, but consumption-oriented; so higher prices tended to reduce
peasant consumption rather than increase production.

We have seen that, in the 1470s and 1480s in Moscow uezd, evidently
recently colonised arable lands might be taken in 'because they have come to
the monastery lands' and that the Grand Prince was also involved in exchanges to improve layout.[3] In the 1540s in Toropets a new clearance was
'let into the field as arable' to a village of post-horse men; similar incorporations took place on a large estate held by service where active land exploitation
was taking place.[4] In the 1560s in Kazan' a crown settlement had an incorporated third field.[5] Evidently these late fifteenth–sixteenth-century attempts
to rationalise layout and extend the unit of arable land took place on the
estates of lords, monasteries and better-off peasants such as post-horse men;
probably, like the slightly earlier spread of the three-field layout itself, such
rationalisation was generally restricted to such levels at this time. The Kazan'

[1] See p. 15.   [2] Appendix 1.   [3] Pp. 243, 123.   [4] Pp. 166 and 192 above.
[5] P. 204 above.

evidence, however, suggests that incorporations sometimes took place at a somewhat lower social level.[1] In Moscow uezd we have no such evidence for the 1560s. In any case, not many attempts at rationalisation are known, though there were somewhat more in the 1580s.[2] It is noticeable, however, that even then it was usually heritable or ecclesiastical estates which showed this concern for a more rational layout.

The three-field layout, occasionally with incorporated land, was worked on such estates by villeins (or, possibly, by enserfed peasants); this enabled exactions and obligations to be closely controlled on all supervised estates. The importance of the growth of demesne arable land and a system of close control on estates should not be exaggerated; even an important Moscow monastery continued towards the end of the sixteenth century to rely on both demesne arable land and a variety of other forms, including peasants' own holdings, for its income.[3]

It was also only on the advanced estates of the tsar and of ecclesiastics that changes from gathering to cultivation (represented by orchards and fish ponds) were found in the sixteenth and early seventeenth centuries. Control was mainly exercised over appropriation rather than production.

The local diminution of agricultural activity in what Goehrke has called the Great Waste Period (1580–1620) was in large part the result of physical destruction and excessive demands associated with warfare.[4] The data we have, especially those from Toropets, suggest that it was larger settlements and those on roads which particularly suffered; precisely those which might have contributed most to growth and development.[5] The hamlets and isolated households might be compelled to move if they survived, but their simple technology and relative independence from the market enabled them to re-establish themselves elsewhere as long as land was available.[6] It was this largely autarkic nature of the household farm which allowed the Russians to survive so many disasters, albeit at a low level, and to recover from them fairly rapidly (as in the 1590s and in the 1620s–1630s). This type of peasant economy ensured survival for many who would otherwise have perished; but it did not enable peasants to co-operate effectively in resisting the encroachments of the state; its simple, largely single-cell, structure prevented much specialisation and put a brake on general economic development. In amoeba fashion, at the micro-level, the farm households replicated themselves, colonising the vast East European plan; but on the macro-level of society as a whole, only the violence of the state reduced the entropy of these basic constituent units and formed them into a more complex organism.

Increased concern for control of its constituent parts by the Moscow state is demonstrated by a number of features. Lists of mainly military officers were compiled from the 1470s; the office dealing with allocations of military posts

[1] P. 208 above. [2] P. 125 above. [3] Appendix 2.
[4] Goehrke, *Die Wüstungen*. [5] Pp. 168, 122 above. [6] See p. 168 above.

Razryadnyi prikaz) emerged in the first half of the sixteenth century.[1] The growth of Moscow's power is also signalled from the late fifteenth century by the issuing of largely administrative laws (*Sudebniki*) intended to apply over the wide areas acquired with the help of the armed forces.[2] The first attempt at a more systematic, codified law book took place only in 1649.[3]

From the mid sixteenth century attempts were made by Moscow to establish a uniform system of weights and measures throughout the country. Units of measure bearing a government imprint were despatched to various localities.[4] Nevertheless, even in the seventeenth century, local measures, different from those of Moscow, continued to exist. Such units were often measures of capacity; a type of measure frequently used in farming. Thus, Staden's statement that Ivan IV 'achieved, throughout Russia, throughout his realm, one faith, one weight, one measure' was an overstatement; but it reflected the intention in the Moscow state.

A common coinage for the Moscow state was introduced with the monetary reform of the 1530s. The Moscow mint was established and local centres were closed down.[5] Thus, the uniform regulation of coining by the Moscow state came later than the earliest inquisitions, but somewhat earlier than some other measures aimed at uniformity and the elimination of local custom. It is also remarkable that, despite the convulsions of the sixteenth century, the coinage remained of stable weight from 1535 to 1606. Table 23 shows the number of coins minted from a given quantity of silver during the sixteenth and much of the seventeenth centuries.

TABLE 23. *Moscow dengas coined per small grivenka of silver*

	Number of dengas	Index	Silver per denga (gm)
–1535	520	100	0.392
1535–1606 or 1610	600	115	0.340
During intervention	700	135	0.291
1613	800	154	0.255
1620s–1640s	850	163	0.240
1640s–1680s	900	173	0.227

Source: PSRL, IV.296; Kamentseva & Ustyugov, *Russkaya metrologiya*, 143–60, has a convenient summary of the debate on coining. The small grivenka, half a Russian pound, weighed 204 gm.

[1] *Razryadnaya kniga 1475–1598 gg.*
[2] *Sudebniki XV–XVI vekov.* Translations in Dewey, *Muscovite Judicial Texts*, 7–21; 45–74; Vernadsky, *Source book*, 118–19, 134–7 (excerpts).
[3] *Sobornoe ulozhenie 1649 g.*
[4] Ustyugov, *IZ*, 19.311, 317–19.
[5] Kaufman, 'Serebryanyi rubl' ', 53, 65; Spassky, *The Russian monetary system*, 109–12.

In order to link Moscow with its dependent territories literacy and numeracy were required. The clerks of the departments numbered only 50 or 60 early in the seventeenth century, but had roughly doubled in number by the middle.[1] They elaborated a new style of language: the early laconic, often obscure, brief records were replaced by long, repetitive formulae, the restricted code of the royal bureaucracy of the sixteenth and the seventeenth centuries. In the same period Russian bead-boards were developed to cope with the complicated calculations involved in dealing with land areas and with fractions.[2] By the seventeenth century then, the clerks and junior clerks had to be specialists, but it was still possible for an officer of inquisition occasionally to be unqualified. Ten years after Foka Ratmanov Durov had recorded Tot'ma he acknowledged, when errors were discovered, that 'he did not know about sokha-units nor how to calculate them'.[3]

It was in the late fifteenth century that, as we have seen, the first registers of inquisition (*pistsovye knigi*) were compiled; the earliest extant ones, from the early sixteenth century, relate to Novgorod and Toropets, areas newly incorporated into the Moscow state. A system of standard allocations in return for service developed in the mid sixteenth century.[4] In terms of size – actually, of their estimated wealth, since any items of income were notionally counted as chets of arable land – estates held by service in the sixteenth century showed regional variations (table 24). It is not unexpected that the Moscow area had a greater proportion of larger or wealthier estates than Kazan' and still more

TABLE 24. *Allocations by size (estates held by service) (percentage)*

		Chets of good quality arable land					
		1–14	15–49	50–99	100–349	350–799	800 or more
Moscow*	1584–6		1.8	13.9	28.6	21.4	34.3
Toropets	1539–40	0	0.3	6.4	44.0	49.3	0
Kazan'	1565–8	0.1	3.7	12.8	27.1	44.9	11.4

Source: Rozhkov, *Sel'skoe khozyaistvo*, 438–40, 443.
* Heritable estates in Moscow showed a somewhat different pattern

| | | 0.5 | 11.9 | 17.1 | 39.0 | 19.9 | 11.6 |

The higher proportion of smaller sized allocations is presumably due to official policy as well as to the impact of partible inheritance.

[1] Bogoyavlenskii, *IZ*, I.228–9. In the same period the number of departments increased from about 40 to about 45. For each clerk there were 2 or 3 junior clerks.
[2] Spasskii, 'Proiskhozhdenie i istoriya russkikh shchetov', 321ff, especially 331–7, on fractions and the need to calculate land areas in terms of units of money. See also Ryan, 'John Tradescant's Russian abacus', *Oxford Slavonic Papers*, n.s., v.83–8.
[3] Veselovskii, *Soshnoe pis'mo*, II.34–6.
[4] *Tysyachnaya kniga 1550 g.*, ed. Zimin. See p. 125 above; Smith, *Enserfment*, 20–2; Blum, *Lord and peasant*, 141–2.

than Toropets. If we consider estates of over 350 chets, however, Moscow and Kazan' had about 56 per cent of their allocations in this broad category. Moscow and Kazan' also had about 16 per cent of estates with less than 100 chets. Toropets shows quite a different pattern with 93.3 per cent of allocations in the intermediate range of 100–799 chets. One hundred chets (say, 5–10 peasant households) was probably the minimum required to maintain a military servitor and enable him to perform his service.[1] The highest ranks had land allowances of 200 chets or more.[2] Thus, Toropets appears to have been in a position to maintain a higher proportion of its tenants holding by service; Kazan' and especially Moscow had a greater proportion of both larger and smaller estates. In this sense Toropets may be regarded as more characteristic of an earlier stage of state development, before control of military servitors and food supplies had been made more centralised.

The registers of inquisition also became somewhat more regularised, especially in the seventeenth century.[3] The aftermath of the Time of Troubles demanded a whole series of surveys from 1613 onwards. These developed particularly after 1620, and the techniques then established lasted at least till the 1670s.[4] The officers were issued with detailed instructions and with measuring ropes; they had 'to measure the arable land and to write in the books, measuring in one field, and in two at the same rate'. Yet the continuing pressure of the forest and of the old symbiosis of the three central fields and numerous outfields is felt even in the mid seventeenth century.[5] The officers sometimes made use of maps in the case of land disputes; this experience evidently underpinned the production in 1627 of probably the earliest Russian major cartographic work.[6]

The system of tenures held by service involved not only the provision of military forces for the state, but also the reduction, often the elimination, of peasant self-administration in the central areas, at least at manorial centres. In physical layout, the closes were giving way to enlarged arable areas around the lord's settlement. In social terms, the free peasant became a serf (or, sometimes, especially in the late sixteenth century, a villein). In economic terms, the estimation of the surplus 'according to strength' (a system which might have acted as a brake on enterprise) later became taxation according to the number of tenements. Sokha, scythe and axe were implements of production; in the

---

[1] See p. 150 above on Stashevskii's evidence for the comparatively well-endowed Moscow servitors in the 1630s having about 21 peasants and labourers.
[2] Smith, *Enserfment*, 21. Rozhdestvenskii, *Sluzhiloe zemlevladenie*, 333.
[3] See especially *APD* and Sedashev, *Ocherki*, especially 107–9.   [4] Sedashev, *Ocherki*, 23.
[5] Sedashev, *Ocherki*, 45–6, cites an instruction to the officers about to carry out an inquisition in Tot'ma in 1645. The forest was, of course, more dominant there than in Moscow, but the instruction does not seem at all exceptional.
[6] *Kniga Bol'shomu chertezhu*, ed. Serbina. The map itself is not extant. In a volume by B. A. Rybakov, *Russkie karty Moskovii kontsa XV–nachala XVIv.*, an attempt is made to place the origin of Russian map making in the reign of Ivan III.

three-fold fourteenth–fifteenth-century formula they designated functionally specific work areas with no precise indication of limits. Later, charters indicate boundaries with some attempt at precise description. The registers of inquisition focus not on production, but on appropriation. At first, they indicated appropriation vaguely 'according to strength'; later, they specified amounts of income in grain or money, and petty income in livestock products and the produce of gathering, for the lord; and taxation, often at least estimated in money terms, for the state. Yet, paradoxically, the peasant tenement became the basis for calculating state taxation in the sixteenth and seventeenth centuries when its role as the main production unit was declining somewhat.[1]

The concern of lords to control the constituent parts of their estates more effectively, to ensure food supplies and income, is demonstrated most strikingly by the appearance of account books, books of income and expenditure. The earliest one known is dated 1531. The manner in which the account books of this period are drawn up suggests an early stage in their development.[2] While primarily concerned with any deals involving money, some registered even the quantities of hay, oats and other feed to be consumed by livestock.[3] Such accounts, however, were virtually restricted to monastic and a few boyar estates. There is a sixteenth century work extant which shows concern for production. This is a translation of *Opus ruralium commodorum* of Crescentiis, probably from a Polish fifteenth-century version.[4] A work dealing with the running of a great household, including many aspects of consumption, has also survived.[5] From the late sixteenth century monasteries also kept harvest and threshing books, as well as other registers dealing with their income.[6] The main concern of the majority of estates was not control of primary production, except indirectly through control of labour, but control of appropriation and consumption.

This brings us to the problem of obligations to lord and state; the peasant had to meet a range of obligations, both rents and exactions for his lord and state taxation. From the mid sixteenth century a number of peasant contracts made with lords are known (*poryadnye*).[7] These were notes of loans or assistance given to peasants in need who in return became dependent and under-

---

1 Veselovskii, *Soshnoe pis'mo*, I.233–4.

2 Mazdarov, *Istoriya razvitiya bukhgalterskogo ucheta v SSSR*, dates the adoption of accounting procedures, a form of single entry record keeping, by the Russian state and business enterprises to 1650–75.

3 Man'kov, *Tseny*, 15–17; Shchepetov, *IZ*, 18.112.

4 *Naziratel'*, esp. 69–74.

5 'Domostroi', *Vrem. MOIDR*, kn. I, 1849, otd II.1–116. A French version is available: M. E. Duchesne, *Le Domostroi*, Paris, 1910.

6 Some were published in *RIB*, XXXVII and *Kniga klyuchei i Dolgovaya kniga Iosifo-Volokolamskogo monastyrya*. Man'kov gives extracts from such books in *Materialy po istorii krest'yan v russkom gosudarstve XVI veka*.

7 D'yakonov, *Ocherki iz istorii sel'skago naseleniya*, 754; also his 'Akty otnosyashchiesya k istorii tyaglago naseleniya'; *PRP*, IV.92.

took to bear the burden of obligations after a specified time. In the mid sixteenth century, apart from such individual contracts, there were generalised charters calling on peasants to obey their lord (especially *poslushnye gramoty*), to fulfil various obligations and, sometimes, 'to pay the quit-rent which he imposes on you'.[1]

It seems impossible, however, to estimate the burden of obligations with any degree of accuracy.

In general, undoubtedly, the artisan quarters were often, but not always, taxed more heavily than the uezds; undoubtedly, lands held by lords paid much less to the state treasury than the sokha-units of black and court lands. Further than these general remarks it is impossible to go. The fact is there is no means of evaluating at all satis-factorily the burden of produce and labour rents paid by peasants to their lords, and without that it is useless and wrong to compare their tax liability to the state with the whole mass of taxes and dues paid by the black and court peasants.[2]

The difficulty is, in fact, greater than this; the evidence not only does not normally allow us to work out the burden per tenement, per area or per man, but we also know little of the amounts of produce and money actually handed over. We do not know how far, or, often, whether, interchange between pay-ment in produce and in money took place.

In the seventeenth century, according to Got'e, there were zones round Moscow distinguished by differing forms of rent: first labour rents, then rents in grain and, finally, money rents.[3] If this were so, it would be reasonable to expect differing effects on peasant holdings. Labour rents are likely to have had an adverse effect on the viability of the holding, since labour demands would occur at peak work periods, themselves somewhat inflexible owing to climatic conditions. In such circumstances, peasants would be prepared to pay high money rents in lieu. Grain rents would not interfere in this way with the normal work cycle of the tenement. Money rents, however, raise the problem of how peasants in an outer zone would obtain money. Distance might make marketing grain or other perishables more costly and difficult, and farms doing so would be exposed to market fluctuations. Thus, one might expect some development of crafts and trades and of off-farm activities in the outer-most zones. The evidence seems insufficient to support these speculations in detail, but Russian proverbs stress the importance of the axe (i.e. the forest) as the basis for money income.

Part of any peasant surplus might be traded; but statutes distinguished peasant trade 'among themselves' within an estate or locality, as well as imports (which were not subject to any impost), from trade elsewhere (which, was).[4] This suggests that the concept of trading for profit was still differenti-ated from local exchanges for consumption, perhaps in the form of bartering,

[1] *PRP*, IV.87; Koretskii, *Zakreposhchenie*, 23.   [2] Veselovskii, *Soshnoe pis'mo*, I.153.
[3] P. 150 above. Cp. von Thünen, *The isolated state*.   [4] Appendix 2, §15.

even in a Moscow monastic estate towards the end of the sixteenth century. In the Code of Laws of 1649 the focus had shifted from peasant consumption needs to the demands of lordly appropriation, but lords could not easily adjust rates of appropriation to market prices; so market-oriented estates favoured villeinage or other forms of close labour control.[1]

Grain and grain products evidently in considerable quantities were marketed in Moscow in the sixteenth century. Many town traders, artisans, hired men and some servitors with inadequate government grants had to buy food on the market.[2] The early-seventeenth-century assizes of bread and the data on the grain-based grades also imply much trade and a new need for control of food supplies for Moscow.[3] Unfortunately, we know virtually nothing of the quantities involved or of the sources of these supplies. Much grain not consumed by the producer will still not have passed through the market. Man'kov considered that the peasant was the main supplier in the sixteenth century; he was clearly the main producer, either on his own holding or as labour on the demesne arable.[4] But Man'kov failed to develop his statement that the process of grain becoming a commodity was most developed in the central zone of the state and, in consequence, involved the serf peasantry considerably in the circulation of the grain market.[5]

State taxation increased greatly in the sixteenth and seventeenth centuries A summary of available information relating to the sixteenth century is given in table 25, but these figures have to be treated with caution. Not all taxes were exacted at any one time or place; many were for special purposes, or were applied only in certain areas; some were commutations of labour obligations. Rozhkov concluded that

however inexact all these calculations are, we can say that money payments to the treasury changed as follows: in the course of the 16th century in the first two decades they were no more than 5 rubles from 800 chets, in the 1520s–1540s they reached 8 rubles, in the 1550s–1580s 42 rubles and, finally, at the end of the 1580s and beginning of the 1590s 151 rubles.[6]

He regarded the rise as mostly taking place in the second half of the century. Even if Rozhkov's calculations were overestimates, the increased use of money as a basis for estimating taxation, the conversion of labour obligations to money and the raising of some *ad hoc* money taxes, even on a limited scale, would in themselves have had considerable impact on the farm unit on which we have concentrated. Its predominantly autarkic nature depended both on available land and on a largely in-kind economy. Money could not be grown; it had to be won largely by means of crafts and trades. At a simple level such activities could be accommodated within the household unit and the usual

---

[1] Chapter IX, §§3–5 for example.   [2] Bakhrushin, *Nauchnye trudy*, I.35.
[3] Pp. 143–4; appendix 3, and table 14, p. 153.   [4] Man'kov, *Tseny*, 35.   [5] *Ibid.* 42.
[6] Rozhkov, *Sel'skoe khozyaistvo*, 223. The 80c-chet sokha-unit applied to lands held by service; peasant lands had a 500-chet unit.

TABLE 25. *Sixteenth-century tax rates*
(Moscow dengas per 800-chet sokha-unit)

	1510	1525	1500–1550	1551 to about 1580	1580
1 Treasurer's, clerk's and junior clerk's dues			200	200	
2 Maintenance for royal representative or volost head			256	256	
3 Tribute			280	4,000	
4 Post-horse monies			950	2,000	16,000*
5 Prisoner's monies			†	400	1,080
6 Service by sokha-units and defence work	Labour duties, but sometimes commuted, especially in second half of century.				
7 Taxes for defence work and siege works				800	1,360
8 Firearm monies				328	
9 Saltpetre monies				260	3,400
10 Siege works due				128	★
11 Allowance for junior clerks, carpenters and smiths					400
12 Assistance for post-horse men					1,520
13 Travel money for post-horse men					5,320
14 Maintenance of German prisoners					1,160
TOTAL	972	2,517	1,686	8,372	30,280

Source: Rozhkov, *Sel'skoe khozyaistvo*, 221–33. See Hellie, *Enserfment*, 125–6; Veselovskii *Soshnoe pis'mo*, I.414; Milyukov, *Spornye voprosy*, 90.

Note: The 800-chet sokha-unit applied to lands held by service; court lands and the lands of black peasants had a unit of only 500 chets.

★ Items 4 and 10.

† Not a regular tax in the first half of the sixteenth century.

calendar of activities, often using the labour of women and children. Any further development of such activities, however, meant work outside the tenement; to a great extent it involved not farming (which produced small surpluses), but gathering, trade and even some sale of labour off-farm.

The Toropets material gives some information on the burden of money payments per tenement around 1540; it also demonstrates the difficulties of interpretation. Defects in the text oblige us to calculate a minimum (1080) and a maximum number (1339) of tenements on the black peasant lands; they paid a total of 75 rubles 120 dengas in money.[1] This averages out at 11–14 dengas a tenement, roughly the price of a side of meat or a quarter of a chet of rye.[2] Individual entries, however, vary from a minimum of 4–6 dengas to a maximum of 44 dengas. Perhaps the former amount was reduced because of the obligations of the post-horse men; the latter very high amount, however, also excluded tribute. Rozhkov, on the basis of much other, independent data, estimated that money taxes in the Moscow state rose from about 2 dengas per chet in the 1540s to 27 or more by the 1580s.[3] Thus, according to this estimate, the average Toropets tenement had $5\frac{1}{2}$–7 chets. The average obtained from our limited data was in fact 9 chets per tenement, 7 chets per man.[4] Given the inherent difficulties, this is remarkable confirmation of Toropets farm-size and of the comparatively low level of money payment in the first half of the sixteenth century.

The situation had radically changed by the time of Giles Fletcher's visit to Russia in 1588–9.

Besides the taxes, customs, seizures, and other public exactions done upon them [i.e., the commons] by the emperor, they are so racked and pulled by the nobles, officers, and messengers sent abroad by the emperor in his public affairs, especially in the *iamy* (as they call them [i.e., the post-horse service]) and through fair towns, that you shall have many villages and towns of half a mile and a mile long stand all uninhabited, the people being fled all into other places by reason of the extreme usage and exactions done upon them.[5]

The tendency for money taxation to rise became even more marked in the early part of the seventeenth century, as a result mainly of war and invasion, though some taxes remained in kind (see table 26). The tax in grain intended to support musketeers (*streletskii khleb*) (100 chets of grain per sokha-unit in 1614) was commuted to money only among black peasants and in the artisan quarters of towns. The conversion rate was adjusted to halve this tax in 1619, as a result of peace with Poland, but when a renewal of war threatened in 1622 the old rate was restored. 'When in peacetime, between the two wars with Poland, the government halved the rate for posthorse monies and reduced the musketeers' grain even more, these two new taxes still exceeded the old artisan quarter taxation 10 times and uezd taxation 20–30 times'.[6]

In the first twenty to twenty-five years of its existence the musketeers'

1 'Toropetskaya kniga', ff. 131(43)v, 143(55)v, 126(38)v, 128(40), 158(70)v–159(71), 160(72)v, 161(73)v–162, 167(79)v, 169(81), 171(83), 171(83)v, 172(84), 173(85), 175(87), 181(93)v, 191(103)v, 205(117)v, 214(126)v; 158(70)v has been omitted from our calculation, since the number of tenements is not known.

2 Man'kov, *Tseny*, 104, 140.   3 See above.   4 Table 15, and p. 162.

5 Giles Fletcher, *Of the Russe Commonwealth*, in Berry and Crummey, *Rude and Barbarous Kingdom*, 170.

6 Veselovskii, *Soshnoe pis'mo*, I.238.

TABLE 26. *Seventeenth century tax rates* (per 800-chet sokha-unit)
(money in rubles, grain in chets, unless otherwise indicated)

			Musketeers' grain tax					
			(in kind)		(money in lieu)			
	Events	Post-horse monies	rye	oats	Private lands	Black peasants: towns of Novgorod office	Black peasants: towns of Ustyuga office	Exceptional taxes
c. 1600*	Time of Troubles–	10						
1613	Mikhail Romanov elected tsar							
1614		105?(35?)	100†		175 or 150	250		
1615			100† from some towns			150	171?	120–150
1616		280	80	40				
1617		350	from other towns					
1618	Peace with Poland		200	200‡	800	150	198½,200	
1619		800			200	150	90	
1620			100 yufts		120§	90	90	
1621		584			100	100		
1622	Threat of war		200 yufts		150§	200	200	
1623–6		468			75§ 70‖	60	90	
						White Sea towns		
1627–9					80§ 76‖		90	
1630		400			85§ 80‖		95	
1631–3	Smolensk campaign		100 yufts		80§		95	
1634	Peace with Poland						96	
1635–6		534					120	
1637–8			200 yufts				240	
1639–40			100 yufts		280¶		120	
1641		726						
1642								
1643		784?	700 yufts		672		168	
1644–53		784						
1654–60								

Sources: Veselovskii, *Sem' sborov* and *Soshnoe pis'mo*, 1.162–7, 170–5, 188, 414–17; Milyukov, *Spornye voprosy*, 91.

* Around 1600 heritable lands and those held by service were rated on average at 13r. 9 altyns 2 dengas (10r. post-horse monies, 2r. prisoners monies, 1r. 9 altyns 2 dengas maintenance for royal representative). Court lands were rated at 33–70r. plus grain or money in lieu, and black peasant lands, which had a smaller sokha-unit of 500 chets, at 48–200r.

† Rye, grits and oatmeal, estimated as equivalent to 65–70 *yufts*; Veselovskii, *Soshnoe pis'mo*, 1.189.

‡ Or 320 chets of rye, 40 of grits and 40 of oatmeal.

§ From distant towns.

‖ From nearby towns.

¶ From Polish towns.

grain tax was fairly uniform, and relatively moderate, throughout the state; but, probably in 1641, possibly somewhat earlier, the rates for different types of sokha-unit became differentiated. Artisan quarters, lands held by service, heritable and monastic estates were assessed at four times the rate for the black lands of the White Sea area and also, probably, for court lands.[1] In 1641 or 1642 the musketeers' grain tax was increased to 700 chets of rye and 700 of oats, converted at 192 dengas a *yuft'*.[2] The seven-fold increase in tax rates in grain did not mean a seven-fold increase in the burden of taxation, if we take into account the changes in the conversion rates of money in lieu and the differentiated rates for different types of sokha-unit.

Tribute and rent monies (*dannye i obrochnye den'gi*) or chetvert incomes (*chetvertnye den'gi*) came to subsume many other commuted taxes; these developed over time to a payment of 120–150 rubles per sokha-unit. The post-horse tax was not commuted for court and black peasants who continued to meet this obligation in the form of labour and the provision of horses. In the period 1614–60 the sokha-unit of black peasants was evidently rated in the range of 90–228 or 250 rubles (18,000–50,000 dengas).[3] Over this period the value of money declined at least $12\frac{1}{2}$ per cent (see table 23), nevertheless the erratic tax rates show a marked tendency to rise substantially, more than doubling in the case of the black peasants on whom the heaviest total tax burden fell. Tax-payers fled; the government was obliged to allow time for arrears to be met, or even to forego their collection altogether.

Man'kov made the point that 'it is very important to know, for example, how much grain a sixteenth-century peasant had to take to market in order to pay his quit-rent to his lord and buy himself some coarse cloth [*sermyaga*], an axe or a number of other essentials'.[4]

In practice it is not easy to make such estimates, especially if you attempt to restrict yourself to using prices from one area. Table 27 shows the prices in Moscow of livestock, axes (the prices of which are for the country as a whole), rye and oats. The paucity of data prevents any certainty in our conclusions and stresses the need for caution. Rye generally cost around 40 to 50 dengas a chet throughout the century, except during the 1570s and the late 1580s.

Livestock show a different pattern of prices. Geldings cost about 300 dengas until around 1590, through with wide fluctuations about a figure of 400 dengas in the 1580s; in the final decade of the century they cost around 1000 dengas. Horse prices were higher, but followed a similar pattern, rising from 200–400 dengas in the first half of the century to a peak above 2000 around 1580, then fluctuating above 1000 dengas until the early 1590s and apparently rising to about 1800 dengas in the last years of the century. A tentative explanation of stabler grain than livestock prices is that rye was a fairly stable crop, grown by virtually every tenement and was readily marketed. Prices could rise steeply when warfare, or other accidents

---

[1] *Ibid.* I.187.   [2] *Ibid.* I.162–7.   [3] *Ibid.* I.188.   [4] Man'kov, *Tseny*, 101.

TABLE 27. *Prices of livestock, axes, rye and oats, Moscow* (Moscow dengas)

	Horse (loshad')	Geld-ing	Mare	Horse (kon')	Colt	Bull	Cow	Ram	Axe*	Rye	Oats
										per chet	
1500		342									
1506				456							
1540	600										
1553				198					6–10		
1556										22	
1557										40	
1564					1248						12
1569										23	30
1570										200	100
1576		300									
1580				2178							
1583										40	20
1585		534		1188							
1586		318						24		45	28
1587		402	258				246	14		50	
1588		498		1290			258		15–20	50	
1589		300				198				90	
1591		1002		1020							
1592		1182									27
1593		708	714							30	21
1594										18	
1595		990									
1598		990		1782	1986						
1599											21
1600										40	19

Source: Man'kov, *Tseny*, 104–13, 122–33; 183–8. Table 12, p. 143 above.

* Data for Moscow is inadequate, so approximate prices derived from all available evidence are shown.

dislocated the market, but recovery was rapid; the widespread cultivation of rye, however, helped to minimise fluctuations in price due to storage and transport problems. Livestock prices rose three-fold or more in the course of the century. The rise in the second half of the century is probably due to greatly increased demand for horses, both for the army and for agriculture, after the destruction of the early 1570s and the later period of desertions. But livestock have a longer reproduction cycle than grain, so recovery was slower than with rye. There is a hint of recovery in the early 1590s, but by the end of the century there was evidently a further scarcity of horses. We lack material to calculate any net import of animals from the steppe zone.

Thus, while it seems impossible to assess the total burden of obligations on a peasant tenement in the sixteenth century, we can point to certain changes. Between the 1540s and the 1580s state taxation rose from the equivalent, in money terms, of less than 1 chet of rye to about 5 chets; the cost of a horse rose from about 10–12 chets to 22–45 chets; the cost of an axe, however, remained steady at the equivalent of $\frac{1}{2}$ or less of a chet of rye. There were no doubt considerable regional variations in the amounts and forms of obligations to lords. We have no hint of any reduction of such obligations, however; overall, therefore, it seems clear that in the half century from about 1540 the burden on peasants rose considerably and any increase in the amounts demanded in money meant the involvement of the farm in a new, more extended network of economic relationships.

The axe was the essential prerequisite for forest farming; in terms of grain it did not increase in price in the sixteenth century. Outlays in real terms on this item remained both steady and at a low level. With axe alone the main tillage and other implements could readily be made. Metal tips for the sokha were not essential, and for farming on the ash of closes may not even have been particularly advantageous. They doubtless became an advantage once three regular fields were being worked. The house, however, was sufficiently highly valued, especially where building timber came to be in short supply, to justify carting it to a new location when peasants moved.[1] Capital inputs in the basic peasant household production unit thus remained small as late as the sixteenth century; but at this time the constraints influencing the unit grew considerably and diminished or even eliminated its former capacity for largely independent activity. Nevertheless the relatively rapid recovery of the Russian economy from the 1620s onwards has ultimately to be ascribed to the continuing viability of this unit, its ability to ensure something more than mere subsistence for the overwhelming majority of the population.

It may be objected that there is no reason for the historian to concern himself with the dimension of peasant everyday life on which this work has focussed. After all, it may be alleged, it makes little impact on the historical record; it is social anthropology or ethnography rather than history. In terms of these objections in a sense there can be no peasant history, and perhaps this book should not have been attempted as history. On the other hand, Russian history in the period dealt with here would be so distorted were its peasant dimension unrepresented that it would be nonsense. To restrict peasant appearances on the historical scene to those significant but exceptional

---

[1] *Sudebniki XV–XVI vekov*, 27 (1497, §57), 172–3 (1550, §88); Smith, *Enserfment*, 80, 82, 92–3. *APD*, I, no. 120 (1623). Olearius commented on partly assembled houses being available at little expense in Moscow: Baron, *Travels*, 112.

occasions when mass disturbances irrupt on to the political stage is to compound the distortion of the documents. It is at least as historically important to depict the usual life of the mass of society as to focus on such cataclysmic events as the nominally peasant risings.

The tenement had been the primary cell, relatively little specialised, of the manor or estate certainly until the three-field organisation and close labour control had been effected and, in most cases, for long after. The manors and estates, in their turn, formed a level of administrative organisation (consisting of somewhat more differentiated and developed cells) of the state as a whole. Moreover, the series of increasingly complex units – tenement, hamlet, village, manor, estate, uezd, state – also displays increasingly complex networks of communications. At its simplest this appears to be due to the increasing number of units in the network at each stage, and the consequent rise in the number of links. Three units require three links for all to be in contact; increase the number of units to twenty and 190 links are needed.[1] In a community where decision making is not mediated by representatives but is face-to-face, it is probable that all units would be in contact, at least occasionally. This crudely indicates the likely density of communication links in a Moscow area hamlet and village. In addition, of course, the number of links has to be increased not merely as the number of units at each level in the series increases, but also because links are also required between levels; though, since Muscovy was a highly hierarchical organisation, this vertical type of link was not multi-stranded: relatively few officials linked different levels.

The change in the nature of the communications transmitted at different levels, however, is even more important. At the tenement, hamlet and village level the direct, relatively ephemeral nature of the communications in a universe with short spatial and temporal horizons, and with relatively few constituent units, meant that oral communication met most needs. Word of mouth was sometimes supplemented by such aids as tally sticks, personal marks and, perhaps calendars, but literacy was not an essential skill.

Law books, registers, lists, maps and survey calculations required literate clerks, at times professionally literate and numerate persons; moreover, the attempts by the state's officials to regularise the multiplicity of local custom and tradition they encountered gave rise to many court cases and, consequently, a great increase in the number of written documents. The question is, however, to what extent those documents which have survived reflect the total reality of the time.

Our documents, even if they are a valid sample of all those which then existed, adequately reflect little beyond that part of the network based on literacy. The lawsuits sometimes refer to peasants, sometimes a peasant may

---

[1] The formula for determining the number of such links (or more accurately, pairs of links – since they can operate in either direction) is $\frac{1}{2}n(n-1)$, where n is the number of units.

even be party to a lawsuit; the registers list peasant tenements household by household (though without describing them in detail). Both deeds and registers, however, and virtually all other documents, were concerned with the hierarchical situation of peasants; income and justice (itself a source of income for the lords) were virtually the only mediators between government and peasant. Peasant life, however, mostly comprised horizontal interactions between peasants. This peasant life was scarcely of interest to, or dealt with by, the clerks and the officers of inquisition; for the officials the peasant tenement, the hamlet, the village, and sometimes the manor and estate, were black boxes: they were units whose internal structure and activity were not defined or investigated; only the appropriation of their output of men, services, produce and money, was of concern. How the output was achieved did not matter. The peasant tenement and the peasant community, for their part, were viable economic and social units. They could operate without the state. The state, however, could not operate without them.

Yet the factors which enabled the peasant tenements and communities to prove so viable had a negative side, at least in the eyes of officers and monarch. Working abundant land in closes, with a simple technology, sometimes with slash and burn, generally meant peasants achieved an adequate real income. It provided little surplus for outsiders and gave them little means to exert any detailed control; these possibilities only came to be realised with the development of estates, a three-field organisation and serfdom. The peasant tenements and communities experienced little internal stimulus to long-term accumulation or to growth other than by replication of the existing units. The extensive colonisation of the vast East European plain is at once the major historical achievement and also determinant of Russian peasant culture in these centuries; intensive development and growth came with the exploitation and violence of the Moscow state. The tenement was able to succeed without the state as long as land was available for colonisation, whether resulting from the replication of peasant tenements or from peasant flight. Its very success, in turn, both diminished the effectiveness of state control and thereby stimulated further efforts of the state to achieve control. It generated a very low level of demand and thus contributed to a slow rate of economic growth. In 1649 Russian colonisation of Siberia reached the Pacific coast – flight could go on no further east – and the Code of Laws completed the legal enserfment of the peasantry. The means had been elaborated whereby, only half a century later, Peter the Great launched Russia on a quarter century of reforms and modernisation.

APPENDICES

I

Three-field layout

In order to try and establish what was the situation with regard to three-field layout, especially in relation to peasant lands, we will take all explicit references to three fields to be found in the fifteenth- and early-sixteenth-century documents available to us and see what picture emerges.

1. In April 1454 a monastery in Zvenigorod uezd was granted a privilege on the basis of which at some later date the monastery's title to the wastes of Karinskoe village, Negodnevo and Somovo, was confirmed; the peasant claimants lost their case 'because those wastes from of old, for sixty years, have been subject to Karinskoe village, a third field...'[1] Vasko, who spoke 'on behalf of all the Andreevskoe peasants', regarded the wastes as Andreevskoe land; Karinskoe and Andreevskoe were more than five miles apart.[2]

The view of 'all the Andreevskoe peasants' (a commune of some sort) was that the disputed lands were wastes. The judicial view was that the wastes were a third field of a village about five miles distant, and that it had been subject to that village for sixty years, perhaps as a third field.

2. On 27 August 1474 an exchange of lands was concluded between Prince Ivan Vasil'evich and the Trinity monastery of St Sergius, evidently in order to achieve a more compact disposition of land.[3] After listing the lands granted to the monastery, the prince added: 'and since my Ploshchevo lands have gone as third field to their village of Vyakhorevo I have also given them that land'.

Here we have evident concern for organised layout and an implication that the monastic village's cultivated area had been extended to take in the prince's 'lands'.

3. Between 1485 and 1490 a land dispute in the Sol' Galich area was settled. A witness for the Trinity monastery of St Sergius stated: 'I, lord, remember for sixty years; that Matkovo manorial settlement, lord, was held by Oksin'ya, wife of Ontufii Oberuchev, and her son Iev Ontuf'ev lived with his mother in that Matkovo manorial settlement and, lord, that Kashino waste is the third field of Matkovo towards the river, and that is boyar land, lord.'[4] The peasant claimants had stated that they had been granted the waste by the reeve (*starosta*) 'and the peasants', i.e. that this was black land, administered by peasant officials in effect acting for the prince.

Here, again, a waste is now regarded by the monastery as a third field. The peasants, again perhaps a commune, regarded it as a waste wrongfully tilled by the monastery for twenty-five years or more.

[1] *ASEI*, III, no. 54a. [2] *Ibid.* 488. [3] *ASEI*, I, no. 424 and comments on p. 622.
[4] *ASEI*, I, no. 523.

4. About 1488–90 the Trinity monastery had a land dispute with peasants in Nere-khta volost, Kostroma uezd. The monastery overseer claimed that 'that land, lord, is our monastery's, the third field of Poemech'e village, but now, lord, on that land the Grand Prince's peasants, Fedotko and Mikhal' Zhirovkin and Mikitka Feodotov have set up three hamlets, a tenement in each hamlet'.[1] The peasants had been settled there by a prince's official establishing settlements free of obligations. A witness claimed the land had been the monastery's third field for seventy years.

The three peasant tenements established were each regarded as a hamlet which, as we have seen, indicated a fully viable peasant farming unit. They had only been established a year before, so it seems that monastic control of its third field was fairly regular, though not sufficiently constant to prevent farms being set up on it.

5. Between 1490 and 1498 there was a land dispute in Goretov stan, Moscow uezd. A peasant complained that Vasko Us, a peasant of the Simonov monastery, had tilled from a distance for eleven years land of the Grand Prince's Soninskoe manorial settle-ment. Vasko, however, claimed that 'that land, lord, is the Simonov monastery's, of Ivanovskoe village, which Semenko calls the Grand Prince's Soninskoe waste; but, lord, that is not a waste, that, lord, is the third field of the monastery's Bolkoshino hamlet, beyond the river. And my father, Malafei, lord, tilled that land fifty years ago. And I, lord, have been tilling that land eleven years. And before this, lord, before me, Ivashko Dubnev lived on that same Bolkoshino land and, lord, tilled that Zarech'e [lit. beyond the stream] for Bolkoshino four years, the third field'.[2] Evidence of both parties showed that at some periods the land had been used as hayfield.[3]

Here the prince's 'land' allegedly tilled from a distance for eleven years is claimed as a third field. It was a third field fifteen years before (about 1475–80). Its use as arable seems, in any event, to have been intermittent, as it was mown at least from time to time. There is no direct suggestion that it was a third field when tilled fifty years previously.

6. Between 1495 and 1511 the Metropolitan of All Rus' sent an official to remeasure lands in the villages of the Emperor Constantine and Helena monastery, Vladimir, because the archimandrite had complained 'that their monastery peasants of Suralomy and Dobroe Selo till much arable for themselves, but till little monastery arable. And you are to ride to those villages with archimandrite Matfei and to remeasure the land in all three fields, and you are to give the peasants in those villages in all three fields five desyatinas each, and you are to order them to till a sixth desyatina for the monastery'.[4] Provision was made for deviations from this norm, the obligations to be exacted in proportion to the amount tilled.

The rate of five desyatinas, plus a sixth to be worked for the monastery, is pre-sumably in each of the three fields and may be compared with the norm of five desya-tinas per vyt mentioned in no. 13 below.

7. In 1497 Prince Danil Aleksandrovich issued a charter confirming certain grants of land made to the Saviour of Kamenny monastery in an area near the Yarenga by his father and grandfather. These included 'the Agapitov waste, a third field to the hamlet of Podol'sk'.[5] Of all the items listed one was a 'corner', one was 'hayfield land' (navolok

[1] *ASEI*, I, no. 540.   [2] *ASEI*, II, no. 404, 414.   [3] *Ibid.* 415.   [4] *AFZ*, I, no. 205.
[5] *DAI*, I, no. 21, 17.

zemlyu) and forty were simply 'land'. The charter also refers to wastes 'which are not recorded in the monastery's documents' where the prince's official had been establishing peasants.

8. In 1497–8 there was a land dispute in Shutkin stan, Yur'ev uezd, between peasants and the Trinity monastery of St Sergius. Three peasants alleged that 'the monastery peasants mow that site heavily'. A peasant witness claimed that his grandfather had told him 'that a bee-man of the Grand Prince had been settled on that Medvezh'e site; and, lord, Ivan Shilo began to till it from the battle of Suzdal', as I remember'.[1] A witness for the monastery, however, stated that 'the Medvezh'e site is the monastery's and, lord, Anna wife of Ivan Vasil'evich Olfer'ev gave it. And, lord, it was the third field of that same Mikhailova hamlet...'

As in no. 5 above there is no suggestion that the original cultivator, Ivan Shilo, regarded the site as a third field.

9. In 1497–8 three brothers, peasants of the Kukhmyrev family, received a ten-year relaxation from certain obligations in Gorokhovets volost, Suzdal' uezd. A peasant reeve stated that 'the Saviour's servant, Nechai, says that we established Derevnishche clearance on Machkovo village's turnip patch, but we, lord, established that Derevnishche clearance on the Grand Prince's Gorokhovets volost land, on the third field by the Dubrovki site on the river Lyulekh...and the arable, lord, on that site they tilled two fields, and the third field, lord, they tilled on this side of the river Mortka, at the place at Derevnishche where that clearance is now established'.[2] The hamlet itself was apparently moved across the Mortka as the river Lyulekh was washing away the first hamlet; but the new site was also subsequently abandoned 'because they had no luck', and then resettled by one of the original holders' sons.

Here we see that a clearance might be on a turnip patch and that two fields might be on a site. A third field was elsewhere and the hamlet itself was later resited there. This document is especially interesting because it is the only one which clearly shows peasants with three fields. It may be significant that the peasants were three brothers, sons of Senka Kukhmyrev who had been a witness to a court case in the 1480s and thus probably a substantial peasant.[3]

10. A note in a Gospel of a grant in Moscow uezd dated 1 May 1498, made by Grand Prince Vasili Ivanovich, specifies the lands for the maintenance of the church as '6 chets in a field and the same in the other two, and hay 20 of our big ricks, and on the other side of the river Pruzhenka the Popova waste and the Bobriki waste, in which are 12 chets of tilled arable land in a field and the same in the other two, and of tillable forest 17 desyatinas, and of marsh forest 3 desyatinas, and 21 ricks of hay on the Pruzhenka'.[4] A further note records that Popova waste had 2 chets of medium quality church arable lying fallow and 10 chets overgrown with forest in each field, 20 ricks of hay on the Pruzhenka and 7 desyatinas of tillable forest.[5] Here, the three fields may well be used as a notational device; there is no clear indication of real fields, only of the presence of arable land, wastes and forest. The formula 'in a field and the same in the other two'

---

[1] *ASEI*, I, no. 615. Ivan (Vasil'evich Olfer'ev) Shilo was a member of the Kuchetskii boyar family; see Veselovskii, *Issledovaniya*, 427–8. The battle of Suzdal' was in 1445.

[2] *ASEI*, II, no. 489. [3] *ASEI*, II, no. 483.

[4] *ASEI*, III, no. 52. [5] *Ibid.* p. 487.

is regularly used in sixteenth-century registers of inquisition. It would be odd for abandonment to fallow and to forest to occur evenly in the three fields.

11. In 1498–9 metropolitan Simon authorised a transfer of lands in Moscow uezd from one servitor to another. The lands taken from Turikovo village were in two fields: 'in one field 20 desyatinas of the Metropolitan's Petrovskii lands, and in the other field 13½ desyatinas of the Metropolitan's Likhoradovo lands, and in Likhoradovo hamlet which had been held by Never, son of Ivan Yur'ev, the Metropolitan's grant in all three fields was 21 desyatinas of land; and the whole arable and the hay which was mown on the cut forest was 76 desyatinas. And the Metropolitan's Sel'tsi volost head was ordered to administer that land and to till it for his sovereign, the Metropolitan.'[1]

The amounts of arable here do not agree with those given in no. 12, but there were two Likhoradovo hamlets.[2]

12. In 1498–9 also in Sel'tsi volost, Moscow uezd, the land of the metropolitan's Likhoradovo hamlet was measured. 'And in that hamlet there are 30 desyatinas of arable in all three fields and 20 ricks of hay in those same desyatinas'.[3] This, again, may well be a notational record in rounded figures rather than a statement about the real field layout.

13. In 1498–9 a register of an inquisition relating to the Opol'e stan of Vladimir uezd recorded the Grand Prince's Simizinskie villages. 'And the Grand Prince's arable in all three fields was 16 desyatinas a field, and they sowed two chetverts of rye on a desyatina and twice as much of the spring crop. And that arable was given the peasants on quit-rent. And the total peasant arable was 97 desyatinas a field. The Grand Prince had no hay: peasant hay was 600 ricks. In the settlement three sokha-units of arable.'[4] The fourteen peasant tenements were taken as 19½ vyts 'and the peasant is allocated five desyatinas a vyt', i.e. a total of 97½ desyatinas or just over 20 desyatinas in all fields per tenement.

The Grand Prince's Zhelezovo settlement was also recorded. 'And the Grand Prince's arable in all three fields was 11 desyatinas a field, and they sowed two chetverts of rye on a desyatina and twice as much of the spring crop, and the arable was given the peasants on quit-rent. And the total peasant arable in all three fields is 36 desyatinas a field. Peasant hay is 310 ricks; the priest has 30 ricks. In the settlement two-and-a-half sokha-units of arable.' The nine peasant tenements (one was a widow's) were rated as 7 vyts, i.e. a total of 35 desyatinas or nearly 12 desyatinas in all fields per tenement.

Here, too, we have an estimation in terms of conventional three-field allocation. It is not clear how the holdings were organised in reality: was the prince's area separate or intermingled with that of the peasants?

14. At the end of the fifteenth century a land dispute took place in Biserovo, a village of the metropolitan in Kolomna uezd. 'Standing on the metropolitan's land of Biserovo village, on the rye field [na pole na rzhishche] at the Popovo site, Ondreiko Pelepelkin said: That site, lord, is the Popovo land of the Grand Prince and the metropolitan's [people], lord, have taken over the site fifty years ago. And Konstantin asked the metropolitan's village overseer, Senka: Answer. And Senka said: That

---

[1] AFZ, I, no. 51.   [2] AFZ, I, no. 46b.   [3] AFZ, I, no. 52.   [4] AFZ, I, no. 166, 152.

Popovo site, lord, is the metropolitan's third field at Biserovo village .[1] Some old-established peasants confirmed this version. 'As we remember, lord, that Biserovo village, Popkovo is the third field of Biserovo's arable by the village; and our fathers told us, lord, and we remember that from of old Popkovo has been subject to Biserovo, and Biser gave that village to metropolitan Fotii together with Popkovo and that site, lord, is the third field of that village.'

Ondreiko lost his claim because he had failed to take suit against the metropolitan's people for fifty years and had not petitioned the prince. Again, there is a difference between the peasant speaking of a site, or simply land, and the monastery having a third field; again, too, there is no explicit statement that the disputed land had been a third field fifty years earlier.

15. On 20 January, 1501 a land dispute in Yaroslavl' uezd was settled. The peasant reeve claimed, on behalf of all the Shakhovskaya volost peasants, that 'that Malaya Khinovka hamlet is black land of the Grand Prince's Shakhovskaya volost; and that was forest; but, lord, those monks from Bol'shaya Khinova hamlet cut that forest and established that hamlet [subject] to that hamlet; and it is about fourteen years ago that they established that hamlet'.[2] For their part the monks claimed that 'that Malaya Khinovka hamlet is ours, the Tolgskii Immaculate Virgin monastery's; and that was forest, lord, of Bol'shaya Khinova in the third field; and we, lord, cut that forest from Bol'shaya Khinova and established that hamlet on the third field'. The waste of Bol'shaya Khinova, they added, had been granted the monastery by a prince fifty years ago; 'and, lord, abbot Ivonya settled peasants on that waste, into that Bol'shaya Khinova hamlet thirty years ago. And after that abbot, the old Ivonya, lord, another abbot Ivonya settled peasants on the third field by that Bol'shaya Khinova hamlet into the second hamlet, Malaya Khinovka; that was fourteen years ago.' The peasant reeve conceded that for fourteen years 'we, lord, have made no petition to the Grand Prince or his boyar about this land against those monks and have sent no court officials to summon them; and we have been silent, lord'.[3]

Here we have a clear timetable of monastic forest colonisation. Fifty years ago (c. 1450) a waste had been given to the monastery; thirty years ago (c. 1470) peasants were settled 'on that waste'; fourteen years ago (c. 1487) the monks cleared the forest and established a new hamlet 'in the forest. . . on the third field'. The third field seems here to have been an administrative division of the manor, not necessarily a regular arable field of equal area to two others.

16. About 1501–2 a land dispute was settled in Minskii stan, Kostroma uezd, between peasants and the metropolitan's rural overseer, Vanya. The peasants, representing all the peasants of the stan, claimed the monastery had established hamlets, in one case forty years ago, in others ten, on black lands deserted due to plague.[4] Vanya stated that 'those lands, lord, Poddubnoe hamlet and the Demidkovo, Fedorkovo, Perepechino, Osotovo, Sukhorevo wastes and Karpovo, Katunino and Dmitreitsovo are of old the Metropolitan's lands of Kulikovo village; and what, lord, the peasants called Lyutogo hamlet which they say stands on Katunino land, that, lord, is Andreeva hamlet, not

---

[1] *AFZ*, I, no. 114.   [2] *ASEI*, III, no. 221, 239.   [3] *Ibid.*, p. 240.
[4] It does not seem possible to date the plague referred to; there were serious outbreaks in 1465–7, 1478 and 1486–7, for example. Vasil'ev & Segal, *Istoriya epidemii*, 38.

Lyutogo, and does not stand on Katunino land; and, lord, they called a waste Ievtsovo, but that, lord, is one land of Dmitreitsovo and there was a drying barn there, lord; and, lord, they called a waste Ignatovo, but that, lord, is not a waste, but a new clearance (*rospash'*) and it is known as the Nosilovskaya meadow; and, lord, they called a waste Mikulkino, but that, lord, is the third field of Tryastinskaya hamlet, but not Mikulkino and that Tryastinskaya hamlet stood, lord, but, lord, they moved that hamlet from there to the place where the church of St Nicholas the Great now stands, and all those lands are the metropolitan's, of Kulikovo village'.[1] In their rebuttal, the peasants stated that 'as for that Mikulkino waste, lord, which Vanya calls one Tryastinskaya land, if you, lord judge, will follow us, we will show you where Tryastinskaya hamlet stood. And they led (him) to the rye field, to the waste.' The decision went against the peasants because 'Poddubnoe hamlet and Andreevo are entered in the register of Mikhail Volynskii as lands of the metropolitan and on Ignatovo waste they sought traces of a hearth [*pechishche*] but failed to find it; and Mikulkino waste is by the metropolitan's Tryastinskaya hamlet in the middle of the field, and Demidkovo, Fedorkovo, Perepechino, Osotovo, Sukhorevo, Dmitreitsovo, Ievtsovo, Karpovo, Katunino and Lyutogo hamlet, those lands too are in the middle of the metropolitan's lands, and the Grand Prince's lands do not come up to those lands'.[2]

Here, too, the peasants referred to lands and wastes, not to a third field. It seems at least possible that the monastery lands had become locationally distinct from the Grand Prince's due to internal colonisation during the forty years mentioned. The confusion possible over placenames in a forest area of small dispersed settlements, compounded by the fact that, as here, hamlets might be moved to new locations, is well illustrated.

17. Between 1503 and the 1540s an extract was made from a register of inquisition relating to holdings of the Trinity monastery of St Sergius in the Radonezh area. In the case of Glinkovo village it noted that 'two fields of that village are in Kinela, in Pereyaslavl' uezd, and the third field is in Radonezh, and on the Radonezh side are': twenty-nine tenements are then listed.[3] Three other hamlets of this village, with fourteen tenements, are added.

This village, then, had at least 29 peasant tenements, possibly 43 or more, on its third field.

18. In 1504 Grand Prince Ivan Vasil'evich sent a report to his son Yuri on the bounds of the lands granted to Yuri in Dmitrov, Ruza and Zvenigorod.[4] In part this reads: 'And from the mouth of the river Vyaz'ma by the river Moskva downstream towards Vyaz'ma village of the two children of Ondrei Mikhailov Ovtsyn; on the right the Vyaz'ma village, and the third field of that village is beyond the river Moskva, and meadows, a small lake and trapping runs are in Zvenigorod.'[5]

Several settlements of Andrei Povadin are mentioned, and also lands of other members of this family. In one case, 'on the right a village, a clearance of Ondrei Povadin, and two fields of that village are in Dmitrov, in Mushkova'.[6]

---

[1] *AFZ*, I, no. 258, 222.   [2] *Ibid.* 226.   [3] *ASEI*, I, no. 649, 567.
[4] On the date of this document see Zimin, *PI*, VI.320, 324.
[5] *DDG*, no. 95, 381. Ovtsyn was a minor noble line, but Andrei Mikhailov is not mentioned by Veselovskii, *Issledovaniya*, 458–9.
[6] *DDG*, no. 95, 388.

In another part of this report, the Lytkino hamlets are mentioned: 'Fofanovo hamlet, Bechmanovskoe hamlet, the third field of that hamlet has gone over beyond the river Radomlya'.[1]

Another section reads: 'And from the little pond by the pits to Khlyabovo gully and by the Khlyabovo gully down to the river Chernaya Gryaz', on the right is Khlyabovo village of Ignat Chertov and on the left, also Ignat's land, is the third field of Khlyabovo in Dmitrov, in Vyshgorod and his monastery church of the Resurrection.'[2]

The document several times specifies the administrative subordination of lands which 'have gone over that boundary'.

We have here, then, three landed proprietors, minor lords, with three fields on their lands. The third field is apparently in each case located away from the other two and in one case has been moved to a new location.

19. On 18 April 1505 a report on a land dispute in Vol'skaya volost', Sheksna uezd, was drawn up. The peasant case of 'Okish Olyunov and his fellows' against Gridya, the steward of the boyar owner, Dmitri Vasil'evich Shein, was that 'he let in, lord, the Aleshino hamlet into the third field, to Dmitri's village, to Lavrentievo; and that hamlet, lord, has been empty since the great plague; and, lord, on that field fifty chetverts of rye are sown'.[3] Gridya replied that 'that, lord, is the third field of Dmitri Vasil'evich's Lavrentievo village; and on that field of Lavrentievo village, lord, a tenement stood, but it was carried off, lord, to another field'.

The boyar's third field here seems to have amounted to 25 desyatinas. This seems to be a high figure for a single tenement (see nos. 6, 13 above); taken in conjunction with the fact that a steward was in charge, it may suggest that it was worked by peasants paying labour rent or by servile dependants.

Clashes between peasant close farming and some sort of three-field organisation are represented in a quarter of the documents cited (nos. 1, 3, 14, 15, 16); expansion incorporation or reorganisation of fields is also shown (e.g. nos. 2, 16, 19). It is also clear that a field might be a waste (no. 7), or have tenements on it (nos. 15, 17, 19) or even be moved (no. 18). Even in this central area of the Moscow state, then, three-field organisation appears not to have been uniformly established by the early sixteenth century save on some, mainly monastic and princely, lands.

[1] *Ibid.* 389.   [2] *Ibid.* 392. Mikula Chertov also had villages in this area.
[3] *ASEI*, I, no. 658, 582.

A monastic statute

The following charter, granted by the Patriarch Iov on 5 February 1590 to the abbot of the Novinskii monastery, gives a picture of an important late-sixteenth-century monastic estate and the range of obligations imposed on its peasants.[1]

The monastery had been founded by the metropolitan Fotii in 1430; it was located west of Moscow, on the east bank of the Presnya, by the confluence of the Presnya and the Moskva, near Kudrino village.[2] By the sixteenth century the monastery's lands in this area stretched from what is now the Arbat to the Krasnaya Presnya; the monastery was, thus, important in terms both of its links with the highest levels of the church and its location.

For ease of reference only, sections of the charter have been numbered; in what follows figures in brackets refer to these numbered sections.

The charter shows that the monastery held lands in a number of locations. The monastery itself was on the outskirts of Moscow, but there were holdings in Moscow uezd, in Ruza, Dmitrov (3), Romanov (21), Kostroma (23) and Kashin (24) uezds. It had artisan quarters, free settlements (2, 27), manorial settlements, villages and their dependent hamlets (e.g. 2 and 3). One group of settlements is specified as 'distant villages' (21).

The various categories of monastic clergy, monks and servants, and also rural servants, and peasants are mentioned, as well as hired men (2, 26, 27, 28). The servants and, in particular, the peasants who 'begin to live in the villages and hamlets' (3) appear to be the basis of the monastic economy. If a peasant holding becomes vacant, the remaining peasants in the settlement are responsible for cultivation and for ensuring payment of taxes, dues and labour obligations (17). There is concern to fill any such vacant peasant holdings and to ensure control of labour mobility (14, 16, 17). The estate is run by officials some of whom live in the monastery's tenements in the villages (4, 8, 26). These officials are obliged to defend the labour force in cases involving parties other than the monastery (28). The importance of the administration, and the income derived from it (especially from justice), is shown by the large number of sections concerned with it (2, 7, 8, 9, 10, 11, 12, 13, 14, 15, 16, 19, 26, 27, 28).

Agricultural production was apparently organised in two ways. First, there was production by the peasant tax unit (vyt) (3, 24); second, the monastery had land which was cultivated 'compulsorily' (21, 23). In the latter case peasants also had to pay a quit-rent and, sometimes, certain other dues. Unfortunately the charter gives no clear picture of the arable or hay fields, but these might be contiguous (13). The obligations imposed on the peasants included not only cultivation and mowing, but also carting, threshing, cutting and supplying both firewood and building timber, the erection of

---

[1] *AFZ*, III, razdel 1, no. 23.   [2] See Veselovskii, *IGAIMK*, 139.70.

buildings and fences, repairing mills (3, 4, 5, 17, 21, 23, 24). Peasants had to supply draught horses for certain purposes (22) and a range of supplies, including grain, dung (3), textile fibres and a range of livestock products (20, 21, 23, 24). There is no mention of local fish. Some of these obligations might be commuted by money payments. Supplies were delivered to Moscow once a year (24).

One section (15) shows that peasants engaged in trade in livestock and agricultural produce; an interesting distinction is made between such peasant trade 'among themselves' within the volost, i.e. local trade, and that which involved sales outside the volost; purchases into the volost were treated similarly to local trade.

Peasants had certain communal responsibilities, mainly to fulfil all obligations; there seems to be some distinction between the volost, which is responsible for setting up the drying barn (7), and the village (17, 18) and hamlet (17). Probably the volost in this context was a district or area like a parish or manor which might include more than one settlement.

In the free settlements, where there was remission of certain obligations, notably labour rent, deals in tenements and buildings took place and men were hired (27).

This charter, then, shows a great estate with holdings in a number of different areas, drawing in supplies from farming both on peasant units and on monastic demesne. The central complex of the monastery, together with nearby free settlements, was economically more advanced than the rural settlements from which it drew its raw and some processed materials; the processing of raw materials, artisan production and trade were concentrated here.

1.   By God's mercy I, the humble Iov, Patriarch of Moscow and All Russia, have granted abbot Varsonofei who is in Christ, and the brethren, or whatever other abbot there shall be in our age-old patriarchal Novinskii monastery of the Immaculate Mother of God and of the Miracle workers Petr [Peter], Aleksei and Ivan [John] in accordance with the former grant of Makarii, metropolitan of all

2.   Russia. Whatever black [i.e. monastic] clergymen and monks, monastic servants and rural servants, rural priests, peasants in artisan quarters, villages and hamlets begin to live with him in that monastery, my public and court week-men shall not impose sureties nor indicate terms for nor ride in for any cause to the abbot's monks, monastic and rural servants, rural priests, and peasants of artisan quarters, villages and hamlets. The abbot himself, or whoever he may indicate, shall have cognizance of and judge those monks and black monastic priests, monks and servants, peasants of artisan quarters and rural servants, rural priests, and peasants of villages and hamlets. If anyone shall take suit against the abbot Varsonofii, himself, or whatever abbot there shall be after him in that monastery, I, Iov, Patriarch of Moscow and of All Russia, or whoever I shall indicate, shall judge.

3.   As to their monastery [8–10 letters missing] in our patriarchal Moscow uezd and in other towns, Chizhovo manorial settlement and hamlets, and in it the church of the Blessed Miracle-Worker Sergii, and Tsareva hamlet; Telepnovo village and hamlets in Moscow uezd, in it the church of the Universal Elevation of the Holy Cross, and (subject) to that village is Ivashkovo hamlet in Ruza uezd; Mikhailovskoe village and hamlets in Moscow uezd, and in it the church of the Archangel Michael; in Dmitrov uezd the manorial village of Kartsovo and hamlets, in it the church of the Resurrection of Christ; Nivki village and hamlets also in Dmitrov

251

uezd, in it the church of St Simon; in Dmitrov uezd, Poddano village and hamlets; and whatever servants and peasants begin to live in the villages and hamlets, the peasants are to till for [the monastery] per vyt one-and-a-half desyatinas for rye; to cart as much dung as there shall be in any village to the monastery's arable land; to mow as much hay as there shall be in any village and cart it to the monastery and the monastery's yard; and the peasants are to provide every sort of grain stocks for the monastery as ordered per vyt and cart them to the monastery. And they are also to thresh the monastery's rye for the monastery, two chetverts a vyt.

4.   They are also to cart to the monastery from a vyt three loads of firewood and a three-sazhen beam of timber to the cell, when the abbot buys beams or firewood, or from the monastery's coppices to its building. If in any year the abbot does not order the peasants to cart firewood, the peasants give the monastery an altyn a vyt per load. The peasants of the villages are to cut firewood and beams after tillage in the autumn or in spring before the arable or after tillage in the spring, three days a year. The firewood and beams they cut in three days they are to stack in heaps and to cart the firewood and beams to the monastery in the winter in accordance with the charter; in winter, they are to cart, also by vyts, beams and firewood to the monastery's tenements in the villages for the whole year; and they are to put up buildings for themselves in the villages, and make the monastery's tenements in monastic villages, in the monastery's forest and where the abbot makes any purchases.

5   If the abbot and the brethren in any village or Moscow want a mill [a line missing] the mill timber to the mill, to cart where the abbot makes a purchase, similarly in spring before the arable or in spring after tillage, or in autumn after tillage.

6   If any peasant [6–8 letters missing] is disobedient in any monastery matters, the abbot shall order the steward to take a grivna for the monastery treasury from whoever is disobedient and send him to work for the monastery.

7   If any peasant [6 letters missing; probably 'owes the mo'] nastery grain without fraud, grain is not to be exacted from that peasant. The drying barn is to be set up by the volost. If the monastery's silver [due] passes from one peasant on to another, the clerk of the treasury transcribes the contract; the treasurer and the clerk receive two dengas per ruble. If an exaction is imposed on anyone in the year, the treasurer and the clerk have four dengas per ruble of the exaction.

8   If a peasant makes a claim against a peasant before the abbot, or before the monks in chapter, or before the town or rural stewards, two or three of the best men of the artisan-quarter peasants, approved by the latter, are to sit on the case with the abbot and the monks in chapter. Three or four good village men, approved by the latter, are to sit on the case with the steward in the villages. Whatever peasants the villagers approve [5 or 6 letters missing] to sit on the case and the steward is to send the abbot their names, after copying them, over a priest's signature.

9.   The customary due for the abbot, monks and stewards [something unclear] per ruble [6–8 letters missing] ten dengas. The court investigator is to have a local allowance of two dengas, for a true [copy] twice that, and to have a denga per

10. versta for travel money. Whatever poor people begin to live at the churches on the monastery's land and go to law between themselves, the monastery steward shall judge them and take from them for the case a customary due [unclear word] two

dengas a ruble; and for local allowance and a true copy a denga; when he submits
11. the copy of the report he takes ten dengas for travelling. If any peasants undergo trial by battle on the basis of the report to the Patriarch's house-steward, the
12. steward takes a grivna from the man killed, and the court investigator [a denga]. If the steward of any village at all sends his list to the prior as a report with the court investigator, the court investigator takes for travelling [two words unclear] ten dengas. The Moscow court investigator again takes sureties according to the copy from the two litigants to place themselves every day [6–8 letters missing] and before the Patriarch's house-steward and he takes two dengas for the sureties.
13. If any peasant in those villages [6–8 letters missing] names written in the document, tills over a boundary or mows across one or if anyone who is involved in trampling takes a steward, the steward takes from the guilty party an altyn as a boundary forfeit
14. [boran] and the court investigator has two dengas. Whoever takes a wife in the volost, the court investigator is to take an altyn as *ubrus* payment. If anyone gives a daughter in marriage out of the volost, the steward takes a grivna for the abbot as marriage payment [*vyvodnaya kunitsa*].
15. Whoever sells a horse or cow, or a tub [*odon'e*] of rye, a load of corn [*zhito*], or a building, or a tub of oats, a stack of hay or five sheep, or five pigs, out of the volost, the steward is to have a declaration payment of two dengas. But whoever buys or exchanges among themselves in their volost or whoever buys from outside the volost, from such the steward does not have a declaration payment. Whoever does not make his declaration before the day, the steward takes from him a default
16. payment of four altyns one-and-a-half dengas. If anyone goes into another volost, or from tenement to tenement in a village or in hamlets to make a living, the steward takes two dengas as a declaration payment. If anyone goes out of a volost from arable land, or goes into one, the steward is to have a denga as a declaration payment. If anyone passes from a tenement into a tenement to make a living and does not make a declaration to the steward before that day, the steward is to take four altyns one-and-a-half dengas default payment.
17. The abbot is to summon peasants to empty vyts and the stewards are to have sureties on them with notes so that they may be good people. If any peasant goes, withdrawing from his volost at the date, the vyt is to be tilled by the peasants of that village; they are each to give the taxes of the Tsar and Grand Prince and the monastery's dues and to perform the labour obligations until there is a peasant on that vyt, and the abbot shall not enter that vyt. And in all monastery villages and hamlets the peasants among themselves are to [half a line unclear] tenements,
18. buildings and erect fences. If any peasant goes out of the volost from the villages because the tenement has rotted, the remaining peasants of the villages are to set up that tenement at their own expense because they have protected one another against the tenement rotting and the peasant vyt holders should live, the gates [unclear].
19. The peasants should not keep an evil man among themselves, they should seek properly among themselves for an evil man and if any bad hearsay arises about anyone they should send out those bad people so that the peasants should not have fines for those evil people.
20. In those villages, too, the peasants by Moscow and of Dmitrov have to give the monastery a quit-rent at Easter, on St Peter's Day and at Christmas of twenty eggs each. On the same date, at Christmas, they are each to give two grivenkas of

butter, a cheese and a sheepskin. If a sheepskin or cheese is not agreeable, they are
to give an altyn for a sheepskin and two dengas a cheese. And in those villages the
peasants are to spin [unclear] what monastic spinning flax and hemp is grown for
21. table cloths and wool from the monastery's sheep for cloth. The distant villages of
our monastery, Spaskoe village and hamlets, Romanov uezd, in R[1–2 letters
missing]zhevets in the Sheksna area, in it the church of the Transfiguration of the
Saviour and the heated church of the Nativity; the peasants are to till compulsorily
fifteen full desyatinas each, and the quit-rent they pay from each vyt to the
monastery is a chet of rye, two chets of oats, a sheepskin and at Christmas thirty
eggs, a cheese and two grivenkas of butter.
22. If the steward goes to Beloozero to buy fish or salt for his stores the peasants
provide him with seven changes of horse and to Moscow for salt and fish they
provide him with two.
23. In Kostroma uezd, Kulikovo village and hamlets, in it the church of the Miracle
of the Archangel Michael and another church, heated, called Pyatnitsa; and
Tresino village, in it the church of Nicholas the Miracle-Worker, four desyatinas
of monastery arable land which the peasants till and reap compulsorily, they mow
the monastery's hay and cart the dung from the monastery's yard to the arable land.
They give a quit-rent to the monastery, from each vyt four chets of rye; four chets
of oats and for cartage they each give five altyns. And from the monastery's ten
mills they give a rent of six rubles and a grivna to the monastery treasury, and a
sheepskin from each vyt, and at Christmas, thirty eggs, a cheese and two grivenkas
24. of butter. In Kashin uezd, Koshelevo village and hamlets, in it the church of the
Holy Miracle-Worker Dmitrii, they each till one-and-a-half desyatinas of arable
land for rye for the monastery and the same amount of spring-sown crops and mow
whatever hay there is. They travel to Moscow with stores from each vyt once a
year as ordered on St Dmitrii's day or St Filip's eve. They give a quit-rent to the
monastery at Christmas of two grivenkas of butter, a cheese, thirty eggs and a
sheepskin.
25. Thus have I [granted] the abbot Varsonofii, or whatever prior there may be
after him, the villages and hamlets recorded in this charter; and the abbot is to have
per vyt from the peasants ten dengas for himself for livery.
26. If any monastery servants begin to live in the villages, in office (na prikazekh),
those stewards and court investigators are to have from each vyt a denga as entry
money, and in autumn the rural overseer has nine dengas and the court investigator
an altyn according to custom. In autumn the steward and court investigator is to
have a chicken and a lamb's fleece, but if a chicken or lamb's wool is not agreeable,
they are to take for the chicken and for the wool a denga a vyt, for both of them.
27. As to the monastery's free settlement at Moscow, by the monastery on the
Prisnya, and another Novinskii free settlement, if any peasants in those settlements
sell a tenement between themselves or exchange one, or sell a building elsewhere,
or take a hired man making a livelihood to themselves, from an exchanged tenement the declaration payment to the court investigator is two dengas, and from a
shared building half a denga; from a hired man the declaration payment is a denga.
Whoever registers a tenement gives the customary ruble due according to the old
decree. The abbot takes an impost of two dengas per merchant as of old. If anyone
sells a tenement or exchanges one, or sells a building from a tenement, or takes a

hired man, and does not make a declaration before the day, the court investigator is to take for that a default payment of four altyns one-and-a-half dengas. The abbot and the brethren are to have from those settlements, in accordance with the charter of instruction of the Tsar and Grand Prince, for all the monastery's labour obligations, each year on St Peter's Day, five altyns per tenement; they are not to give more than that money nor to perform any labour for the monastery.

28. Also the monastery's stewards who are in office in the villages and settlements are to travel to town with peasants if they are falsely accused, or for any cause, and to work for them, to give sureties for them and defend them in all matters. If the monastery's servants do not begin to travel to town for the peasants and to work for them and defend them from other parties, they will be in disgrace with me, Iov, Patriarch of Moscow and All Russia.

29. I may give some charters on other charters, but on this charter there is no charter in conflict.

The charter was granted the 5th of February in the year 7090 and eight [1590]. Ivan Grigorov.

3

An Assize of Bread

The following is a translation of a document published by P. Polevoi in *Russkaya vivliofika* (1883), 1.68–82. Its text coincides with the first item in *Vrem. MOIDR* (1849), 4. materialy, 1–8, but the heading differs. In the translation the schedule has been tabulated. The text of the publication is evidently corrupt in places; it nevertheless gives a vivid picture of the detail in which price and weight control of bread products was attempted in seventeenth-century Moscow.

The 30th November, 7132, in accordance with the decree of the Sovereign Tsar and Grand Prince Mikhail Fedorovich of All Russia, a memorandum to the town stewards, Vasili Artemov and Bogdan Beketov and their comrades, and to the sworn-men for the weight of bread, to the tax-liable man of Ustyug Half-hundred, Dorofeiko Ivanov and his comrades. They are to go in the Kremlin, in Kitaigorod, in Bely Kamenny gorod and in the wooden town, by streets and alleys and by the small markets and they are to weigh the loaves of bolted and coarse flour and the rubbed out rolls and soft round loaves at the loaf-makers and roll-makers and the dealers in loaves and rolls, and the Patriarch's, metropolitans', bishops', princes', boyars', and monastic yard-men and the musketeers, gunners and any people and food sellers, wherever they find loaves and rolls; and they are to weigh truly, in accordance with their oath to the Sovereign Tsar and Grand Prince Mikhail Fedorovich of All Russia, not to favour a friend, nor take vengeance on an enemy, nor to have any bribe or reward from that, nothing from anyone by any means at all; and where they find loaves and rolls and they are not in accord with the weight according to this decree of the Sovereign, the stewards and sworn-men are to bring those people to the Land Department and a fine is to be imposed on those people; in the first case half a poltina a man, but if they cheat anyone in weight a second time, then the fine is a poltina, and if they cheat in weight a third time, then the fine to be exacted is two rubles four altyns one and a half dengas.

If anyone stops loaves and rolls from being weighed anywhere, the stewards and sworn-men are to take away those loaves and rolls from them with witnesses. If they take them away from anyone, then a fine is to be imposed on them of two rubles four altyns one and a half dengas also. And the stewards and sworn-men are to look and convincingly acquaint themselves that the loaves of bolted and coarse flour and the rubbed out rolls and round rolls have been fully baked and that there were no thickeners or additive at all in the loaves, by any means at all. If any loaf-makers and roll-makers fail to bake their loaves and rolls fully, so that they might be heavy in weight, or add a thickener or any other additive, the sworn-men are to cut open those loaves and rolls of theirs and if the loaf or roll is not fully baked for weight, or if there is any additive in anyone's loaves, a fine of half a poltina is to be exacted from those people. A schedule is given to them, attached to this memorandum of instruction, showing how much loaves of bolted and coarse flour and rolls should weigh at a given price, according to how much loaf-makers and roll-makers pay for flour, and how many grivenkas there should be in a loaf or roll.

Schedule laid down for maintenance [*kharch*] for a quarter of flour and the weight of rolls and loaves in a quarter of flour.

For a quarter of rolls:	altyns	dengas
yeast	2	2
salt		6
firewood		8
cartage and carriage		8
bolting		4
labour		10
stall	2	2
candles and broom		2
TOTAL FOR A QUARTER	11	—

For a quarter of loaves from bolted flour:	altyns	dengas
ferment		3
salt	1	
firewood		8
cartage to the house and from it		8
bolting		4
tax and trading		10
candles and broom		1
distribution to the stall	2	
	9	—
[should be	8	4]

Cost of a quarter of flour		Number per quarter of		Weight, in grivenkas and zolotniks* of loaf costing:		
altyns	dengas	altyn loaves	4 denga loaves	two dengas gr. zol.	four dengas gr. zol.	one altyn gr. zol.
8	2	18		5 15	10 18	16 30
9	2	19		5 24	10 35	15 48
10	2	20		4 69	9 54	14 64
11	–	21		4 64	9 32	14
12	–	22		4 38	8 90	13 32
13	–	23		4 12	8 48	12 60
14	–	24		4 3	8 18	12 15
15	4	25	1	4	8	11 69
16	4	25	1	3 84	7 48	11 30
17	4	27	1	3 54	7 3	10 87
18	4	28	1	3 48	7	10 48
19	4	29	1	3 3	7 6	10 12
20	4	30	1	3 8	6 4	9 42
21	4	31	1	3 3	6 6	9 33
22	4	32	1	3	6	9 16
23	4	33	1	2 84	5 72	8 66
24	4	34	1	2 66	5 48	8 51
25	4	35	1	2 72	5 48	8 24
26	4	36	1	2 54	5 24	8
27	4	37	1	2 51	5 24	7 72
29	4	39	1	2 48	5	7 48
30	4	40	1	2 32	4 64	7 24
31	4	41	1	2 24	4 60	6 93

* A grivenka weighed a Russian pound (410 gm), a little less than a pound avoirdupois; the ninety-sixth part (*zolotnik*) weighed 4.267 gm.

Cost of a quarter of flour		Number per quarter of		Weight, in grivenkas and zolotniks of roll costing:		
altyns	dengas	altyn rolls	2 denga rolls	two dengas gr. zol.	four dengas gr. zol.	one altyn gr. zol.
16	2	26	1	4 16	6 80	10 48
17	2	28	1	3 32	6 64	10
18	2	28	1	3 24	6 48	9 87
19	2	29	1	3 8	6 16	9 54
20	2	30	1	3 19	6 44	9 21
21	2	31	1	2 78	5 84	8 87
22	2	32	1			8 69
23	2	33	1	2 72	5 54	8 30
24	2	34	1	3 12	5 48	8 12
25	2	35	1	2 48	5 33	7 81
26	2	36	1	2 42	5 15	7 57
27	2	37	1	2 32	4 84	7 33
28	2	38	1		4 72	7 15
29	2	39	1	2 24	4 93	7 3
30	2	40	1	2 16	4 52	6 68
31	2	41	1			6 69
32	2	42	1			6 51
33	2	43	1			6 38
34	2	44	1			6 24
35	2	45	1	2	4 9	6 9
36	2	46	1			6 36
37	2	47	1	1 90	3 81	5 84
38	2	48	1	1 93	3 78	5 72
39		49	1	1 72	3 60	5 60
40	2	50	1	1 80	3 64	5 48

4

Chets of good quality arable land per tenement and per person (Kazan' uezd)

	A. Lords with both peasants and people, i.e. serfs and slaves				B. Lords with peasants only				C. Lords with slaves only				D. Lords with neither peasants nor people	
	Tenements		Persons		Tenements		Persons		Tenements		Persons		Tenements	Persons
	Total inc. lords	*Exc. lords and those of non-tillers, handicraft workers etc.*	*Total inc. lords*	*Exc. lords, non-tillers, handicraft workers etc.*	*Total inc. lords*	*Exc. lords and those of non-tillers, handicraft workers etc.*	*Total inc. lords*	*Exc. lords, non-tillers, handicraft workers etc.*	*Total inc. lords*	*Exc. lords and those of non-tillers, handicraft workers etc.*	*Total inc. lords*	*Exc. lords, non-tillers, handicraft workers etc.*	*Total inc. lords*	*Total inc. lords*
I					5.0	5.0	5.0	5.0	5.0	5.0			16.0	16.0
	12.8	21.3	12.8	21.3	7.8	13.2	7.3	13.0	18.2	38.5			19.3	17.5
					2.72	5.4	2.2	5.6	10.9	21.8			12.2	12.5
II							18.5							
					18.8	37.5	18.8						9.0	
													10.6	
													5.6	
III													4.0	13.3
													5.0	5.0
IV					8.3	9.7	8.3	9.7					24.0	24.0
					13.6	17.0	13.6	17.0						
V													4.0	4.0
VI	3.0	3.0	3.0	3.0	3.0	42.0	3.0	42.0	12.0	12.0	12.0	12.0	21.0	21.0
	13.4	21.1	13.4	16.9	10.8	11.3	10.8	10.7	15.0	31.8	13.9	31.8	19.5	18.3
	4.1	11.7	4.1	5.8	3.0	6.0	3.0	6.2	8.8	25.4	8.6	25.4	17.4	17.4
VII						4.0	4.0							
						6.9	7.0							
						2.4	2.5							
VIII						13.0	13.0							
					15.0	7.7	7.5							
						8.3	8.3							
IX						11.0	11.0							
					7.8	3.7	2.8							
						3.4	2.7							

Category of estate holder and sources

I   Near Kazan', junior boyars, mostly old Kazan' lords holding before the 1565–8 inquisition. A. 107v; B, 103v, 109–109v, 112v, 115v, 119; C, 104, 105v, 108, 110, 113; D, 104, 105, 106–106v, 107, 108, 110, 111–12, 113, 114–15, 116v, 117.

II   Small shares of settlements, held by widows and minors. B, 121, 121v; D, 120v, 122, 123, 123v, 124, 124v, 125.

III   Musketeer officers. D, 125v, 126v.
IV   Interpreters. B, both 127v; D, 128.
V   Artisan quarter prisoners at Laishev. D, 129.
VI   New allocations. A, 149v, 153v, 166; B, 136, 136v, 138, 138v, 139, 140, 140v, 142, 142v, 143, 144, 144v, 145v, 147, 148, 149v, 151, 152v, 157, 165, 173–5, 176v, 177, 178, 179, 179v, 180v, 181v, 183, 185, 186, 186v, 187v, 188v, 189; C, 154, 155, 158v, 161, 163, 167, 169, 169v, 170, 172, 172v; D, 154v, 156v, 157v, 158, 159, 159v, 160, 161v, 162, 162v, 163, 164, 165v, 166v, 168, 168v, 170v, 171, 172.
VII Estates refused and not taken up. 191–191v, 195.
VIII Archbishop. 202–205v, 208–209v.
IX   Transfiguration and Ilantov monasteries. 212v–214v, 218v.

Notes:
Figures in italic are raw data.
Other figures are shown in the sequence: no. of entries: mean size(chets); standard deviation – for the given category.

5

Registers of the new grants of estates
(Kazan' uezd)[1]

Registers of the new grants of estates held by service in Kazan' uezd recorded and
carried out by the chamberlain Nikita Vasilevich Borisov and Dmitri, son of Dmitri
[sc. Andrei] Kikin, and their fellows in October 1567. They recorded and measured the
villages and hamlets of the 1565–7 grants of estates held by service, the Sovereign
Tsar's and Grand Prince's villages and hamlets, wastes and sites in Kazan' uezd which
were of the court according to the 1563–4 record of Semen Narmanski; no income was
noted according to his record in those court villages and hamlets, other than arable
land; the peasants of live vyts in the villages and hamlets tilled desyatinas for the
Sovereign and Grand Prince, a desyatina of rye per vyt and a desyatina of the spring-
sown crop.

In October 1567 the Kazan' officers of the inquisition, the chamberlain Nikita
Vasilevich Borisov and Dmitri, son of Andrei Kikin, with their fellows, in accordance
with the charter of Ivan Vasilevich, Sovereign and Grand Prince of All Russia, and with
the list of appointments, recorded and measured the Sovereign's court villages, hamlets
and wastes of their former 1565–7 record and allocation; the lands held by service by
old, retired attendants; escheated empty lands which of old had been Tatar, Chuvash
or Mordva; that is lands which were to be distributed in return for service, but Semen
Narmanski had not recorded those lands as of the court villages and they were not
granted to anyone.

And having recorded and measured all those lands as estates to be held by service
they distributed them to the princes and junior boyars whom the Sovereign sent to
Kazan' to his heritable estate for their livelihood; and they allocated them to them all
equally. How many chets they were allotted as half an allotment, arable land per
hundred chets, long fallows, and land overgrown and tillable oak woods; that is the
record of these registers under estates held by service in the lists. But the income of
lords holding by service from the peasants is not noted, apart from the fact that the
peasants are to till desyatinas of arable for the lords, a desyatina a vyt, as it was for the
Sovereign. The tillable good quality land is recorded by desyatinas per vyt and from
those desyatinas the amount which the peasants have to till for the lord is noted; and
the long fallow land, and land overgrown with oak woods are not noted in vyts, and
the tillable dense woods and untillable black forest is recorded in desyatinas as an
appurtenance to villages, hamlets and wastes and in estimated verstas, other than bee-
tree forests, and the rivers, apart from beaver runs; because the appurtenances in forest
and the beaver runs in the rivers are, in Kazan' uezd, all Tatar and Chuvash and there

[1] TsGADA, fond 1209, no. 152, f. 134v–135v.

were no bee-forests and beaver runs held by peasants; but if, in any villages or hamlets, there were bee-forests or beaver runs in rivers prior to the capture of Kazan' and those bee-forests and beaver runs have not been recorded in the new record and have not been distributed as estates to be held by service; from those bee-forests and beaver runs the prisoners give tax in kind [*yasak*] to the Sovereign's treasury in Kazan' according to the old registers; and in these registers, after the lands to be held by service, are recorded lands to be held by service, those which are empty, wastes and sites and surplus lands which remain over from the tenants and also the Sovereign's rented mills; all this is recorded genuinely, item by item.

Abbreviations

AAE	*Akty sobrannye v bibliotekakh i arkhivakh Rossiiskoi imperii arkheografi-cheskoi ekspeditsieyu, Arkheograficheskaya Kommissiya*
AAG	*Afdeling Agrarische Geschiedenis*
AE	*Arkheograficheskii ezhegodnik*
AFZ	*Akty feodal'nogo zemlevladeniya i khozyaistva XIV–XVI vekov*
AGR	*Akty, otnosyashchiesya do grazhdanskoi raspravy drevnei Rossii*
AHR	*Agricultural history review*
AI	*Akty istoricheskie*
AOIA	Arkhangel'skii oblastnoi istoricheskii arkhiv
APD	*Akty pistsovago dela . . .*
ASEER	*American Slavic and East European Review*
ASEI	*Akty sotsial'no-ekonomicheskoi istorii severo-vostochnoi Rusi kontsa XIV–nachala XVI v.*
AZR	*Akty, otnosyashchiesya k istorii Zapadnoi Rossii*
BAN	Biblioteka Akademii Nauk
ChOIDR	*Chteniya v Obshchestve istorii i drevnostei rossiiskikh pri Moskovskom universitete*
CPHB	*Conventus primus historicorum Balticorum Rigae*
DAI	*Dopolneniya k Aktam istoricheskim*
DDG	*Dukhovnye i dogovornye gramoty velikikh i udel'nykh knyazei XIV–XVI vv.*
DOOS	*Dopolnenie k opytu oblastnago Velikorusskago slovarya*
DRV	*Drevnyaya rossiiskaya vivliofika*
EHR	*Economic History Review*
ERMA	*Eesti Rahva Muuseumi Aastaraamat*
EzhAI	*Ezhegodnik po agrarnoi istorii Vostochnoi Evropy*
GBL	Gosudarstvennaya biblioteka imeni Lenina
GIM	Gosudarstvennyi Istoricheskii Muzei
GVN	*Gramoty Velikogo Novgoroda i Pskova*
IGAIMK	*Izvestiya Gosudarstvennoi akademii istorii material'noi kul'tury*
IKDR	*Istoriya Kul'tury Drevnei Rusi*
Ist. SSSR	*Istoriya SSSR*
IZ	*Istoricheskie zapiski*
KS	*Kratkie soobshcheniya o dokladakh i polevykh issledovaniyakh*
LZAK	*Letopis' zanyatii Arkheograficheskoi Komissii*
MIA	*Materialy i issledovaniya po arkheologii SSSR*
MISKh	*Materialy po istorii sel'skogo khozyaistva i krest'yanstva SSSR*

ABBREVIATIONS

MIZ	*Materialy po istorii zemledeliya SSSR*
NGB	*Novgorodskie gramoty na bereste*
NIL	*Novgorodskaya pervaya letopis' starshego i mladshego izvodov*
NPG	*Novye Pskovskie gramoty XIV–XV vekov*
NPK	*Novgorodskie pistsoviya knigi*
OOS	*Opyt oblastnogo Velikorusskago slovarya izdannyi vtorym otdeleniem Imperatorskoi Akademii Nauk*
PI	*Problemy istochnikovedeniya*
PIMK	*Problemy istorii material'noi kul'tury*
PKMG	*Pistsovyya knigi Moskovskago gosudarstva*
PL	*Pskovskie letopisi*
PRP	*Pamyatniki russkogo prava*
PSRL	*Polnoe sobranie russkikh letopisei*
RANION	Rossiiskaya assotsiatsiya nauchnoissledovatel'skikh institutov obshche-stvennykh nauk
RIB	*Russkaya istoricheskaya biblioteka, izdavaemaya Arkheograficheskoyu komissieyu*
RIZh	*Russkii istoricheskii zhurnal*
SA	*Sovetskaya arkheologiya*
SAI	*Svod arkheologicheskikh istochnikov*
SE	*Sovetskaya etnografiya*
SF	*Studia Fennica*
SGGD	*Sobranie gosudarstvennykh gramot i dogovorov*
SGKE	*Sbornik gramot Kollegii ekonomii*
SRIO	*Sbornik imperatorskago russkago istoricheskago obshchestva*
SRL	*Scriptores rerum Livonicarum*
TrGIM	*Trudy Gosudarstvennogo istoricheskogo muzeya*
TrIE	*Trudy Instituta Etnografii im. Miklukho–Maklaya*
TrIL	*Trudy Instituta lesa*
TrLOII	*Trudy Leningradskogo otdeleniya Instituta istorii*
TrMIRM	*Trudy Muzeya istorii i rekonstruktsii Moskvy*
TrMORVIO	*Trudy Moskovskago otdeleniya russkago voenno-istoricheskago obshchestva*
TrODRL	*Trudy otdela drevnerusskoi literatury*
TsGADA	Tsentral'nyi Gosudarstvennyi arkhiv drevnykh aktov SSSR
Uch. zap	*Uchenie zapiski*
ULS	*Ustyuzhskii letopisnyi svod*
VE	*Voprosy ekonomiki*
VI	*Voprosy istorii*
VIMK	*Voprosy istorii mirovoi kul'tury*
VIRA	*Voprosy istorii religii i ateizma*
VMU	*Vestnik Moskovskogo universiteta*
Vrem.MOIDR	*Vremennik Moskovskago Imperatorskago Obshchestva Istorii i drevnostei Rossiiskikh*
ZhMNP	*Zhurnal Ministerstva narodnago prosveshcheniya*

Bibliography

PRIMARY SOURCES

Akty pistsovago dela, materialy dlya istorii kadastra i pryamogo oblozheniya v Moskovskom gosudarstve, tt. I–II, vyp. I, ed. S. Veselovskii. M, 1913.

Artsikhovskii, A. V. *Novgorodskie gramoty na bereste (iz raskopok 1952g.)*. M, 1954.

Artsikhovskii, A. V. & V. I. Borkovskii. *Novgorodskie gramoty na bereste: (iz raskopok 1953–1954gg.)*. M, 1958; *(iz raskopok 1955g.)*. M, 1958; *(iz raskopok 1956–7gg.)*. M, 1963; *(iz raskopok 1958–61gg.)*. M, 1963.

Artsikhovskii, A. V. & M. N. Tikhomirov. *Novgorodskie gramoty na bereste (iz raskopok 1951g.)*. M, 1953.

Baron, *see* Olearius.

Berry, L. E. & R. O. Crummey. *Rude and barbarous kingdom*. Madison, 1968.

Chaev, N. S. 'Severnye gramoty XV v.', *LZAK 1927–8*, XXXV; 1929, pp. 121–64.

Dewey, H. W. *Muscovite Judicial Texts 1488–1556*, Michigan Slavic Materials, no. 7. University of Michigan Department of Slavic Languages and Literatures, Ann Arbor, 1966.

Dewey, H. W. 'The 1497 Sudebnik', *ASEER* (1956), 15.3.325–38.

Dioptra ili zertsalo mirozritel'noe. GIM, no. 2709.

'Domostroi blagoveshchenskago popa Sil'vestra, soobshch. Dm. Pavlov Golokhvastovym', *Vrem.MOIDR* (1849), I.II, materialy, pp. I–IV, 1–116.

D'yakonov, M. A. *Akty, otnosyashchiesya k istorii tyaglago naseleniya v Moskovskom gosudarstve*, vyp. I, *Krest'yanskie poryadnye*. Yur'ev, 1895.

Fletcher, G. *Of the Russe Commonwealth 1591*, facsimile edition of Hakluyt Society, first series, vol. 20. Cambridge, Mass., 1966.

Hakluyt, R. *The principal navigations, voyages, traffiques and discoveries of the English nation*, 12 vols., Glasgow, 1903–5.

von Hamel, J. *England and Russia: comprising the voyages of J. Tradescant the elder, Sir A. Willoughby, R. Chancellor, Nelson and others to the White Sea, etc.* (translated from the German by J. S. Leigh). London, 1854.

von Herberstein, S. *Rerum Muscoviticarum Comentarii*, 3 parts. Vienna, 1549.

Horsey, Sir J. *The travels of Sir Jerome Horsey*. Hakluyt Society, first series, no. 20. London 1856.

Howes, R. C. (trans. and ed.). *The testaments of the Grand Princes of Moscow*. Ithaca, N.Y., 1967.

Iosafovskaya Letopis', M. 1957.

Izbornik slavyanskikh i russkikh sochinenii i statei,, vnesennykh v khronografy russkoi redaktsii, ed. Andrei Popov. M, 1869.

James, R. Manuscript dictionary published in B. A. Larin, *Russko-angliiskii slovar'-dnevnik Richarda Dzhemsa 1618–1619 gg*. L, 1959.

Khrestomatiya po istorii SSSR XVI–XVII vv., ed. A. A. Zimin. M, 1962.

Kniga Bol'shomu chertezhu, ed. K. N. Serbina. M-L, 1950.

Kniga kazanskogo uezdu pomesnykh zemel detei boyarskikh kazanskikh starykh zhiltsov pisma i mery Dmitreya Andreeva syna Kikina s tovarishchi leta 7074. TsGADA f. 1209, no. 152, folios 103–256.

Kniga klyuchei i Dolgovaya kniga Iosifo-Volokolamskogo monastyrya XVI veka, ed. M.N. Tikhomirov & A. A. Zimin. M-L, 1948.

Kniga ob izbranii na tsarstvo Velikago Gosudarya, Tsarya i Velikago Knyazya Mikhaila Fedorovicha. M, 1856.

Margaret, capitaine. *Estat de l'empire de Russie et Grande Duché de Moscovie*. Paris, 1607.

Massa, I. *Skazaniya Massy i Gerkmana o Smutnom Vremeni v Rossii...s prilozheniem portreta Massy, plana Moskvy, 1606 g., i dvortsa Lzhedimitriya I*, ed. with notes, by E. E. Zamyslovskii, 2 parts. SPB, 1874.

Materialy po istorii krest'yan v Rossii XI–XVII vv., ed. V. V. Mavrodin. L, 1958.

Materialy po istorii krest'yan v Russkom gosudarstve XVI veka, ed. A. G. Man'kov. L, 1955.

Materialy po istorii Tatarskoi ASSR, Pistsovye knigi goroda Kazani 1565–8 gg. i 1646 g., Trudy istoriko-arkheograficheskogo instituta, Materialy po istorii narodov SSSR, vyp. 2. L, 1932.

Melnikov, S. *Akty istoricheskie i yuridicheskie i drevniya tsarskiya gramoty Kazanskoi i drugikh sosedstvennykh gubernii*, 1. Kazan', 1859.

Mukhamed'yarov, Sh. F. 'Maloizvestnaya pistsovaya kniga Kazanskogo uezda 1602–1603 gg.', *Izv. Kazanskogo filiala AN SSSR, seriya gumanitarnikh nauk* (1957), 2.191–6.

Mukhamed'yarov, Sh. F. 'Tarkhannyi yarlyk Kazanskogo khana Sakhib-Gireya 1523 g.', in *Novoe o proshlom nashei strany*, 104–9, M, 1967.

Naziratel', ed. S. I. Kotkov. M, 1973.

Novgorodskaya pervaya letopis' starshego i mladshego izvodov, ed. A. N. Nasonov. M-L, 1950.

Novye pskovskie gramoty XIV–XV vekov, ed. L. M. Marasinova. M, 1966.

Olearii, Adam. *Opisanie puteshestviya v Moskoviyu i cherez Moskoviya v Persiyu i obratno*, with introduction, translation and index by A. M. Lovyagin. SPB, 1906.

[Olearius], trans. & ed. S. H. Baron. *The travels of Olearius in seventeenth-century Russia*. Stanford, 1967.

Pamyatniki sotsial'no-ekonomicheskoi istorii Moskovskogo gosudarstva XIV–XVII vekov, t. 1. M, 1929.

Perry, J. *The state of Russia under the present Czar*. London, 1716.

Pistsovyya knigi Moskovskago gosudarstva, ed. N. V. Kalachov. SPB, 1872.

Plan sela Mikhailovskogo i ego okrestnostei s nadpis'yu: 'Zvenigorodskogo uezdu', TsGADA, f. 27, no. 484, ch. III, d. 68.

Plan sela Sofarina na reke Talitse, ego zemel' i okrestnostei. TsGADA, f. 27, no. 484, ch. III, d. 99.

Printz, Daniel of Bucchau. 'Moscoviae ortus, et progressus', *SRL*, 2.687–728.

Radzivilovskaya ili Kenigsbergskaya letopis', tt. 2. SPB, 1902.

Raskhodnaya kniga Patriarshago Prikaza kushan'yam, podavavshimsya patriarkhu Adrianu i raznago china litsam s sentyabrya 1698 po avgust 1699g., ed. A. A. Titov. SPB, 1890.

Razryadnaya kniga 1475–1598gg. M, 1966.

Severnye gramoty, see Chaev.

Shumakov, S. A. *Obzor "Gramot Kollegii ekonomii"*, 4 vyp. M, 1899–1917.

Shumakov, S. A. *Sotnitsy (1537–1597gg.) gramoty i zapisi (1561–1696gg.)*. M, 1902.

Skazanie Avraamiya Palitsyna, RIB XIII.473–524. SPB, 1892.

Sobornoe ulozhenie 1649 g., ed. M. N. Tikhomirov & P. P. Epifanov. M, 1961.

Spisok s pistsovoi mezhevoi knigi goroda Sviyazhska i uezda; pis'ma i mezhevaniya N. V. Borisova i D. A. Kikina (1565–67), ed. A. Yablokov. Kazan', 1909.

Spisok s pistsovykh knig po gor. Kazani s uezdom, ed. K. I. Nevostruev. Kazan', 1877.

Sudebniki XV–XVI vekov. Zakonodatel'nye pamyatniki russkogo tsentralizovannogo gosudarstva XV–XVII vekov, t. 1. M-L, 1952.

Toropets. Materialy dlya istorii goroda XVII i XVIII st. M, 1883.

'Toropetskaya kniga 1540/41g.' ed. M. N. Tikhomirov & B. N. Florya, *AE za 1963*, 277–357. M, 1964.

Tysyachnaya kniga 1550 g. i dvorovaya tetrad' 50-kh godov XVIv. ed. A. A. Zimin. M-L, 1950.

Vernadsky, G. *et al. A source book for Russian history from early times to 1917*, 3 vols. New Haven, London, 1972.

Veselovskii, S. B. *Sem' sborov zaprosnykh i pyatinnykh deneg v pervye gody tsarstvovaniya Mikhaila Fedorovicha*. M, 1909.

Vosstanie I. Bolotnikova, dokumenty i materialy. M, 1959.

Weber, F. C. *The present state of Russia*, 2 vols. London, 1722–3.

Zakon Sudnyi lyudem, kratkoi redaktsii, ed. M. N. Tikhomirov. M, 1961.

SECONDARY SOURCES

Adelung, F. von. *Kritisch-literärische Übersicht der Reisenden in Russland bis 1700*, 2 Bd. SPB, 1846.

Aakjaer, S. *Maal og Voegt: Nordisk kultur*, 30. Copenhagen, Stockholm, 1936.

Abramovich, G. V. 'Neskol'ko izyskanii iz oblasti russkoi metrologii XV–XVIvv.', *PI*, XI.364–90.

Agrarnaya istoriya severo-zapada Rossii, vtoraya polovina 15 nachalo 16 v. L, 1971.

Alekseev, L. V. *Polotskaya zemlya*. M, 1966.

Andriyashev, A. M. *Materialy po istoricheskoi geografii Novgorodskoi zemli, Shelonskaya pyatina po pistsovym knigam, 1498–1576 gg.: I, Spiski selenii; II, Karty pogostov*. M, 1913–14.

Antropov, V. and V. *Rozh' SSSR i sopredel'nykh stran (Prilozhenie 36-e k Trudam po prikladnoi botanike, genetike i selektsii)*, with English summary. L, 1929.

Aristov, N. Ya. *Promyshlennost' drevnei Rusi*. SPB, 1866.

Arkheologicheskie pamyatniki Moskvy i Podmoskov'ya (TrMIRM, vyp. 5), ed. A. P. Smirnov. M, 1954.

Artsikhovskii, A. V. *Drevnerusskie miniatyury kak istoricheskii istochnik*. M, 1944.

Atkinson, J. A. & J. Walker. *Picturesque representation of the manners, customs and amusements of the Russians*, 3 vols. London, 1803–4.

Bakhrushin, S. V. 'Knyazheskoe khozyaistvo XV i pervoi poloviny XVI v.', in his *Nauchnye trudy*, t. II, 13–45. M, 1954.

Bakhrushin, S. V. *Nauchnye trudy*, tt. I–IV. M, 1952–9.

Banfield, E. C. *The moral basis of a backward society*. Glencoe, 1958.

Baranov, F. I. & M. D. Bessonova. *Derevyannye rybolovnye lovushki, ikh ustroistvo i primenenie*. M, 1944.

Bennett, R. & J. Elton. *History of corn-milling*, 4 vols. London, 1898–1904.

Bezhkovich, A. S., S. K. Zhegalova, A. A. Lebedeva, & S. K. Prosvirkina (eds.). *Khozyaistvo i byt russikh krest'yan, pamyatniki material'noi kul'tury; opredelitel'*. M, 1959.

Bloch, M. 'Avènement et conquête du moulin à eau', *Annales d'Histoire Économique et Sociale*, 1935, VII, no. 36, 358–63.

Blomkvist, E. E. 'Krest'yanskie postroiki russkikh, ukraintsev i belorusov', in *Vostochnoslavyanskii etnograficheskii sbornik* (*TrIE, n.s.*, XXXI), 3–458. M, 1956.

Blum, J. *Lord and peasant in Russia from IX–XIX century*. Princeton University Press, 1961.

Bogolepov, A. M. *Prichiny neurozhaev i goloda v Rossii v istoricheskoe vremya*. M, 1922.

Bogoyavlenskii, N. A. *Drevnerusskoe vrachevanie v XI–XVII vv., istochniki dlya istorii russkoi meditsiny*. M, 1960.

Bogoyavlenskii, S. K. *Prikaznye sud'i XVII v*. M–L, 1946.

Bond, E. A. (ed.) *Russia at the close of the sixteenth century*. Hakluyt Society, first series, no. 20. London, 1856.

Borovik, E. A. *Rybo-promyslovye ozera Belorussii*. Minsk, 1970.

Boserup, E. *Woman's role in economic development*. London, 1970.

Bratanić, B. 'Some similarities between ards of the Balkans, Scandinavia, and Anterior Asia, and their methodological significance' in *Selected Papers of the Fifth International Congress of Anthropological and Ethnological Sciences*, 221–8. Philadelphia, 1957.

Brillat-Savarin, J. A. *The philosopher in the kitchen*, translated from the French by A. Drayton. Harmondsworth, 1970.

Brown, J. C. *Forests and forestry*. Edinburgh, 1884.

Budovnits, I. U. *Monastyri na Rusi i bor'ba s nimi krest'yan v XIV–XVI vekakh*. M, 1966.

Burnashev, V. P. (compiler). *Opyt terminologicheskago slovarya sel'skago khozyaistva, fabrichnosti, promyslov i byta narodnago*, 2 tt. SPB, 1843–4.

Bussov, C. [i.e. Conrad Bussow] *Conradi Bussovi et Petri Petrei chronica Moscovitica*. *Rerum Rossicarum scriptores exteri*, t. 1, ed. A. A. Kunik. SPB, 1851.

Chechulin, N. D. *Goroda Moskovskago gosudarstva v XVI veke*. SPB, 1889.

Cherepnin, L. V. *Novgorodskie berestyanye gramoty kak istoricheskii istochnik*. M, 1969.

Cherepnin, L. V. *Obrazovanie russkogo tsentralizovannogo gosudarstva v XIV–XV vekakh*. M, 1960.

Chernetsov, Alexey V. 'On the origin and early development of the East-European plough and the Russian sokha', *Tools and tillage*, II.1.34–50.

Chernetsov, A. V. *Pakhotnye orudiya Drevnei Rusi, Avtoreferat dissertatsii na soiskanie uchenoi stepeni kandidata istoricheskikh nauk*. M, 1963.

Chicherin, B. N. *Oblastnye uchrezhdeniya v Rossii v XVII veke*. M, 1856.

Chicherov, V. I. *Zimnii period russkogo zemledel'cheskogo kalendarya XVI–XIX vekov; ocherki po istorii narodnykh verovanii* (*AN SSSR. TrIE, n.s. t.* 40). M, 1957.

Dal', V. *Tolkovyi slovar' zhivago velikorusskago yazyka*, 2-oe izd., tt. I–IV. SPB, 1880–2.

Danilova, L. V. *Ocherki po istorii zemlevladeniya i khozyaistva v Novgorodskoi zemle v XIV–XV vv*. M, 1955.

Devyatnadtsatyi Mezhdunarodnyi geograficheskii kongress v Stokgol'me. M, 1961.

Dolgorukov, P. D. & D. I. Shakhovskoi. *Melkaya zemskaya edinitsa*. SPB, 1902.

D'yakonov, M. A. *Ocherki iz istorii sel'skago naseleniya v Moskovskom gosudarstve, 16–17 vv.* SPB, 1898.

D'yakonov, M. A. *Ocherki obshchestvennogo i gosudarstvennogo stroya drevnei Rusi.* M-L, 1926.

Dyuvernua, A. *Materialy dlya slovarya drevne-russkago yazyka.* M, 1894.

The Economist guide to weights and measures. London, n.d.

Evliya Efendi, trans. J. von Hammer. *Narrative of travels in Europe, Asia and Africa in the 17th century.* vol. I, pt. I, 1834, pt. 2, 1846; vol. 2, 1850. London.

Efimenko, Aleksandra. 'Krest'yanskoe zemlevladenie na krainom severe', *Russkaya mysl'*: 1882, IV.183–217; V.48–74; 1883, VI.161–91; VII.245–74.

Elnett, Elaine. *Historic origin and social development of family life in Russia.* Columbia University Press, 1926.

Fasmer, M. *Etymologicheskii slovar' russkogo yazyka,* 4tt. M, 1964–73.

Fat'yanov, A. S. 'Opyt analiza istorii razvitiya pochvennogo pokrova Gor'kovskoi oblasti' in *Pochvenno-geograficheskie issledovaniya i ispol'zovanie aerofotos"emki v kartirovanii pochv,* 3–171, M, 1959.

Fedotov, G. P. (compiler and ed.). *A treasury of Russian spirituality.* New York, 1965.

Fekhner, M. V. *Torgovlya Russkogo gosudarstva so stranami Vostoka v XVI veke, TrGIM,* vyp. 27. M, 1952.

Fennell, J. L. I. (ed. and trans.). *Prince A. M. Kurbsky's history of Ivan IV.* Cambridge, 1965.

Filologicheskii sbornik. Smolensk, 1950.

Galakhov, N. N. *Izuchaite griby.* M, 1968.

Galeski, B. *The basic concepts of rural sociology.* Manchester, 1972.

Galton, Dorothy. *Survey of a thousand years of Beekeeping in Russia.* London, 1971.

Gedymin, A. V. & A. T. Kharitonycheva. 'Old cartographic material used in landscape study', *20th Intl. Geographical Cong., Abstracts of papers,* no. 1991/1992, London, 1964.

Gieysztor, A. 'W sprawie początków trójpolówki w Polsce i krajach sąsiednich' in *Prace z dziejów Polski feodalnej,* 71–9. Warszawa, 1960.

Gnevushev, A. M. *Ocherki ekonomicheskoi i sotsial'noi zhizni sel'skago naseleniya Novgorodskoi oblasti, posle prisoedineniya Novgoroda k Moskve,* t. I. Kiev, 1915.

Goehrke, C. *Die Wüstungen in der Moskauer Rus', Studien zur Siedlungs-, Bevölkerungs- und Sozialgeschichte, Quellen und Studien zur Geschichte des Östlichen Europa,* Bd. I. Wiesbaden, 1968.

Goldschmidt, A. *Die Frühmittelalterlichen Bronzetüren,* Bd. II, *Die Bronzetüren von Nowgorod und Gnesen.* Marburg, 1932.

Gorskaya, N. A. 'Zernovoe zemledelie tsentral'nykh oblastei russkogo gosudarstva vo vtoroi polovine XVI v.', Kandidatskaya dissertatsiya. M, 1955.

Gorskii, A. D. *Bor'ba krest'yan za zemlyu na Rusi v XV-nachale XVI veka.* M, 1974.

Gorskii, A. D. *Ocherki ekonomicheskogo polozheniya krest'yan severo-vostochnoi Rusi XIV–XV vv.* M, 1960.

Got'e, Yu. V. *Ocherk istorii zemlevladeniya v Rossii.* Sergiev Posad, 1915.

Got'e, Yu. V. *Zamoskovnyi krai v XVII v., Opyt issledovaniya po istorii ekonomicheskago byta Moskovskoi Rusi.* M, 1906.

Gowers, Sir E. A. (ed.) *A dictionary of modern English usage,* 2nd edn. Oxford University Press, 1965.

Gurlyand, I. Ya. *Yamskaya gon'ba v Moskovskom gosudarstve*. Yaroslavl', 1900.

Hakulinen, L. 'Finnisch Luokka "Klasse" ' *SF* (1938), III. 45–63.

Hajnal, J. 'European marriage patterns in perspective' in *Population in history*, ed. D. V. Glass and D. E. C. Eversley. London, 1965.

Hanse, Downing Street und Deutschlands Lebensraum. Berlin, 1940.

Haudricourt, A–G. 'Contribution à la géographie et à l'ethnologie de la voiture', *La Revue de Géographie Humaine et d'Ethnologie* (1948), I. 1.54–64.

Haudricourt, A–G. 'De l'origine de l'attelage moderne', *Annales d'histoire économique et sociale* (1946), 8. 515–22.

Haudricourt, A–G. 'Lumières sur l'attelage moderne', *Annales d'histoire sociale* (1945), 8. 117–8.

Haudricourt, A–G. & M. J–B. Delamarre. *L'homme et la charrue à travers le monde* (with maps). Paris, 1955.

Hecker, J. F. C. *The epidemics of the middle ages* (*The Black Death in the fourteenth century* [*revised by H. E. Lloyd*]; *The Dancing Mania – Appendix*), 2 parts, translated from the German by B. G. Babington, etc. London, 1833–5.

Hellie, R. *Enserfment and military change in Muscovy*. Chicago, London, 1971.

Hensel, W. *Słowiańszczyzna wczesnośredniowieczna*. Wydanie trzecie, Warszawa, 1965.

Howison, W. 'An account of the manufacture of turpentine from the *Pinus Sylvestris*, as practised by the Native peasantry of the interior of the Russian Empire', *Transactions of the Highland Society of Scotland*, 1820.

Ikonnikov, V. S. *Opyt russkoi istoriografii*, 2tt. Kiev, 1891–1905.

Indova, E. I. *Dvortsovoe khozyaistvo v Rossii, pervaya polovina XVIII veka*. M, 1964.

Iskusstvo Rusi epokhi Rubleva (*XIV nachala XVI vv.*), *Katalog*. M, 1960.

Issledovaniya istochnikov po istorii russkogo yazyka i pis'mennosti, ed. L. P. Zhukovskaya and N. I. Tarabasova. M, 1966.

Issledovaniya po lingvisticheskomu istochnikovedeniyu, ed. S. I. Kotkov. M, 1963.

Istoricheskii arkhiv, ed. B. D. Grekov *i dr.*, tt 1–10. M, 1936–54.

Istoriko-arkheologicheskii sbornik Artemiyu Vladimirovichu Artsikhovskomu, ed. D. A. Avdusin, V. L. Yanin. M, 1962.

Istoriya Kul'tury Drevnei Rusi, ed. N. N. Voronin *i dr.*, 2 tt. M–L, 1948–51.

Istoriya Moskvy v shesti tomakh, t. I, *Period feodalizma XII–XVIIvv.* M, 1952.

Jablonskis, K. 'Apie vergus Didžiojoje Lietuvos kunigaikš tystėje XVI amžiaus pradžioje', *Praeitis*, I. 304–17. Kaunas, 1930.

Jablonskis, K. 'Die offizielle Urkundensprache des litauischen Grosfürstentums als Kulturgeschichtliche Quelle', *CPHB*, 269–75.

Jablonskis, K. *Lietuviški žodžiai senosios Lietuvos raštiniu kalboje*, I. Kaunas, 1941.

Janoušek, D. Emanuel. 'Poznámky k metodě studia vývoje živočišné výroby na feudálním velko statku v 17. a 18. stoleti', *Historie a musejnictví, Sborník Československé Akademie Zemědělských Věd* (1958), 3. XXXI. 11–34. Summaries in Russian, German.

Kahan, A. 'Natural calamities and their effect upon the food supply in Russia', *Jahrbücher fur Geschichte Osteuropas* (1968), 16. 353–77.

Kalima, Jalo *Die Ostseefinnischen Lehnwörter im Russischen, Suomalais-ugrilaisen Seuran Toimituksia XLIV*. Helsinki, 1919.

Kamentseva, E. I. & N. V. Ustyugov. *Russkaya metrologiya*. M, 1965.

Karamzin, N. M. *Istoriya gosudarstva rossiiskago*, 3-e izd., 12 tt. SPB, 1830–1.

Kashtanov, S. M. 'Iz istorii polednikh udelov', *Trudy Moskovskogo gosudarstvennogo istoriko-arkhivnogo instituta*, II, 269–96. M, 1958.

Kashtanov, S. M. *Ocherki russkoi diplomatiki*. M, 1970.

Kashtanov, S. M. *Sotsial'no-politicheskaya istoriya Rossii kontsa XV – pervoi poloviny XVI v.* M, 1967.

Kaufman, I. I. 'Serebryanyi rubl' v Rossii ot ego vozniknoveniya do kontsa XIXv.', *Zapiski Numizmaticheskago otdeleniya Imp. russkago arkheologicheskago obshchestva* t. II, vyp. I–II, 1–268. SPB, 1910.

Kerblay, B. H. *L'isba d'hier et d'aujourd'hui*. Lausanne, 1973.

Khavskii, P. *Drevnosti Moskvy, ili ukazatel' eya topografii i istorii*. M, 1854.

Khoroshkevich, A. L. *Torgovlya Velikogo Novgoroda s Pribaltikoi i Zapadnoi Evropoi v XIV–XV vekakh*. M, 1963.

Khozyaistvo i byt, see Bezhkovich.

Kirikov, S. V. 'Istoricheskie izmeneniya zhivotnogo mira nashei strany v XIII–XIX vekakh', *Izv. AN SSSR, seriya geograficheskaya: 1952*, 6, 31–48; 1953, 4, 15–27; 1955, 1, 32–40; 1958, 1.

Kirikov, S. V. *Izemeneniya zhivotnogo mira v prirodnykh zonakh SSSR (XIII–XIXvv.), lesnaya zona i lesotundra*. M, 1959.

Kirikov, S. V. *Izemeneniya zhivotnogo mira v prirodnykh zonakh SSSR (XIII–XIXvv.), stepnaya zona i lesostep'*. M, 1959.

Kirikov, S. V. *Promyslovye zhivotnye, prirodnaya sreda i chelovek*. M, 1966.

Kizevetter, A. A. *Mestnoe samoupravlenie v Rossii IX–XIX st.* M, 1910.

Klyuchevskii, V. O. *Boyarskaya Duma drevnei Rusi*, 5-e izd. P, 1919.

Klyuchevskii, V. O. *Opyty i issledovaniya*. Pg, 1918.

Klyuchevskii, V. O. *Sochineniya*, tt. 1–8. M, 1956–9.

Kobrin, V. B. 'Iz istorii zemel'noi politiki v gody Oprichniny', *Ist. arkhiv* (1958), 3. 152–60.

Kochin, G. E. *Sel'skoe khozyaistvo na Rusi v period obrazovaniya Russkogo tsentralizovannogo gosudarstva, konets XIII–nachalo XVI v.* M–L, 1965.

Kolchin, B. A. *Novgorodskie drevnosti, reznoe derevo*. Arkh. SSSR, SAI E1–55. M, 1971.

Kolycheva, E. I. *Formirovanie klassa krepostnykh krest'yan v Rossii XVI veka*. M, 1971.

Kolycheva, E. I. *Kholopstvo i krepostnichestvo (konets XV–XVIv.)*. M, 1971.

Konstantinov, N. A. 'Narodnye reznye kalendari', *Sbornik muzeya antropologii i etnografii* (1961), 20. 84–113.

Kopysskii, Z. Yu. *Ekonomicheskoe razvitie gorodov Belorussii XVI–XVIIvv.* Minsk, 1966.

Koretskii, V. I. *Zakreposhchenie krest'yan i klassovaya bor'ba v Rossii vo vtoroi polovine XVI v.* M, 1970.

Korolev, D. A. *Russkii kvas*. M, 1963.

Kosven, M. O. *Semeinaya obshchina i patronimiya*. M, 1963.

Kovalev, S. A. *Sel'skoe rasselenie; geograficheskoe issledovanie*. M, 1963.

Krest'yanstvo i klassovaya bor'ba v feodal'noi Rossii (TrLOII, vyp. 9). L, 1967.

Krzhizhanovskii, G. M. *Energeticheskie resursy SSSR*. M, 1937.

Kudryavtseva-Molodchikova, L. P. *Gribnaya byl'*. M, 1956.

Kul'tura i byt tsentral'no-promyshlennoi oblasti. M, 1929.

Kushner, P. I. i dr. (eds.) *Russkie, istoriko-etnograficheskii atlas: [I] Zemledelie. Krest'yanskoe zhilishche. Krest'yanskaya odezhda*. M, 1967. [II] *Iz istorii russkogo narodnogo zhilishcha i kostyuma*. M, 1970.

Kutepov, N. I. t. 1, *Velikoknyazheskaya i tsarskaya okhota na Rusi s X po XVI vek*. SPB, 1896. t. 2, *Tsarskaya okhota na Rusi tsarei Mikhaila Feodorovicha i Alekseya Mikhailovicha XVII v*. SPB, 1898. [Title of vol. 1 is *also* title of 4 vols, 1896–1911. Only first two vols have been used.]

Lamb, H. H. *The changing climate*. London, 1966.

Lappo-Danilevskii, A. S. *Organizatsiya pryamogo oblozheniya v Moskovskom gosudarstve*. SPB, 1890.

Laptev, M. *Kazanskaya guberniya, Materialy dlya geografii i statistiki Rossii*. SPB, 1860.

Lebedev, A. N. *Klimat SSSR*, vyp. 1, *Evropeiskaya territoriya SSSR*. L, 1958.

Lefebvre des Noettes, R. J. E. C. *L'attelage. Le cheval de selle à travers les âges* (with plates), 2 vols. Paris, 1931.

Lekarstvennye rasteniya dikorastuyushchie. Minsk, 1965.

Lesostep' i step' russkoi ravniny. M, 1956.

Levasheva, V. P. 'Sel'skoe khozyaistvo', in *Ocherki po istorii russkoi derevni X–XIIIvv*. *TrGIM*, vyp. 32. M, 1956.

Łowmiański, H. *Studja nad początkami społeczeństwa i państwa Litewskiego*, 2 tt. Wilno, 1931. (Rozprawy wydziału III Towarzystwa Przyjaciół Nauk w Wilnie. tom 5, 6).

Luk'yanov, P. M. *Istoriya khimicheskikh promyslov i khimicheskoi promyshlennosti Rossii do kontsa XIX veka*, tt. 1–5. M–L, 1948–61.

Lyubavskii, M. K. *Istoricheskaya geografiya Rossii v svyazi s kolonizatsiei*. M, 1909.

Lyubimova, E. L. & E. M. Murzaev. 'Place-name evidence of former geographic conditions on the Russian plain', *20th Intnl. Geographical Cong., Abstracts of papers*, no. 1993/1994. London, 1964.

Lyubomirov, P. G. *Ocherki po istorii russkoi promyshlennosti*. M, 1947.

MacAdam, R. 'Ancient water-mills', *Ulster J. of Archaeology* (1856), 4.6–15.

McConnell, Primrose. *Notebook of agricultural facts and figures for farmers and farm students*. London, 1883.

Maisterov, L. E. & S. K. Prosvirkina 'Narodnye derevyannye kalendari', *Istoriko-astronomicheskie issledovaniya*, vyp. 6. M, 1960.

Makovetskii, I. V. *Arkhitektura russkogo narodnogo zhilishcha, sever i verkhnee Povolzh'e*. M, 1962.

Makovetskii, I. V. *Pamyatniki narodnogo zodchestva russkogo severa, po materialam kompleksnoi ekspeditsii Inst-a istorii iskusstv AN SSSR i GIM-a*. M, 1955.

Makovetskii, I. V. *Pamyatniki narodnogo zodchestva verkhnego Povolzh'ya*. M, 1952.

Makovskii, D. P. *Pervaya krest'yanskaya voina*. Smolensk, 1967.

Makovskii, D. P. *Razvitie tovarno-denezhnykh otnoshenii v sel'skom khozyaistve russkogo gosudarstva v XVI veke*. Smolensk, 1963.

Malowist, M. 'The problem of the inequality of economic development in Europe in the later middle ages', *EHR*, 2nd series (1966), XIX. 1. 15–28.

Man'kov, A. G. *Tseny i ikh dvizhenie v russkom gosudarstve XVI veka*. M–L, 1951.

Matuszewski, Jozef 'Początki nowożytnego zaprzęgu konnego', *Kwartalnik Historii Kultury Materialnej*: 1953, 1–2.78–111; 1954, 4.637–68.

Mavrodin, V. V. *Obrazovanie edinogo russkogo gosudarstva*. L, 1951.

Mazdorov, V. A. *Istoriya razvitiya bukhgalterskogo ucheta v SSSR*. M, 1972.

Medvedev, A. F. *Ruchnoe metatel'noe oruzhie, luk i strely, samostrel, VIII–XIVvv. Arkh. SSSR, SAI* E1–36. M, 1966.

Mendeleev, D. I. *Sochineniya*, tt. 1–25. L–M, 1946–54.

Merder, I. *Istoricheskii ocherk russkago konevodstva i konnozavodstva.* SPB, 1868.

Merkulóva, V. A. *Ocherki po russkoi narodnoi momenklature rastenii, travy, griby, yagody.* M, 1967.

Merzon, A. Ts. *Pistsovye i perepisnye knigi XV–XVIIvv.* M, 1956.

Milyukov, P. *Spornye voprosy finansovoi istorii Moskovskago gosudarstva.* SPB, 1892.

Min'ko, L. I. *Znakharstvo, istoki, sushchnost', prichiny bytovaniya.* Minsk, 1971.

Mongait, A. L. *Ryazanskaya zemlya.* M, 1961.

Moora, H. 'Zur älteren Geschichte des Bodenbaues, bei den Esten und ihren Nachbarvölkern, in *Congressus secundus internationalis fenno-ugristarum Helsingiae habitus* 23–28. *viii.* 1965, pars II, 239–57. Helsinki, 1965.

Morozov, G. F. *Uchenie o lese.* SPB, 1912.

Moszyński, K. *Kultura ludowa Słowian, cz.* I. Kraków, 1929.

Mukhomediyarov, F. B. *Prosteishie rybolovnye orudiya i ikh primenenie.* Irkutsk, 1944.

Needham, J., Lu Gwei-Djen 'Efficient equine harness; the Chinese inventions', *Physis* (1960), II. 2. 121–62.

Nenquin, J. A. E. *Salt,* Dissertationes archaeologicae gandenses vol. 6. Bruges, 1961.

Nilsson, N. M. P. *Primitive time reckoning (Skrifter utgivna av. Humanistika Vetenskapssamfundet i Lund.* no. 1). Lund, 1920.

Nosal', M. A. & I. M. Nosal'. *Lekarstvennye rasteniya i sposoby ikh primeneniya v narode.* Kiev, 1959.

Novoe o proshlom nashei strany; pamyati akademika M. N. Tikhomirova. M, 1967.

Novosel'tsev, A. P., V. T. Pashuto & L. V. Cherepnin. *Puti razvitiya feodalizma.* M, 1972.

Obozov, N. A., A. T. Savel'ev, O. V. Belevtseva & I. K. Fortunatov. *Pobochnye pol'zovaniya v lesakh SSSR.* M, 1971.

Ocherki po istorii Moskovskogo kraya. M, 1962.

Ocherki po istorii russkoi derevni X–XIIIvv., ed. B. A. Rybakov: *TrGIM,* vyp. 32, M, 1956; *TrGIM,* vyp. 33, M, 1959.

Ocherki russkoi kul'tury XIII–XV vekov, ed. A. V. Artsikhovskii *i dr.* 2 parts. M, 1969.

Opisanie dokumentov i bumag khranyashchikhsya v Moskovskom arkhive Min. Yustitsii: kn. 1, SPB, 1869; kn. 2, SPB, 1872; kn. 3, SPB, 1888.

Opyt oblastnago Velikorusskago slovarya Izdannyi vtorym otdeneniem Imperatorskoi AN. SPB, 1852.

Opyt oblastnago Velikorusskago slovarya Izdannyi vtorym otdeleniem Imperatorskoi AN. Dopolnenie. SPB, 1858.

Osnovy zhivotnovodstva, 4-oe izd. M, 1957.

Ostrovskaya, M. 'Volost' i ee zakreposhchenie' *Trud v Rossii,* kn. 2, 132–146. L, 1924.

Paneyakh, V. M. *Kabal'noe kholopstvo na Rusi v XVI veke.* L, 1967.

'Peasant art in Russia', *The Studio* [special issue], no. 320, 1912.

Peretyatkovich, G. *Povolzh'e v XV i XVI vekakh.* M–L, 1946.

Pervaya vseobshchaya perepis' naseleniya Rossiiskoi Imperii, 1897g., ed. N. A. Troinitskii, 89 vyp. SPB, 1899–1905.

Pershin, P. N. *Zemel'noe ustroistvo dorevolyutsionnoi derevni, t.* I, *Raiony: Tsentral'nopromyshlennyi, Tsentral'no-chernozemnyi i Severo-zapadnyi.* M, Voronezh, 1928.

Picheta, V. I. *Belorussiya i Litva XV–XVI vv.* M, 1961.

Pisarev, I. Yu. *Narodonaselenie SSSR; sotsial'no ekonomicheskii ocherk.* M, 1962.

[Platonov, S. F.] *Sbornik statei po russkoi istorii, posvyashchennyi S. F. Platonovu.* P, 1922.

Poboinin, I. I. 'Toropetskaya starina. Istoricheskie ocherki goroda Toroptsa s drevneishikh vremen do kontsa XVII veka', *ChOIDR* (1897), 1. 1–92; (1899), 3. 93–184; (1902), 2. 185–375.

Pochvenno-geograficheskie issledovaniya i ispol'zovanie aerofotos"emki v kartirovanii pochv. M, 1959.

Pochvenno-geograficheskoe raionirovanie SSSR (v svyazi s sel'skokhozyaistvennym ispol'-zovaniem zemel'), ed. P. Z. Letunov. M, 1962.

Podobedova, O. I. *Miniatyury russkikh istoricheskikh rukopisei; k istorii russkogo litsevogo letopisaniya.* M, 1965.

Podvysotskii, A. *Slovar' oblastnago Arkhangel'skago narechiya v ego bytovom i etnograficheskom primenenii.* SPB, 1885.

Pokrovskii, N. N. *Aktovye istochniki po istorii chernososhnogo zemlevladeniya v Rossii XIV–nachala XVI v.* Novosibirsk, 1973.

Polezhaev, K. V. *Moskovskoe knyazhestvo v I polovine XVI veka.* SPB, 1878.

Polosin, I. I. *Sotsial'no-politicheskaya istoriya Rossii.* M, 1963.

Ponomarev, N. A. *Voznikovenie i razvitie vetryanoi mel'nitsy.* M, 1958.

Popov, A. *Izbornik slavyanskikh i russkikh sochinenii i statei, vnesennykh v khronografy russkoi redaktsii.* M, 1869.

Popov, I. S. *Kormovye normy i kormovy tablitsy,* 13–e izd. M, 1955.

Porokhova, O. G. 'Slova *lyada, lyadina* i *niva* v russkikh narodnykh govorakh', in *Leksika russkikh narodnykh govorov,* 175–91. M–L, 1966.

Pospelov, E. M. *Toponimika i kartografiya.* M, 1971.

Presnyakov, A. E. *Obrazovanie velikorusskogo gosudarstva.* Pg, 1920.

Presnyakov, A. E. 'Votchinnyi rezhim i krest'yanskaya krepost'', *LZAK,* vyp. 1 (xxxiv), 174–92. L, 1927.

Prokof'eva, L. S. *Votchinnoe khozyaistvo v XVII veke, po materialam. Spaso-Prilutskogo monastyrya.* M–L, 1959.

Prozorovskii, D. 'Chai, po starinnym russkim svedeniyam' (iz *Domashnaya Beseda*). SPB, 1866.

Rabinovich, M. G. *O drevnei Moskve: Ocherki material'noi kultury i byta gorozhan v XI–XVI vv.* M, 1964.

Rainov, T. I. *Nauka v Rossii XI–XVII v.* M–L, 1940.

Ränk, G. 'Eesti Arder', *Eesti Rahva Museumi, Aastaraamat,* IV. 19–38. Tartu, 1928.

Rasteniya primenyaemye v bytu, 2–oe izd. (*Sredi prirody,* vyp. 54). M, 1966.

Resursy fauny promyslovykh zverei v SSSR i ikh uchet. M, 1963.

Reznikov, F. I. 'Skotovodstvo v nizov'yakh r. Severnoi Dviny v XVII–XVIII vv.' *MISKh,* IV. 105–38.

Reznikov, F. I. 'Zemledelie v basseine r. Severnoi Dviny v XVII–XVIII vv.', *MISKh,* IV. 67–104.

Romanov, B. A. 'Izyskaniya o russkom sel'skom poselenii epokhi feodalizma', *TrLOII,* 2. 327–476. L, 1960.

Rozhdestvenskii, S. V. *Sluzhiloe zemlevladenie v Moskovskom gosudarstve v XVI veke',* Zapiski istoriko-filologicheskago fakul'teta imp. S. Peterburgskago universiteta, 43. SPB, 1897.

Rozhko, Yu. D. *Yagody i ikh lechebnye svoistva.* Kiev, 1966.

Rozhkov, N. A. *Sel'skoe khozyaistvo Moskovskoi Rusi v XVI veke.* M, 1899.

Rubinshtein, N. L. *Sel'skoe khozyaistvo Rossii vo vtoroi polovine XVIII v.; istoriko-ekonomicheskii ocherk.* M, 1957.

Russia; seu, Moscovia itemque Tartaria. Leyden, 1630.

Russkie, see Kushner.

Rybakov, B. A. *Remeslo drevnei Rusi.* M, 1948.

Sadikov, P. A. *Ocherki po istorii Oprichniny.* M–L, 1950.

Sakharov, A. N. *Russkaya derevnya XVII v., po materialam patriarshego khozyaistva.* M, 1966.

Samokvasov, D. Ya. *Arkhivnyi material.* M, 1905.

Savvaitov, P. *Opisanie starinnykh russkikh utvarei, odezhd, oruzhiya, ratnykh dospekhov i konskago pribora.* SPB, 1896.

Sbornik Arkheologicheskago Instituta, kn. 1–6. SPB, 1878–98.

Sedashev, V. *Ocherki i materialy po istorii zemlevladeniya Moskovskoi Rusi v XVII v.* Izvestiya Konstantinovskago mezhevago instituta, vyp. 2 i 3. M, 1912.

Sel'skokhozyaistvennaya entsiklopediya, Izd. 3, 5 vols. M, 1949–56.

Semenov, P. P. *Geografichesko-statisticheskii slovar' Rossiiskoi imperii,* 5 vols. SPB, 1863–85.

Semenov-Tyan-Shanskii, V. P. *Rossiya,* 11 vols. SPB, 1890–1914.

Semenova-Tyanshanskaya, A. M. 'Izmenenie rastitel'nogo pokrova lesostepi russkoi ravniny v XVI–XVIII vv. pod vliyaniem deyatel'nosti cheloveka', *Botanicheskii zhurnal* (1957), 9. 1398–1407.

Shapiro, A. L. *i dr. Agrarnaya istoriya severo-zapada Rossii, vtoraya polovina XV-nachalo XVI v.* L, 1971.

Shimkin, Demitri B. 'National forces and ecological adaptations in the development of Russian peasant societies' in *Process and pattern in culture,* ed. Robert A. Manners. Chicago, 1964.

Shrewsbury, J. F. D. *A history of bubonic plague in the British Isles.* Cambridge, 1970.

Shubin, V. I. *Griby severnykh lesov.* Petrozavodsk, 1969.

Shul'ga, K. V. *Griby nashikh lesov.* Minsk, 1965.

Sidney, S. *The book of the horse.* London, 1873–5.

Singer, C., E. J. Holmyard, A. R. Hall & T. I. Williams, ed. *A history of technology,* 5 vols. Oxford, 1954–8.

Sinskaya, E. N. *Istoricheskaya geografiya kul'turnoi flory (na zare zemledeliya).* L, 1969.

Sinskaya, E. N. & A. A. Bestuzheva 'Formy ryzhika (Camelina sativa) v ikh otnoshe-niyakh k klimatu, l'nu i cheloveku', *Trudy po prikladnoi botanike genetike i selektsii* (1930–1), 25.2. 98–178 (English summary 179–97).

Sizov, I. A. 'Evolyutsiya kul'turnogo l'na', *Problemy botaniki* (1955), II. 113–66.

Slicher van Bath, B. H. 'The influence of economic conditions on the development of agricultural tools and machines in history' in *Mechanization in agriculture,* ed. J. L. Meij (Studies in industrial economics, II), 1–36. Amsterdam, 1960.

Slicher van Bath, B. H. *Yield ratios, 810–1820 (AAG Bijdragen* 10). Wageningen, 1963.

Slovar' na shesti yazykakh: Rossiiskom, Grecheskom, Latinskom, Frantsuzkom, Nemetskom i Angliiskom. Izdannyi v pol'zy uchashchagosya Rossiiskago yunoshestva. SPB, 1763.

Smirnov, A. P. *Ocherki drevnei i srednevekovoi istorii narodov srednego Povolzh'ya i Prikam'ya, MIA,* no. 28, M, 1952.

Smirnov, I. I. *Vosstanie Bolotnikova, 1606–1607 gg.* L, 1951.

Smith, R. E. F. *The enserfment of the Russian peasantry.* Cambridge, 1968.

Smith, R. E. F. 'Forest cultivation in Toropets (sixteenth century)', *Forschungen zur osteuropäischen Geschichte,* 18. 125–37.

Smith, R. E. F. *The origins of farming in Russia*. Paris, The Hague, 1959.

Solov'ev, S. M. *Istoriya Rossii s drevneishikh vremen*, 15 vols. M, 1959–66.

Sorokin, P. A. & R. K. Merton. 'Social time', *Am. J. Sociol.* (1937), XLII. 615–29.

Sovremennye problemy geografii. M, 1964.

Spasskii, I. G. 'Proiskhozhdenie i istoriya russkikh shchetov', *Istoriko-matematicheskie issledovaniya*, 5. 269–420. M, 1952.

Spasskii, I. G. *Russkaya monetnaya sistema*, 4-oe izd. L, 1970. English translation, *The Russian monetary system*. Chicago, 1967.

Spegal'skii, Yu. P. *Pskovskie kamennye zhilye zdaniya XVII veka*, *MIA*, vyp. 119. M–L, 1963.

Spravochnik predsedatelya kolkhoza, ed. V. A. Chuvikov *i dr.* 2 vols. M, 1956.

Sprinchak, Ya. A. 'Nekotorye voprosy izucheniya vostochnoslavyanskoi leksiki' in *Voprosy leksikologii*, 52–73. Dnepropetrovsk, 1969.

Sreznevskii, I. I. *Materialy dlya slovarya drevnerusskago yazyka*, tt. 3. SPB, 1893–1903.

Stashevskii, E. D. *Opyty izucheniya pistsovykh knig Moskovskago gosudarstva XVI veka*, vyp. 1, *Moskovskii uezd*. Kiev, 1907.

Stashevskii, E. D. *Zemlevladenie Moskovskogo dvoryanstva v pervoi poloviny XVI veka*. M, 1911.

Strutosov 'Okhota v dopetrovskoi Rusi', *Priroda i okhota*, M, 1881, t. II: aprel': 30–50; mai: 1–16.

Talve, Ilmar. *Badstu och torkhus i Nordeuropa*, Nordiska Museets Handlingar 53. Stockholm, 1960.

Tatishchev, V. N. *Istoriya rossiiskaya*, 7 vols. M–L, 1962–8.

Thorner, D., B. Kerblay & R. E. F. Smith (ed.) *A. V. Chayanov on the theory of peasant economy*. Chicago, 1966.

Thünen, J. H. von. *Von Thünen's 'Isolated state'*. Oxford, 1966.

Tikhomirov, M. N. *Drevnyaya Moskva (XII–XV vv.)*. M, 1947.

Tikhomirov, M. N. *Klassovaya bor'ba v Rossii XVII v.* M, 1969.

Tikhomirov, M. N. *Posobie dlya izucheniya Russkoi Pravdy*. M, 1953.

Tikhomirov, M. N. *Rossiya v XVI stoletii*, M, 1962.

Tikhomirov, M. N. *Srednevekovaya Moskva v XIV–XV vv.* M, 1957.

Tkhorzhevskii, S. 'Gosudarstvennoe zemledelie na yuzhnoe okraine Moskovskogo gosudarstva v XVII veke, Kolonizatsiya "polya" i "gosudareva desyatinnaya pashnya" ', *Arkhiv istorii truda v Rossii*, kn. 8, 64–78. SPB, 1923.

Tkhorzhevskii, S. 'Pomest'e i krest'yanskaya krepost', K istorii zemledel'cheskogo truda XVI–XVII vv.', *Arkhiv istorii truda v Rossii*, kn. 11–12, 72–97. SPB, 1924.

Toivonen, Y. H. *Suomen kielen etymologinen sanakirja*, Lexica Societatis Fenno-Ugricae, vol. 12. Helsinki, 1955.

Tret'yakov, P. N. 'Podsechnoe zemledelie v Vostochnoi Evrope', *IGAIMK*, t. XIV, vyp. 1. M, 1932.

Tsalkin, V. I. *Materialy dlya istorii skotovodstva i okhoty v drevnei Rusi*. *MIA*, no. 51. M, 1956.

Tsvetkov, M. A. *Izmenenie lesistosti evropeiskoi Rossii s kontsa XVII stoletiya po 1914 god*. M, 1957.

Tupikov, N. M. *Slovar' drevnerusskikh lichnykh sobstvennykh imen*. SPB, 1903.

Turnerelli, E. T. *Russia on the borders of Asia, Kazan the ancient capital of the Tartar khans*, 2 vols. London, 1854.

Umnikov, N. Z. *Plody, yagody, ovoshchi, zlaki i pryanosti*. Tbilisi, 1959.

Unbegaun, B. O. 'The language of Muscovite Russia in Oxford vocabularies', *Oxford Slavonic Papers* (1962), 10. 46–59.

Ustyugov, N. V. 'Ocherk drevnerusskoi metrologii', *IZ*, 19. 294–348.

Vakhromeev, I. A. *Tserkov' vo imya svyatogo i slavnago proroka Bozhiya Ilii v g. Yaroslavle*. Yaroslavl', 1905.

Vasil'ev, K. G. & A. E. Segal. *Istoriya epidemii v Rossii*. M, 1960.

Vasil'kov, V. P. *S''edobnye i yadovitie griby srednei polosy evropeiskoi chasti SSSR, Opredelitel'*. M–L, 1948.

Vasmer, Max, *see* Fasmer, M.

Veselovskii, S. B. *Issledovaniya po istorii klassa sluzhilykh zemlevladel'tsev*. M, 1969.

Veselovskii, S. B. 'Monastyrskoe zemlevladenie v Moskovskoi Rusi vo vtoroi polovine XVI v.', *IZ* (1941), 10. 95–116.

Veselovskii, S. B. *Selo i derevnya v severo-vostochnoi Rusi XIV–XVI vv. IGAIMK*, vyp. 139. (With French summary). M–L, 1936.

Veselovskii, S. B. *Soshnoe pis'mo: issledovanie po istorii kadastra i pososhnago oblozheniya Moskovskago gosudarstva*, 2tt. M, 1915–16.

Veselovskii, S., V. Snegirev & B. Zemenkov *Podmoskov'e, pamyatnye mesta v istorii russkoi kul'tury XIV–XIX vekov*, 2–oe izd. M, 1962.

Veshnyakov, V. *Krest'yane sobstvenniki v Rossii*. SPB, 1858.

Vikhrov, V. E. 'Ispol'zovanie drevesiny v drevnem Novgorode', *TrIL*, XXXVII. 266–79. M, 1958.

Vikhrov, V. E. & B. A. Kolchin 'Drevesina v khozyaistve i byte drevnego Novgoroda', *TrIL* (1962), II. 142–57.

Vilkuna, K. 'Das Krummholz im Kumtgeschirr', *SF* (1938), 3. 67–82.

Vilkuna, K. *Die Pfluggeräte Finnlands, Sonderdruck: SF*, 16. Helsinki, 1971.

Vilkuna, K. *Die Pfluggeräte Finnlands*, Vortrag auf dem II Finnougristenkongress, in Helsinki am 24 VIII 1965, Helsinki, 1965.

Vinter, A. V. & E. M. Fateev. *Ispol'zovanie energii vetra v sel'skom khozyaistve*. M, 1955.

Voprosy ekonomiki i klassovykh otnoshenii v Russkom gosudarstve XII–XVII vekov, TrLOII, vyp. 2. L, 1960.

Vorob'ev, N. I. *Kazanskie Tatary, Etnograficheskoe issledovanie material'noi kul'tury dooktyabr'skogo perioda*. Kazan', 1953. (With illustrations and a map.)

Voronin, N. N. *K istorii sel'skogo poseleniya feodal'noi Rusi. Pogost, sloboda, selo, derevnya, IGAIMK*, vyp. 138. L, 1935.

Vossoedinenie Ukrainy s Rossiei, dokumenty i materialy, 3 tt. M. 1953.

Wasson, V. P. & R. G. Wasson. *Mushrooms, Russia and History*, 2 vols. New York. 1957.

Weinreich, U. *Languages in contact, Findings and problems etc.* Publications of the Linguistic Circle of New York, no. 1. New York, 1953.

Yakovlev, A. I. *Namestnich'i, gubnye i zemskie ustavnyya gramoty Moskovskago gosudarstva*. M, 1909.

Yuridicheskii slovar', ed. P. I. Kudryavtsev, Izd. 2, tt. 2. M, 1956.

Yurre, N. A. & V. I. Anikin. *Ispol'zovanie lesnykh bogatstv*. M, 1958.

Zaozerskii, A. I. *Tsarskaya votchina XVII v.* M, 1937.

Zasurtsev, P. I. 'Usad'by i postroiki drevnego Novgoroda' in *Zhilishcha drevnego Novgoroda* (*MIA* no. 123), 5–165. M, 1963.

Zelenin, D. K. *Russkaya sokha eya istoriya i vidy*. Vyatka, 1907.

Zemtsovskii, I. *Toropetskie pesni*. L, 1967.

Zimin, A. A. *Reformy Ivana Groznogo*. M, 1960.

Zimin, A. A. *Rossiya na poroge novogo vremeni*. M, 1972.

Zuev, D. P. *Dary russkogo lesa (griby i yagody)*. Sel'khozgiz, M, 1960.

Zverinskii, V. V. *Material dlya istoriko-topograficheskago issledovaniya o pravoslavnykh monastyryakh v Rossiiskoi imperii*, 3 vols. SPB, 1890–7.

Index

account books, 22, 23, 39, 105, 143, 230
accumulation, 3, 240
administration
  on lands of lords, 100–2
  on peasant lands, 179–80
administrative areas (*see also perevara, uezd*), 100
adze, 48, 75
alder, 56, 158, 199
allocations, 30, 33, 81, 95, 101, 107–8, 116, 187,
  188, 215, 216, 217, 226, 228–9, 261
  of forest, 137, 213, 217
allod, 186
animals, 69, 73, 91, 93, 95, 237
  domestic, 46, 67
  for forces and for transport, 34
  size of, 45
  wild, 46, 50, 55, 66, 67, 68, 78
appropriation, 230, 232
arable, 8, 10–40, 121, 150–1, 170, 171, 172, 181,
  186, 187, 190, 192, 203, 205, 209, 217, 223,
  228, 244–9, 250, 262
  area of, 194, 206, 229: per tenement, 85, 151,
    162–5, 215, 260–1; per man, 151, 164–5,
    216, 217, 260–1
  consolidation of, 109–10, 122–3, 125, 166,
    192, 204, 208, 225, 243, 247–8, 249
  demesne, 132, 141, 150, 156, 226, 254
  location of: on estates, 130f, 207; on peasant
    holdings, 134
  valuation of, 214
archaeology, 1, 11, 14–15, 23, 44, 48, 60
  in agreement with MS evidence, 16, 18
ard, 17, 21–3, 53
artisan(s), 148, 149, 180, 232
artisan quarter, 169, 195–6, 209–10, 212, 218,
  231, 234, 250, 261
ash tree, 114
aspen, 43, 51, 72, 114, 158, 199
autarky, 232
axe, 7, 17, 47–8, 72–3, 108, 116, 123, 168, 223,
  229, 231, 236, 237–8

badger, 70
baking, 54, 152, 156
bark, 54
barley, 33, 34, 39, 87, 88, 143, 146
barns, 36
barrels, 54

barshchina, see rent, labour
baskets and punnets, 54
bast, 54, 55, 75
bast shoes, 55, 155
bead-boards, 228
bears, 50, 55, 59, 70, 117, 138, 139
bear garlic/ramson, 60
beaver, 70, 73, 78, 115, 139, 213
beaver runs, 262–3
bees, 74f, 116, 137, 139, 156, 176, 213
bee forest, 139
bee-men, 114, 139
bee products, 200
bee-trees, 75–6, 134, 139, 177, 209, 213, 262–3
bee-tree-marks, 177
beer, 34, 58, 88, 180
bells, 43
Beloozero, 39, 46, 50, 56, 64, 65
berries, 57f, 72, 77, 79, 88, 93, 137, 178
birch, 43, 51, 52, 72, 79, 111, 114, 158, 199,
  212
birch bark, 54, 56, 61, 72
birds, 66, 70, 73
birth rate, 83
bison, 71–2
black lands, 100, 102, 107, 109, 116, 119, 122,
  156, 161, 162, 176, 183, 188, 189, 202, 218,
  223, 231, 236, 247
black peasants, 2, 28, 30, 31, 101, 110, 129, 148,
  161, 224, 234
boar, 70, 72
boats, 53, 65
bobyl', see labourer
bondage, deeds of full, 140
boundaries, 31, 58, 103, 114, 117, 118, 134, 137,
  191, 199, 200, 204, 205, 206, 207, 213, 215,
  223, 224, 230
bow and arrow, 72, 73
boyar, 101, 102, 103, 105, 106, 121, 129, 147,
  148, 149, 160, 183, 187, 188, 192, 218, 224,
  262
bread, 34, 82, 88, 94, 145, 217, 256
  assize of, 152, 232, 256–9
bread man, 210
bread–salt relationship, 74
bread storage, 36
bridges, 50
brigandage, 105, 118